Introduction to Renewable Biomaterials

Introduction to Renewable Biomaterials

First Principles and Concepts

Edited by

Ali S. Ayoub
Archer Daniels Midland Company, Chicago, IL, United States
North Carolina State University, Raleigh, NC, United States

Lucian A. Lucia
North Carolina State University
Raleigh, NC, United States

This edition first published 2018
© 2018 John Wiley & Sons Ltd

All rights reserved. No part of this publication may be reproduced, stored in a retrieval system, or transmitted, in any form or by any means, electronic, mechanical, photocopying, recording or otherwise, except as permitted by law. Advice on how to obtain permission to reuse material from this title is available at http://www.wiley.com/go/permissions.

The right of Ali S. Ayoub and Lucian A. Lucia to be identified as the authors of the editorial material in this work has been asserted in accordance with law.

Registered Office(s)
John Wiley & Sons, Inc., 111 River Street, Hoboken, NJ 07030, USA
John Wiley & Sons Ltd, The Atrium, Southern Gate, Chichester, West Sussex, PO19 8SQ, UK

Editorial Office
9600 Garsington Road, Oxford, OX4 2DQ, UK

For details of our global editorial offices, customer services, and more information about Wiley products visit us at www.wiley.com.

Wiley also publishes its books in a variety of electronic formats and by print-on-demand. Some content that appears in standard print versions of this book may not be available in other formats.

Limit of Liability/Disclaimer of Warranty
While the publisher and authors have used their best efforts in preparing this work, they make no representations or warranties with respect to the accuracy or completeness of the contents of this work and specifically disclaim all warranties, including without limitation any implied warranties of merchantability or fitness for a particular purpose. No warranty may be created or extended by sales representatives, written sales materials or promotional statements for this work. The fact that an organization, website, or product is referred to in this work as a citation and/or potential source of further information does not mean that the publisher and authors endorse the information or services the organization, website, or product may provide or recommendations it may make. This work is sold with the understanding that the publisher is not engaged in rendering professional services. The advice and strategies contained herein may not be suitable for your situation. You should consult with a specialist where appropriate. Further, readers should be aware that websites listed in this work may have changed or disappeared between when this work was written and when it is read. Neither the publisher nor authors shall be liable for any loss of profit or any other commercial damages, including but not limited to special, incidental, consequential, or other damages.

Library of Congress Cataloging-in-Publication Data

Names: Ayoub, Ali S., 1977- editor. | Lucia, Lucian A., editor.
Title: Introduction to renewable biomaterials : first principles and concepts / edited By Ali S. Ayoub, Lucian A. Lucia.
Description: Hoboken, NJ : John Wiley & Sons, 2018. | Includes bibliographical references and index. |
Identifiers: LCCN 2017019395 (print) | LCCN 2017036356 (ebook) | ISBN 9781118698594 (pdf) | ISBN 9781118698587 (epub) | ISBN 9781119962298 (cloth)
Subjects: LCSH: Biomass. | Renewable natural resources.
Classification: LCC TP339 (ebook) | LCC TP339 .I595 2017 (print) | DDC 662/.88–dc23
LC record available at https://lccn.loc.gov/2017019395

Cover design by Wiley
Cover image: © VL2607954/shutterstock

Set in 10/12pt Warnock by SPi Global, Chennai, India

10 9 8 7 6 5 4 3 2 1

Contents

List of Contributors *xiii*
Preface *xv*

1   Fundamental Biochemical and Biotechnological Principles of Biomass Growth and Use *1*
Manfred Kircher
1.1   Learning Objectives *1*
1.2   Comparison of Fossil-Based *versus* Bio-Based Raw Materials *2*
1.2.1   The Nature of Fossil Raw Materials *2*
1.2.2   Industrial Use *3*
1.2.2.1   Energy *3*
1.2.2.2   Chemicals *4*
1.2.3   Expectancy of Resources *8*
1.2.4   Green House Gas (GHG) Emission *8*
1.2.5   Regional Pillars of Competitiveness *9*
1.2.6   Questions for Further Consideration *11*
1.3   The Nature of Bio-Based Raw Materials *11*
1.3.1   Oil Crops *11*
1.3.2   Sugar Crops *13*
1.3.3   Starch Crops *14*
1.3.4   Lignocellulosic Plants *15*
1.3.5   Lignocellulosic Biomass *16*
1.3.6   Algae *16*
1.3.7   Plant Breeding *17*
1.3.8   Basic Transformation Principles *17*
1.3.8.1   First Generation *17*
1.3.8.2   Second Generation *18*
1.3.8.3   Third Generation *18*
1.3.9   Industrial Use *18*
1.3.9.1   Energy *18*
1.3.9.2   Chemicals *20*
1.3.9.3   Biocatalysts *22*
1.3.9.4   Pharmaceuticals *23*
1.3.9.5   Nutrition *24*
1.3.9.6   Polymers *24*

1.3.10	Expectancy of Resources	26
1.3.11	Green House Gas Emission	26
1.3.12	Regional Pillars of Competitiveness	27
1.3.13	Questions for Further Consideration	29
1.4	General Considerations Surrounding Bio-Based Raw Materials	29
1.4.1	Economical Challenges	29
1.4.2	Feedstock Demand Challenges	30
1.4.3	Ecological Considerations	31
1.4.4	Societal Considerations	31
1.4.4.1	Food Security	31
1.4.4.2	Public Acceptance	32
1.5	Research Advances Made Recently	32
1.5.1	First-Generation Processes and Products	32
1.5.2	Second-Generation Processes and Products	33
1.5.3	Third-Generation Processes and Products	33
1.6	Prominent Scientists Working in this Arena	34
1.7	Summary	35
1.8	Study Problems	35
1.9	Key References	36
	References	36
2	**Fundamental Science and Applications for Biomaterials**	**39**
	Ali S. Ayoub and Lucian A. Lucia	
2.1	Introduction	39
2.2	What are the Biopolymers that Encompass the Structure and Function of Lignocellulosics?	39
2.2.1	Cellulose	40
2.2.2	Heteropolysaccharides	43
2.2.3	Lignin	45
2.2.4	The Discovery of Cellulose and Lignin	47
2.3	Chemical Reactivity of Cellulose, Heteropolysaccharides, and Lignin	48
2.3.1	Cellulose Reactivity	48
2.3.1.1	Reactivity Measurements	50
2.3.1.2	Dissolving-Grade Pulps	51
2.3.1.3	Converting Paper-Grade Pulps into Dissolving-Grade Pulps	51
2.3.2	Hemicellulose Reactivity	51
2.3.2.1	Structural Characterization of Hemicellulose	52
2.3.3	Lignin Reactivity	53
2.4	Composite as a Unique Application for Renewable Materials	53
2.4.1	Rationale and Significance	54
2.4.2	Starch-Based Materials	55
2.4.3	Starch-Based Plastics	56
2.4.3.1	Novamont	57
2.4.3.2	Cereplast	58
2.4.3.3	Ecobras	58
2.4.3.4	Biotec	58
2.4.3.5	Plantic	59

2.4.3.6	Biolice	*59*
2.4.3.7	KTM Industries	*59*
2.4.3.8	Cerestech, Inc.	*59*
2.4.3.9	Teknor Apex	*60*
2.5	Question for Further Consideration	*60*
	References	*60*

3	**Conversion Technologies**	*63*
	Maurycy Daroch	
3.1	Learning Objectives	*63*
3.2	Energy Scenario at Global Level	*63*
3.2.1	Why Our Energy is so Important?	*63*
3.2.2	Black Treasure Chest	*64*
3.2.3	Conventional Fossil Resources and their Alternatives	*66*
3.2.3.1	Light Crude Oil (Conventional Oil)	*66*
3.2.3.2	Coal	*66*
3.2.3.3	Natural Gas	*66*
3.2.3.4	Shale Oil (Tight Oil)	*67*
3.2.3.5	Oil Sands, Bitumen Extra Heavy Oil	*67*
3.2.3.6	Shale Gas	*67*
3.2.3.7	Methane (Gas) Hydrates	*67*
3.2.3.8	EROI – How Much Fuel in Fuel?	*68*
3.2.3.9	Environmental Effects of Fossil Resource Utilisation	*69*
3.3	Biomass	*71*
3.3.1	Renewable Energy and Renewable Carbon	*71*
3.3.2	Why Different Types of Biomass have the Properties they Have?	*73*
3.4	Biomass Conversion Methods	*75*
3.4.1	Conversion of Biochemical Energy Perspective	*75*
3.4.2	Overview of Biomass Conversion Technologies	*78*
3.4.3	Thermochemical Conversion of Biomass	*78*
3.4.4	Biomass Combustion	*80*
3.4.5	Gasification	*81*
3.4.6	Pyrolysis	*84*
3.4.7	Conversion of Oily Feedstocks	*86*
3.4.8	Biochemical Conversion of Biomass	*88*
3.4.8.1	Aerobic and Anaerobic Metabolisms	*88*
3.4.8.2	Central Metabolic Pathway under Anaerobic Conditions	*89*
3.4.9	Harvesting Energy from Biochemical Processes	*91*
3.4.9.1	Ethanol Fermentation	*91*
3.4.9.2	ABE Fermentation	*92*
3.4.9.3	Biohydrogen	*93*
3.4.9.4	Biomethane	*94*
3.5	Metrics to Assist the Transition Towards Sustainable Production of Bioenergy and Biomaterials	*95*
3.5.1	EROI – Primary Metrics of Energy Carrier Efficiency	*95*
3.5.2	LCA – Sustainability Determinant	*96*
3.5.3	Environmental Assessment of Bioenergy Production Processes	*97*

3.5.3.1　Impacts Related to Land-Use Change *97*
3.5.3.2　Impacts of Feedstock Cultivation *98*
3.5.3.3　Impacts of Conversion Process *98*
3.5.3.4　Impacts of Product Use *98*
3.5.4　Sustainability Metrics in Biomass and Bioenergy Policies *99*
3.5.5　Renewable and Non-Renewable Carbon – Taxation and Subsidies *99*
3.6　Summary *102*
3.7　Key References *102*
References *103*

4　Characterization Methods and Techniques *107*
Noppadon Sathitsuksanoh and Scott Renneckar
4.1　Philosophy Statement *107*
4.2　Understanding the Characteristics of Biomass *107*
4.3　Taking Precautions Prior to Setting Up Experiments for Biomass Analysis *108*
4.4　Classifying Biomass Sizes for Proper Analysis *109*
4.5　Moisture Content of Biomass and Importance of Drying Samples Prior to Analysis *110*
4.6　When the Carbon is Burned *111*
4.7　Structural Cell Wall Analysis, What To Look For *112*
4.8　Hydrolyzing Biomass and Determining Its Composition *114*
4.8.1　Analyzing Filtrate by HPLC for Monosaccharide Contents *115*
4.8.2　Choosing the HPLC Column and Its Operating Conditions *115*
4.9　Determining Cell Wall Structures Through Spectroscopy and Scattering *116*
4.9.1　Probing the Chemical Structure of Biomass *116*
4.9.1.1　X-Ray Diffraction (XRD) *118*
4.9.1.2　Cross-polarization/Magic Angle Spinning (CP/MAS) ^{13}C NMR *119*
4.9.1.3　Fourier-Transform Infrared Spectroscopy (FTIR) *121*
4.9.1.4　Raman Analysis *122*
4.10　Examining the Size of the Biopolymers: Molecular Weight Analysis *123*
4.11　Intricacies of Understanding Lignin Structure *125*
4.11.1　^{13}C NMR *126*
4.11.2　^{31}P NMR *126*
4.11.3　2D HSQC *128*
4.11.4　Methoxyl Content Determination *132*
4.11.4.1　^{1}H NMR *132*
4.11.4.2　Hydriodic Acid *132*
4.11.4.3　Direct Methanol *132*
4.12　Questions for Further Consideration *132*
References *132*

5　Introduction to Life-Cycle Assessment and Decision Making Applied to Forest Biomaterials *141*
Jesse Daystar and Richard Venditti
5.1　Introduction *141*
5.1.1　What is LCA? *141*

5.1.1.1   History   *142*
5.1.2   LCA for Decision Making   *142*
5.1.2.1   Eco-labels   *143*
5.2   LCA Components Overview   *144*
5.2.1   Goal and Scope Definition   *145*
5.2.2   Inventory Analysis   *145*
5.2.3   Life-Cycle Impact Assessment   *146*
5.2.4   Interpretation   *146*
5.3   Life-Cycle Assessment Steps   *146*
5.3.1   Goal, Scope, System Boundaries   *146*
5.3.1.1   Goal Definition   *146*
5.3.1.2   Scope Definition   *147*
5.3.1.3   Functional Unit   *148*
5.3.1.4   Cutoff Criteria   *148*
5.3.1.5   Problems Set – Goal and Scope Definition   *148*
5.3.2   Life-Cycle Inventory   *150*
5.3.2.1   Preparation of Data Collection Based on Goal and Scope   *151*
5.3.2.2   Data Collection   *152*
5.3.2.3   Data Quality   *155*
5.3.2.4   Coproduct Treatment – Allocation   *157*
5.3.2.5   Relating Data to the Unit Process   *158*
5.3.2.6   Relating Data to the Functional Unit   *159*
5.3.2.7   Data Aggregation   *159*
5.3.2.8   LCI Data Interpretation   *159*
5.3.2.9   Problems Set – Life-Cycle Inventory   *160*
5.3.2.10 Mandatory Elements   *166*
5.3.2.11 Classification   *168*
5.3.2.12 Characterization   *169*
5.3.2.13 Optional Elements   *170*
5.3.2.14 Life Cycle Impact Assessment Interpretation   *173*
5.3.2.15 Problems Set –Life-Cycle Impact Assessment   *173*
5.4   LCA Tools for Forest Biomaterials   *177*
5.4.1   FICAT   *177*
5.4.2   GREET Model   *178*
          References   *178*

6   First Principles of Pretreatment and Cracking Biomass to Fundamental Building Blocks   *181*
      Amir Daraei Garmakhany and Somayeh Sheykhnazari
6.1   Introduction   *181*
6.1.1   What Is Lignocellulosic Material?   *183*
6.1.1.1   Lignocellulosic Materials   *183*
6.1.1.2   Cellulose   *183*
6.1.1.3   Hemicellulose   *185*
6.1.1.4   Lignin   *187*
6.2   What Difference Should Be Considered Between Wood and Agricultural Biomass?   *189*

6.2.1 Intrapolymeric Bonds *190*
6.2.2 Polymeric Inter Bonds *190*
6.2.3 Functional Groups and Chemical Characteristics of Lignocellulosic Biomass Components *191*
6.2.4 Aromatic Ring *191*
6.2.5 Hydroxyl Group *192*
6.2.6 Ether Bond *192*
6.2.7 Ester Bond *192*
6.2.8 Hydrogen Bond *194*
6.3 Define Pretreatment *194*
6.3.1 What Is the Purpose of Pretreatment? *194*
6.4 Steps of Production of Cellulosic Ethanol *195*
6.4.1 Pretreatment *195*
6.4.2 Hydrolysis *195*
6.4.3 What Are the Inhibitors for Biomass Carbohydrate Hydrolysis? *195*
6.4.4 Fermentation *196*
6.4.5 Formation of Fermentation Inhibitors *196*
6.4.6 Sugars Degradation Products *196*
6.4.7 Lignin Degradation Products *197*
6.4.8 Acetic Acid *197*
6.4.9 Inhibitory Extractives *197*
6.4.10 Heavy Metal Ions *197*
6.4.11 Separation *197*
6.5 What Are the Key Considerations for Making a Successful Pretreatment Technology? *198*
6.5.1 Effect of Pretreatment on Hydrolysis Process *199*
6.6 What Are the General Methods Used in Pretreatment? *199*
6.7 What Is Currently Being Done and What Are the Advances? *200*
6.7.1 Steam Explosion *201*
6.7.2 Hydrothermolysis *204*
6.7.3 High-Energy Irradiations *205*
6.7.4 Acid Pretreatment *207*
6.7.5 Mechanism of Acid Hydrolysis *208*
6.7.6 Alkaline Pretreatment *208*
6.7.7 Ammonia Pretreatment *210*
6.7.8 Ammonia Recycle Percolation (ARP) *210*
6.7.9 Ammonia Fiber Expansion (AFEX) *210*
6.7.10 Defects of AFEX Process *210*
6.7.11 Enzymatic Pretreatment *210*
6.7.12 Advantages of Biological Pretreatment *211*
6.7.13 Defects of Biological Pretreatment *211*
6.8 Summary *211*
References *212*

7	**Green Route to Prepare Renewable Polyesters from Monomers: Enzymatic Polymerization** *219*	
	Toufik Naolou	
7.1	Philosophic Statement *219*	
7.2	Introduction *219*	
7.3	Lipase-Catalyzed Ring-Opening Polymerizations of Cyclic Monomeric Esters (Lactones and Lactides) *220*	
7.4	Lipase-Catalyzed Polycondensation *223*	
7.4.1	Dicarboxylic Acid or Its Esters with Diols *224*	
7.4.2	Dicarboxylic Acid or Its Esters with Polyols *225*	
7.4.3	Polyesters from Fatty Acid-Based Monomers *226*	
7.4.3.1	Lipase-Catalyzed Polycondensation of α, ω-Dicarboxylic Acids and Diols *226*	
7.4.3.2	Lipase-Catalyzed Polycondensation of Hydroxy Fatty Acids *227*	
7.4.3.3	Fatty Acids as Side Chains to Modify Functional Polyesters *228*	
7.4.4	Polyester Using Furan as Building Block *229*	
7.4.5	Conclusions and Remarks *230*	
7.4.6	Questions for Further Consideration *230*	
	List of Abbreviations *230*	
	References *231*	
8	**Oil-Based and Bio-Derived Thermoplastic Polymer Blends and Composites** *239*	
	Alessia Quitadamo, Valerie Massardier and Marco Valente	
8.1	Introduction *239*	
8.2	Oil-Based and Bio-Derived Thermoplastic Polymer Blends *240*	
8.2.1	Comparison Between Oil-Based and Bio-Derived Thermoplastic Polymers *240*	
8.2.2	Thermoplastics Blends *246*	
8.3	Thermoplastic Composites with Natural Fillers *252*	
8.3.1	Wood–Plastic Composites *254*	
8.3.2	Waste Paper as Filler in Thermoplastic Composites *260*	
8.4	Conclusion *263*	
8.5	Questions for Further Consideration *264*	
	References *264*	

Index *269*

List of Contributors

Ali S. Ayoub
Archer Daniels Midland Company
ADM Research
Chicago, IL
USA

and

North Carolina State University
Department of Forest Biomaterials
Raleigh, NC
USA

Amir Daraei Garmakhany
Department of Food Science and Technology
Toyserkan Faculty of Industrial Engineering
Buali Sina University
Hamedan
Iran

Maurycy Daroch
School of Environment and Energy
Peking University
Shenzhen
China

Jesse Daystar
Department of Forest Biomaterials
North Carolina State University
Raleigh, NC
USA

Manfred Kircher
KADIB-Kircher Advice in Bioeconomy
Kurhessenstr.
Frankfurt am Main
Germany

Lucian A. Lucia
Department of Forest Biomaterials
North Carolina State University
Raleigh, NC
USA

Valerie Massardier
INSA de Lyon
IMP/CNRS 5223
Lyon
France

Toufik Naolou
Institute of Biomaterial Science and
Berlin-Brandenburg Centre for Regenerative Therapies
Helmholtz-Zentrum Geesthacht
Teltow
Germany

Alessia Quitadamo
INSA de Lyon
IMP/CNRS 5223
Lyon
France

Scott Renneckar
Department of Sustainable Biomaterials
Virginia Tech
Blacksburg, VA
USA

Noppadon Sathitsuksanoh
Department of Chemical Engineering
University of Louisville
Louisville, KY
USA

Somayeh Sheykhnazari
Department of Wood and Paper Technology
Gorgan University of Agricultural Sciences & Natural Resources
Gorgan
Iran

Marco Valente
Department of Chemical and Material Engineering
University of Rome La Sapienza
Rome
Italy

Richard Venditti
Department of Forest Biomaterials
North Carolina State University
Raleigh, NC
USA

Preface

Over the past few decades the ratio of production to new discoveries has gradually fallen and is currently estimated to about three to one. For every discovered barrel of oil, we consume three. At the same time, more and more regions of the world are seeking high-quality lifestyles that are resource intensive. Until relatively recently (about 30 years ago), high consumption of energy was reserved for the developed economies of the "West." Since then, rapid development of other countries such as China, India, and Brazil has resulted in a huge increase in demand for energy sources worldwide. The entire population of OECD countries is estimated as about 1.25 billion people, and their primary energy use as 4.37 toe per capita. When China, India, and Brazil, altogether about 2.75 billion people, approach even conservative "European" levels of fossil resources usage (3.29 toe per capita), an additional supply exceeding current use of all OECD countries will be required. It is difficult to envisage how this demand could be met with nonrenewable resources in the medium to long term. It is therefore evident that resources at our disposal are shrinking fast. Moreover, most of these petroleum polymers are not biodegradable and, thus, cannot be decomposed naturally. Furthermore, the addition of carbon dioxide to the atmosphere at the end of its life cycle has increased the need to use materials from renewable and CO_2-neutral resources. There is more carbohydrate on earth than all other organic materials combined. Carbohydrates are readily biodegradable and tend to degrade in biologically active environments like soil, sewage, and marine locations where bacteria are active. However, the basic construct of biopolymer matrices remains a virtually insurmountable obstacle to the "best laid plans of mice and men" of providing products to compete with petro-based chemicals and associated commodity items. A more robust and precise understanding of the factors that limit the widespread use of lignocellulosic substrates in society is perhaps the most pressing challenge that the emergent bio-economy faces. The goal, therefore, of this book is to elucidate the fundamental physicochemistry and characterization of the biomaterials, emphasize their value proposition for supplanting petrochemicals, tackle the challenges of conversion, and ultimately provide a milieu of possibilities for the biomaterials. The reader will be conversant and knowledgeable of the critical issues that surround the field of lignocellulosic intransigence, possible successful strategies to cope with their inertness, and potential pathways for the successful use of lignocellulosics and starch in the new bio-economy.

Turning the bio-economy into reality is more than a technical issue. From an abstract point of view, it needs scientific and technical push as well as market pull to make the bio-innovation leap. Therefore, the future role of biomass and its life cycle analysis as industrial feedstock to provide fuel and chemicals is discussed in this book with an analysis of the fossil economy, especially the chemical sector. But first and foremost it needs visionary people: devoted scientists, future-oriented entrepreneurs, a supportive political framework and last but not least a willing general public.

Ali S. Ayoub
July 2017
Chicago, USA

1

Fundamental Biochemical and Biotechnological Principles of Biomass Growth and Use

Manfred Kircher

KADIB-Kircher Advice in Bioeconomy Kurhessenstr, Frankfurt am Main, Germany

> *For the first time in history, we face the risk of a global decline. But we are also the first to enjoy the opportunity of learning quickly from developments in societies anywhere else in the world today, and from what has unfolded in societies at any time in the past.*
>
> Jared Diamond, 2005

1.1 Learning Objectives

This chapter discusses about vegetable biomass and its future role as industrial feedstock to provide fuel and chemicals. In the transition phase from the current fossil-based into the bio-based economy, vegetable biomass needs to face up to competition against the fossil benchmark, which is at mineral oil. Therefore, this chapter starts with an analysis of the fossil economy, especially in the chemical sector.

In future, when fossil feedstock inevitably becomes scarce and the bio-economy increasingly unfolds, vegetable biomass must meet the industrial feedstock demand for a growing global population. While further serving the traditional food, feed, and fiber markets, this is no easy challenge. More sustainable carbon sources and applications are another topic of this chapter.

Turning the bio-economy into reality is more than a technical issue. From an abstract point of view, it needs scientific and technical push as well as market pull to make the bio-innovation leap. But first and foremost, it needs people with visionary: devoted scientists, future-oriented entrepreneurs, a supportive political framework, and last but not least a willing general public. These so-called pillars of competitiveness are presented as well.

The learning objectives of this chapter are

1. the significance of carbon in our economy;
2. the fundamental biochemical and biotechnological principles of fossil- and bio-based carbon sources concerning nature, production, and processing; and
3. the complex challenges in making vegetable biomass the dominant sustainable feedstock.

Introduction to Renewable Biomaterials: First Principles and Concepts, First Edition.
Edited by Ali S. Ayoub and Lucian A. Lucia.
© 2018 John Wiley & Sons Ltd. Published 2018 by John Wiley & Sons Ltd.

1.2 Comparison of Fossil-Based *versus* Bio-Based Raw Materials

1.2.1 The Nature of Fossil Raw Materials

The current global economy is very much based on fossil resources to produce energy (electricity, fuel, heat) and organic chemicals. The initial source of these feedstock has been biomass transformed through geological processes into crude oil, natural gas, black coal as well as lignite and peat. What makes these materials valuable for use in energy and chemistry processes is their high energy as well as carbon content (Table 1.1). The most valuable fossil resources are the hydrocarbons that consist only of carbon and hydrogen. Subgroups are, for example, alkanes (saturated hydrocarbons; C_nH_{2n+2}), cycloalkanes (C_nH_{2n}), alkenes (unsaturated hydrocarbons; C_nH_{2n}), and aromatics (ring-shaped molecules) differing in the number of carbon and hydrogen and molecular structure.

Coal, especially black coal, is the oldest fossil resource. Formed from terrestrial plant biomass, it has been consolidated between other rock strata and altered to form coal seams by the combined impact of pressure and heat under low-oxygen conditions over about 300 million years. Black coal is extracted by open-cast mining as well as deep mining (up to a depth of 1500 m). It is composed primarily of carbon.

Fossil oil has been formed over a time period of about 100 million years by the exposure to similar conditions on sedimentation layers of marine organisms such as algae and plankton. Under such conditions, the long-chain organic molecules of the vegetable biomass are split into short-chain compounds forming liquid oil. It accumulates in specific geological formations called crude oil reservoirs.

Some fractions even split down to molecules with only one carbon and become gaseous methane (CH_4). Therefore, oil deposits (and coal mines) always contain methane of more or less similar age. Methane sources covered by nonpermeable geological layers lead to real methane deposits. From such geological formations, the gas can be extracted in the form of natural gas. Natural gas can also be the result of biological catabolic processes degrading biomass. These deposits are also found under nonpermeable geological formations but have been formed over a period of about 20 million years.

As oil and gas generation needs high-pressure conditions the corresponding deposits are highly pressurized. If such sites are drilled, oil and gas escape through the well – a process called primary recovery allowing to exploit 5–10% of the total oil and gas. By

Table 1.1 Composition (%) and heat value (MJ kg^{-1}) (Herrmann and Weber, 2011) of fossil feedstock.

	C	H	N	O	S	MJ kg^{-1}
Natural gas	75–85	9–24	Traces	Traces	Traces	32–45
Mineral oil	83–87	10–14	0.1–2	0.5–6	0.5–6	43
Black coal	60–75	6	Traces	17–34	0.5–3	25–33
Lignite	58–73	4.5–8.5	Traces	21–36	3	22
Peat	50–60	5–7	1–4	30–40	0.2–2	15

pumping (secondary recovery) and more sophisticated methods (tertiary recovery) more oil and gas can be extracted. Obviously exploiting an oil and gas deposit is easy in the beginning but becomes more and more technically complex and costly with time.

Lignite has a similar origin as black coal. It has been exposed to the harsh geological conditions for up to 65 million years and can be extracted by open-cast mining. The carbon content is lower than that in black coal, but extraction costs are in average more beneficial.

Peat is another fossil resource. It is as well formed from terrestrial plants under aplent moor conditions when the biomass decays for several 1000 years under low-oxygen conditions. Peat contains the lowest carbon and highest water share under fossil resources. It is recovered from ground.

All fossil resources have the following common characteristics: (i) they are rich in carbon and energy; (ii) their composition is not very complex and quite homogeneous; (iii) they can be produced at moderate, though growing cost; and (iv) fossil resources can be shipped easily by railway, tankers, and pipelines.

1.2.2 Industrial Use

1.2.2.1 Energy

All fossil feedstocks are characterized by high energy content. By oxidation (adding oxygen) the chemical energy stored in the molecules is released in the form of heat – a process called burning in everyday language. Therefore, fossil feedstock is an efficient and easy material to produce energy. In 1709, it was used for the first time in England for industrial purposes when black coal instead of wood-based charcoal was used for iron melting in a coke blast furnace. Discovering this energy source came just in time to start metal-based industrialization because charcoal production had significantly decimated the area under forests. Since then black coal is one of the most relevant primary energy carriers. In 1859, the Pennsylvania Rock Oil Company drilled the first oil well in Titusville (Pennsylvania, USA). Only 10 years later, John. D. Rockefeller founded the Standard Oil Company in 1870, thus starting the era of multinational companies serving the global energy markets. Gas exploitation followed in 1920 in the United States and in 1960 in Europe. Table 1.2 shows the share of fossil material use in different global regions.

In summary, production of heat, fuel, and electrical power from fossil resources has been the starting point of industrialization and is still today by far the dominant application. Ninety-three percent of oil, 98% of gas and coal, and 100% of peat are going into energy markets (Höfer, 2009a; Ulber *et al.*, 2011b). It is estimated that even in 2040

Table 1.2 Use of fossil feedstock in different global regions (%) (EKT Interactive Oil and Gas Training, 2014).

	North America	Europe	Asia Pacific
Mineral oil	55	40	37
Natural gas	30	38	12
Coal	15	22	51

Table 1.3 Feedstock mix (%) in German chemical industries (2011) (Benzing, 2013).

Naphtha	Natural gas	Coal	Bio-based feedstock
71	14	2	13

mineral oil, natural gas, and coal will serve more than three-fourths of total world energy supply (US Energy Information Administration, 2013). The mix of fossil feedstock differs among global regions dependent on regional resources and trade routes.

1.2.2.2 Chemicals

The cheap and seemingly unlimited availability of fossil resources not only triggered an energy-hungry industrialization but also the innovation leap into today's chemical industry. High carbon content in combination with easy logistics through pipelines and tankers made especially oil and gas an ideal industrial feedstock. Seven percent of the global oil and about 2% of world natural gas consumption go into chemicals demonstrating that fossil oil still dominates the global chemical industry (70–80% of chemicals are derived from oil, 8–10% from gas, 10–13% from biomass, and only 1–2% from coal; compare Table 1.3)

Since ancient times chemicals and biochemicals had been produced from natural reservoirs or from biological resources, respectively. For example, sodium carbonate was imported by Europe from soda lakes in Egypt and Turkey or extracted from water plants. The alkaline solution of soda ash (sodium carbonate) is in fact named after Arabic "al kalja" for the ashes of water plants. In 1771, an alternative method changed the world when Nicolas Leblanc (1742–1806) in France invented the chemical synthesis of sodium carbonate by using coal as the carbon source. This real innovation is today acknowledged as the starting point of chemical industries.

Structural Materials Since the mid-nineteenth century natural product chemistry tried to use biomaterials as a feedstock to organic chemicals. For example, cellulose, the most abundant plant polysaccharide, has been investigated intensively. The fact that in 1846 three German chemists simultaneously but independently invented a method to produce nitrocellulose from cellulose demonstrates how the time was right for such an innovation. It marked the change from biomaterials to bio-based materials. Though highly inflammable, nitrocellulose entered the market with great success because it was able to replace expensive materials such as whale baleens especially used in ladies costumes as well as silk. Later in 1910 viscose was developed from cellulose in Germany as a fiber material that is still in use. Another example of the efforts to gain independence from natural starting materials by developing synthetic materials is the invention of galalith plastic from casein in 1897 again by German chemists. Obviously, there was a market waiting for more materials from modified biological sources, and there were scientists exploring chemistry.

Dyes Whereas the examples mentioned so far represent more structural materials for fibers and tissues, instruments, and housing, the next group demonstrates the boosting power of added-value chemicals. Color design mostly does not directly determine the utility of a product, but it adds value and makes a difference. Since ancient times dyestuffs were produced at considerable expense from plants, animals, and minerals. At the end of the nineteenth century, chemistry paved the way to cheap dyestuffs and a world full of colors for the first time, thus ending the industrial era of dye plants. The synthesis of the red dye Alizarin in 1869 by German chemists Carl Graebe (1841–1927) and Carl Liebermann (1842–1914) replaced the natural dye made from dyer's madder (*Rubia tinctorum*) within a short time period. Alizarin became one of the first products of BASF, founded in 1865 in Mannheim, Germany, by Friedrich Engelhorn (1821–1902). Another red dye, fuchsin, first synthesized in 1858 became the starting point for Hoechst AG, founded by Carl Friedrich Wilhelm Meister (1827–1895), Eugen Lucius (1834–1903), and Ludwig August Müller (1846–1895) in Hoechst close to Frankfurt and only 80 km (50 miles) from Mannheim. In 1878 followed Indigo, another synthetic dye, which was developed by Adolf von Baeyer (1835–1917). Indigo gained industrial relevance at BASF and Hoechst when Johannes Pfleger (1867–1957), chemist at Degussa AG in Frankfurt/Main, improved the process economics significantly. Until the early twentieth century, dye products were dominating commercial chemistry and even the whole industry was called dye chemistry.

Receptive markets and growing chemical science were now joined by entrepreneurs. It is important to understand the significance of these three factors working together. But in the end, industry is made by competent individuals who complement each other, build friendship, realize the business option, and take the chance. The men mentioned here – many friends since university studies – formed such a network that became the starting point of the German chemical industry.

Drugs As of today successful companies use scientific and technical competence to broaden their product portfolio, develop new application fields, and enter profitable markets. In the early twentieth century, the potential of synthetic drugs had been realized and especially the German dye industry started to invest in research and development. Arsphenamine (Salvarsan®), a syphilis drug, developed by the German physician Paul Ehrlich (1854–1915) and the Japanese bacteriologist Sahachiro Hata (1873–1931) in 1910 became a cash cow to Hoechst AG. In 1935 followed Prontosil®, the first sulfonamide developed by Fritz Mietzsch (1896–1958) and Josef Klarer (1898–1953) at Bayer AG in Wuppertal. Noticeably, this chemical group is also used as azodyes demonstrating how competence in a specific field can lead to a spillover invention in a very different application. Gerhard Domagk (1895–1964) discovered the antibacterial effect and received the Nobel Prize in 1939. These examples not only demonstrate how gaining experience in synthetic chemistry in one field (materials, dyes) led to exploring very different markets (pharmaceuticals) but also how chemical industries early integrated microbiological competence.

The pharmaceutical business opened the door for biotechnology in chemical industries when the Scottish bacteriologist Alexander Fleming (1881–1955) explored antibiotics in 1928. He realized that the fungi *Penicillium* secretes the antibiotic penicillin, a discovery that was honored with the Nobel Prize in 1945. Since 1942 in England Glaxo (pharma company; founded in 1873 and originally in the baby food business) and ICI (chemical industry, founded in 1926) but especially in the US Merck & Co (1917; separated from Merck KGaA, a German pharma company founded in 1668) and Pfizer & Co (founded in 1849; biological pesticides) developed fermentative production processes based on the cultivation of *Penicillium chrysogenum*. Companies with very different backgrounds in chemistry, synthetic drugs, and food production got involved in developing early fermentation methods. It should be emphasized that those companies focused on fermentation because there was no technical alternative. Penicillin antibiotics were not available by chemical synthesis. The production of penicillin is therefore seen as the starting point of industrial biotechnology (in contrast to traditional food biotechnology using microbial processes such as yogurt, beer, and wine fermentation).

Drugs added a quite different quality to the chemical industry's product portfolio. This chemical product sector is characterized by extremely high functionality to fight diseases, thus adding real value and commercial profit. In addition, this sector is extremely knowledge based – documented by Nobel Prize–winning research.

Polymers A combination of extensive science and the availability of carbon sources triggered in the 1930s another chemical success story: polymers. Increasing capacities in oil refineries not only provided gasoline and diesel but with naphtha (Table 1.4) also the fraction of long-chain hydrocarbons to be cracked down to methane, ethylene, and propylene. Platform intermediates like these are till today the biggest chemicals by production volume. Their carbon content is the share of carbon of the molecule's molecular mass (g mol^{-1}). Ethylene, for example, consists of two carbon (atomic mass 12 u) and four hydrogen atoms (atomic mass 1 u), which gives a molecular mass of 28 g mol^{-1} and a share of carbon of 85.7% (Table 1.5).

Not only the availability of a cheap and easy-to-handle feedstock pushed chemical industries but also the often highly advantageous stoichiometric product yield. For

Table 1.4 Oil-refinery output from low to high distillation temperature.

25 °C	>	>	>	>	>	350 °C
Refinery gas	Gasoline	Naphtha	Kerosene	Diesel oil	Fuel oil	Residue
Bottled gas	Automotive fuel	Chemical feedstock	Aircraft fuel	Truck fuel, bus fuel	Ship fuel, power generation	Bitumen for road construction

Table 1.5 Global production volume of bulk chemicals (2010) (Davis, 2011) and content of carbon.

Chemical category	Chemical		C (%)	Production (million tons)	C content (million tons)
Olefins	Ethylene	C_2H_4	85.7	123	105
	Propylene	C_3H_6	85.7	75	64
	Butadiene	C_4H_6	88.9	10	9
	Hexane	C_6H_{14}	83.7	5	4
Aromatics	Xylenes	C_8H_{10}	90.6	43	39
	Benzene	C_6H_6	92.3	40	37
	Toluene	C_7H_6	91.3	20	18

example, ethylene (MW 28.05 g mol^{-1}) and propylene (42.08 g mol^{-1}) are available from hexane (86.18 g mol^{-1}) with a yield of 98% kg kg^{-1}.

$$C_6H_{14} \rightarrow 3C_2H_4 + H_2$$
$$C_6H_{14} \rightarrow 2C_3H_6 + H_2$$

Already in 1912 the Chemische Fabrik Griesheim-Elektron (later a production site of Hoechst AG) close to Frankfurt (Germany) tried to find new applications for ethylene, which was produced by oil refineries in big amounts. Finally, the chemist Fritz Klatte (1880–1934) synthesized vinyl chloride from acetylene (C_2H_2; synthesized by dehydrating ethylene) and hydrogen chloride. From 1928 (several companies in the United States; 1930 BASF in Germany) started production and polymerization to polyvinylchloride (PVC) on large scale. PVC became the first synthetic material not starting from any natural building block and a real milestone in chemical innovation, which had been induced by the availability of a new feedstock. Nylon, patented in 1935 by the chemist Wallace Hume Carothers (1896–1937) at E. I. du Pont de Nemours in Wilmington (Delaware, USA), turned out to be the next big step in polymer innovation. The theoretical base of polymer chemistry had been laid at the University of Freiburg (Germany) by Hermann Staudinger (1881–1965) who received the Nobel Prize in Chemistry in 1953. Today, polymers represent the biggest chemical product group in a volume of 241 million tons in 2012 (Statista, 2013). China leads with a market share of 23.9%, followed by Europe (20.4%) and the NAFTA region (19.9%).

With polymers the chemical industry finally left also in the field of bulk chemicals the level of craftsmanship, which had characterized this industry in the beginning. From then on science and fast advance in knowledge (documented in patents) became a primary competitive driver (Table 1.6).

Table 1.6 Milestones in chemical innovation.

	1900	1920	1940	1960	1980
Pharmaceuticals	Salvarsan® Aspirin®		Antibiotics	Birth-control pill	Anti-AIDS protease-inhibitor
Paints and coatings		Acryl lacquer			Water-based lacquer
Adhesives	Phenolic resin			UV-crosslinked adhesives	Solvent-free adhesives
Surfactants				Biologically degradable tensides	Phosphate-free tenside
Polymers	Synthetic rubber Viscose	Nylon	Teflon Styropore		Microfibers
Agrochemicals	Haber–Bosch process				Linking herbicide and plant breeding
Energy			Solar cell		

Table 1.7 Cost of oil production (US$ per barrel) (Birol, 2008).

Near East	North America	Deep sea	Enhanced oil recovery	Arctic
3–14	10–40	32–65	30–82	32–100

1.2.3 Expectancy of Resources

Common sense suggests that fossil resources are limited and will be consumed eventually. From a physical point of view, such a statement sounds simple and is absolutely right. Economically, it is more complex because geological resources differ in cost of exploitation (Table 1.7).

Geological deposits too costly to be explored today may become competitive tomorrow. An example is today's shale gas boom especially in the United States and the earlier oil sand exploitation in Canada. Both deposits remained untouched and were not included in oil statistics for decades but reached competitiveness because the rising oil price allowed more expensive oil production methods. Therefore, we need to differentiate between reserves and resources. Resources define the total volume of fossil feedstock deposited underground, whereas reserves give an idea of what is exploitable today with the state-of-the-art profitable methods. Economists therefore calculate the "static lifetime," which is the time range within which a given feedstock will be available under current economical conditions with current technical means under consideration or the current consumption.

The total resources in fossil oil are estimated to amount to 752 billion tons. Out of this volume, 383 billion tons is known as exploitable with today's technical means at feasible coast; 167 billion tons or 44% has already been delivered since the beginning of industrial oil production. About 4 billion tons is produced annually. Nonetheless, oil resources are of course limited but are not to be expected to run out within a short term. The very same is true for gas and coal (Table 1.8). Static lifetime expectancy is an important issue because as long as fossil feedstock is on the market it will be the competitive benchmark for bio-based raw materials.

1.2.4 Green House Gas (GHG) Emission

Nevertheless, in view of the climate change, we need to ask whether it is wise to use fossil resources completely. Undoubtedly, producing energy from oil, gas, and coal by burning

Table 1.8 Static lifetime (years) of fossil resources (Harald Andruleit, 2011).

	Static lifetime of reserves	Static lifetime of resources
Mineral oil	54	146
Natural gas	59	233
Black coal	114	2712
Lignite	282	4400

Table 1.9 Annual CO_2 emission from various fossil feedstock (million tons; 2012) (Marland, Boden, and Andres, 2007; Olivier et al., 2014).

	Mineral oil	Natural gas	Coal	Sum
CO_2 emission	14,500	6840	13,160	34,500
Carbon content	4000	1900	3500	9400

leads to CO_2 (molecular mass $44\,\text{g}\,\text{mol}^{-1}$), which is emitted into the atmosphere; 27.3% of it is carbon (Table 1.9).

As atmospheric CO_2 reduces global infrared emission into space the consumption of fossil resources has a warming effect on the atmosphere, which is broadly agreed to contribute to man-made (anthropogenic) climate change. Due to the already occurred emission an increase in global temperature by $1.3\,°C$ seems unavoidable in the long run of which $0.8\,°C$ increase is already proven (because of the climate system's inertia it is a slow process). However, to limit global warming to $2\,°C$ CO_2 emission should not exceed a cumulative volume of 750,000 million tons till 2050 (Wicke, Schellnhuber, and Klingefeld, 2012). This is equivalent to only 21 years of current emission activity of 34,500 million tons. Already the common people are affected by the climate change by sea-level rise in Bangladesh, desertification in Spain, and drought in the United States. Climate change is one of the most pressing current issues forcing governments and industries to reduce the consumption of fossil resources.

1.2.5 Regional Pillars of Competitiveness

When looking on the global map of fossil resources, it is interesting to note that the sites of deposits and production (Middle East, North America, Russia) are mostly not identical with the sites of processing (Figure 1.1). For example, Belgium, Germany, and Netherlands are among the five biggest global chemical regions. Because this region depends on importing oil, it is called after its harbors and rivers which, however, not only serve as the logistics backbone but also as production sites: ARRR (Antwerp, Rotterdam, Rhine, Ruhr).

Although it must be considered that the starting point of industrial activities in this region has been the availability of coal and a little fossil oil the ongoing success of its industries does not depend on feedstock directly on site. More relevant is an efficient regional logistics system for high-volume feedstock imports and processed goods exports through railroad, pipeline, and river and sea transport. Other equally relevant pillars of competitiveness are academic research and education facilities, skilled workforce, effective governmental and public administrative institutions, and last but not least public acceptance.

How the integration of these factors leads to the innovation leap of successful industries producing marketable goods, creating jobs, and inducing a real innovation cycle with a continuous product pipeline is demonstrated by the history of chemical industries. In the nineteenth century, Germany's universities trained excellent chemists who often kept lifelong friendship and formed an effective business network. They

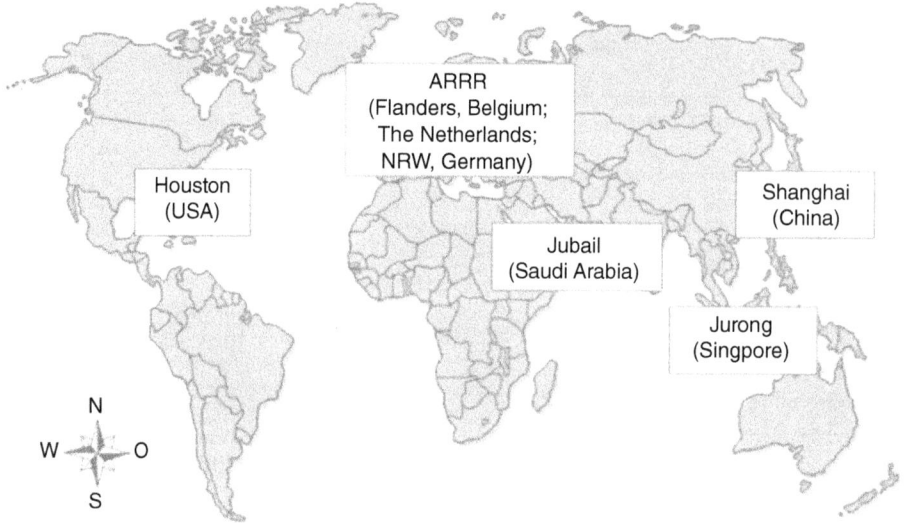

Figure 1.1 Global chemical clusters.

used the new raw material of mineral oil, which was easily available along the river Rhine to develop products for receptive markets like dyestuff and more. With academic excellence, entrepreneurs and investors started production facilities for a society honoring innovation. Nobel Prize and global players in chemical industries were the result. Similar chemical clusters evolved in the United States and Japan (ranking today number 1 and 2). China surpassed Germany a few years ago; its industry grew in the beginning due to beneficial cost but has increasingly gained relevance also because of top-ranking science. Germany's chemical industry still ranks number 4 (Table 1.10). When looking at global regions, the Asian chemical industry is today leading (Table 1.11) especially due to China.

Table 1.10 Chemical industry nation's sales and market share (2013).

	China	United States	Japan	Germany	Brazil
Sales (billion US$)	1245	815	335	250	145
Share (%)	24.9	16.3	6.7	5.5	2.9

Table 1.11 Chemical industry region's sales and market share (2013).

	Asia	EU(27)	NAFTA	Latin America	Africa
Sales (billion US$)	2435	1070	920	260	55
Share (%)	48.7	21.4	18.4	5.2	1.1

1.2.6 Questions for Further Consideration

- What makes fossil feedstock a valuable industrial feedstock?
- What are the most important applications of fossil feedstock? What is their share in fossil feedstock use?
- What are key success factors of leading fossil-based chemical production sites?
- Should fossil feedstock be used till running out? Why not?

1.3 The Nature of Bio-Based Raw Materials

Bio-based raw materials for producing energy and chemicals are provided by agriculture (plant cultivation and animal breeding), forestry, and from marine resources. Plant products and vegetable biomass from agriculture and forestry are most relevant today and will be tomorrow.

Vegetable oil appears in the form of fatty acid esters of glycerol (triglycerides). A typical example is linoleic acid ($C_{18}H_{32}O_2$).

Sugar defines a group of carbohydrates. Monosaccharides include glucose ($C_6H_{12}O_6$), fructose ($C_6H_{12}O_6$), and galactose ($C_6H_{12}O_6$). Disaccharides consist of two sugar molecules such as sucrose ($C_{12}H_{22}O_{11}$; fructose + glucose). Longer chains of sugars are called oligo- or polysaccharides.

Starch is a polysaccharide ($C_6H_{10}O_5)_n$ consisting of α-D-glucose units. It represents one of the most relevant plant reserve molecules stored in special organelles (grain kernel, corn cobs, potato tuber). Most relevant starch crops are wheat, corn, potato, and manioc.

Lignocellulose is the basic material of plant biomass. It is composed of carbohydrate polymers (cellulose (($C_{12}H_{20}O_{10})_{[n]}$) made of glucose dimers, hemicellulose made of D-xylose ($C_5H_{10}O_5$) and L-arabinose ($C_5H_{10}O_5$)) and an aromatic polymer (lignin). The carbohydrate polymer fraction contains different sugar monomers (six and five carbon sugars). Lignocellulose is the most abundant plant material available, for example, from agricultural crops and residuals, forest trees, or steppe vegetation.

Vegetable biomass is characterized by (i) complex polymeric structures and (ii) compound diversity and (in contrast to fossil materials) the presence of oxygen (Table 1.12)

1.3.1 Oil Crops

Oil crops deliver vegetable oil consisting in principle of saturated fatty acids, monounsaturated fatty acids and polyunsaturated fatty acids. Oil crops deposit fatty acids in the seed from which it is extracted. The remaining meal is often rich in protein and used as feed additive.

Soybean (*Glycine max*) delivers oil and protein. The oil content of the seed varies between 14% and 24% and is used for food (cooking oil, salad oil, margarine) and industrial applications whereas protein goes into feed. Linoleic acid ($C_{18}H_{32}O_2$; 49–47% of fatty acids), oleic acid ($C_{18}H_{34}O_2$; 18–25%), and linolenic acid ($C_{18}H_{30}O_2$; 6–11%) are the most relevant fatty acids. The spectrum of fatty acids is the subject of breeding efforts, as especially linolenic acid causes problems concerning oxidation and undesired flavor. After extracting oil from the soybean the bean meal is left. It is rich in protein

Table 1.12 Composition (%) (Michelsen, 1941) and heat value (MJ kg^{-1}) (Herrmann and Weber, 2011) of vegetable biomass and biomass compounds.

	C	H	N	O	S	MJ kg^{-1}
Vegetable biomass (average)	45	6	2	42	Traces	6.8
Wood	50	6	3	41	Traces	14.4–15.8
Peat	55	3	5	37	Traces	4.2
Linoleic acid	77	12	0	11	0	39.1
Glucose	40	7	0	53	0	15.6
Sucrose	43	6	0	51	0	
Starch	44	6	0	50	0	17.5
Lignocellulose	44	6	0	50	0	10–25

(42–47%) and therefore a valuable feed additive. Soybean is cultivated worldwide with highest acreage in the United States and Brazil.

Rapeseed (*Brassica napus*) delivers oil (40–50%) and protein (20–25%). Wild-type (meaning the wild variety) rapeseed contains erucic acid ($C_{22}H_{42}O_2$) with a share of 25–50% among the fatty acids and some glucosinolate (glucoside containing sulfur and nitrogen; a plant defense active against pests). Both compounds have a negative nutritional effect, thus preventing the use of wild-type rapeseed in food and feed applications. Plant breeding reduced the content of both compounds and today's cultivars produce 52–66% oleic acid, 17–25% linoleic acid, and 8–11% linolenic acid. Rapeseed meal contains 33% protein and is a valuable feed component. Rapeseed is cultivated especially in Europe.

Sunflower (*Helianthus annuus*) is an annual crop producing seeds with an oil content of 50%. The main fatty acid components are linoleic acid (55–73%) and oleic acid (14–34%). Extracted meal contains 40–45% and is used both in food and feed applications. Sunflower is especially grown in Russia and Ukraine.

Oil palm (*Elaeis guineensis*) is a palm tree cultivated in plantations. It produces up to 15 years (first time 3 years after planting). Fatty acids accumulate in the fruit pulp as well as in the seed kernel making up 45–50% of the fruit. The pulp fatty acids consist of palmitic acid (44%; $C_{16}H_{32}O_2$), oleic acid (39%) linoleic acid (11%), and some minor fatty acids. Palm kernel is especially rich in saturated fatty acids, mainly lauric acid (48%; $C_{12}H_{24}O_2$) and stearic acid (16%; $C_{18}H_{36}O_2$). Malaysia and Indonesia are leading producers of palm oil.

Jatropha (*Jatropha*) grows as plant, shrub, and tree. Trees that are cultivated for producing vegetable oil grow even on poor soil. Fatty acids are deposited in the seed to up to 30% consisting of 30–52% linoleic acid, 30–44% oleic acid, 15–17% palmitinic acid, and 6–8% stearic acid. Because seeds contain toxic compounds Jatropha has not been domesticated so far and yields are variable. The oil is highly suitable for fuel applications and therefore Jatropha gains increasing industrial interest. Currently, Jatropha oil is especially produced in India, Indonesia, and China. Table 1.13 lists the global consumption of major vegetable oils.

Table 1.13 World consumption of major vegetable oil (2007/2008) (USDA, 2009) and carbon content (75% average assumed).

	Oil (million tons)	C content (million tons)
Palm oil	41.31	30.98
Soybean oil	41.28	30.96
Rapeseed oil	18.24	13.68
Sunflower seed oil	9.91	7.43
Peanut oil	4.82	3.62
Cottonseed oil	4.99	3.74
Palm kernel oil	4.85	3.64
Coconut oil	3.48	2.61
Olive oil	2.84	2.13

1.3.2 Sugar Crops

Sugarcane (*Saccharum officinarum*) is a multiannual grass, which deposits sucrose (70–88%), glucose (2–4%), and fructose (2–4%) in its stalks. Stalks contain 73–76% water and 24–27% solid materials of which 10–16% are soluble. This soluble fraction contains the sugar. It is extracted by repeated chopping, shredding, and washing the stalks. Processed stalks (bagasse) consist of 40–60% cellulose, 20–30% hemicellulose, and 20% lignin. Today bagasse is burned to produce heat and power for the sugar mill. Sugarcane can be harvested up to 10 times before replanting. Up to 10 tons sugar per hectare is produced. Sugarcane is especially grown in Brazil and India (Table 1.14).

Sugar beet (*Beta vulgaris*) is a biennial beet, which accumulates about 20% sucrose in its root. Water content of the beet is about 75–78%. Nine tons of sugar is produced per hectare of sugar beet. Beet leaves can be used as feed. It is cultivated in Europe and Russia.

Table 1.14 The biggest sugar producers, production volume (2012) (USDA, 2013), and carbon content (43% C in sucrose assumed).

Country	Production (million tons)	C content (million tons)
Brazil	38.6	16.6
India	27.4	11.8
Europe	15.6	67
China	14.0	6.0
Thailand	9.9	4.3
United States	8.2	3.5
Mexico	6.6	2.8
Russia	5.0	2.2

1.3.3 Starch Crops

Corn (*Zea mays*) is an annual grass that forms male terminal tassels and female cobs. The cobs develop 8–18 rows with 25–50 maize grains in each row. Grains accumulate starch in up to 62% of fresh matter. The rest consists of 10% protein, 5% fat, and 5% fiber, vitamins, minerals, sugar, and water. Corn grains are at first used as starch source. Corn grain yield reaches 10 tons per hectare. If used as feed special varieties are harvested as whole plant biomass with a yield of up to 50 tons fresh biomass per hectare (15–20 tons per hectare dry mass). The biggest corn grower is the United States providing about 32% of global harvest (Table 1.15).

Potatoes (*Solanum tuberosum*) grow perennially and accumulate starch in tubers. The fresh matter content is about 15% starch (79% amylopectin, a highly branched polymer of glucose; 21% amylose, a helical polymer made of α-D-glucose). Top yields in Germany reach 6 tons of starch per hectare. Potato was at first grown for food, but later for industrial purposes as well. After starch processing, the remaining potato pulp contains hemicellulose, cellulose, protein, and pectin. It is used in enzyme production, fungi cultivation, and as fertilizer and feed additive. A special variety depositing only amylopectin for industrial markets has been bred recently by traditional breeding. A similar variety developed by genetic engineering did not enter the European market because of lack of public acceptance. Potato is grown in Europe, Russia, and China (Table 1.16).

Wheat, barley, and rye (Gramineae family) are grasses depositing starch (28% amylose, 72% amylopectin) in the grain. Starch makes 58% of the grain (fresh matter content) with 15% water plus cellulose, hemicellulose, and lignin of the seed hull. Up to 4 tons of

Table 1.15 Most important corn-producing nations 2012 (Statista, 2014) and carbon content (43% C in sucrose assumed).

	Production (million tons)	Starch content (million tons)	C content (million tons)
United States	274	170	75.4
China	208	129	57.3
Brazil	71	44	19.5
EU	59	37	16.2
Mexico	22	14	6.1

Table 1.16 Most important potato producing nations 2009 (Landesverband der Kartoffelkaufleute Rheinland-Westfalen, 2013).

	Production (million tons)	Starch content (million tons)	C content (million tons)
China	69	10.3	4.6
Russia	36	5.4	2.4
India	34	5.1	2.3
Ukraine	20	3.0	1.3
United States	20	3.0	1.3

Table 1.17 Global starch crop production 2013 (FAO, 2014) (except potato; 2012) and theoretical starch and starch–carbon content according composition given in the text.

	Production (million tons)	Starch content (million tons)	C content (million tons)
Corn	1016	630	279.8
Potato	368	55	24.4
Wheat	713	413	183.6
Barley	144	84	37.1
Rye	16	9	4.1

starch per hectare is produced with wheat. Seed hull and straw are the most important by-products of cereal processing. Wheat is the most important crop and grown worldwide. Barley (*Hordeum vulgare*) and rye (*Secale cereale*) are especially grown in Europe and Russia, whereas sorghum (*Sorghum*) is cultivated in the United States, Africa, and Asia. Triticale (hybrid of wheat and rye) is grown similar to rye (Table 1.17).

Starch crops give an overview of the different characteristics of plants and the wide range of applications. Fresh potatoes come with about 85% of moisture but with more than 80% of starch, about 10% protein, and 5% fiber in dry matter. In contrast, corn kernels are relatively dry (15% moisture) come with a little less starch in dry matter (80%) but contain more protein and lipids. Lack of moisture and the high protein and lipid content make corn a more competitive industrial feedstock. About 80% of global starch production is corn based. The product portfolio available from corn is shown in Table 1.18.

1.3.4 Lignocellulosic Plants

Miscanthus (*Miscanthus*) is a grass related to sugarcane and millet. It reaches up to 4 m in height and dry mass yield of 15–20 tons, even up to 30 tons per hectare. It is harvested like grass for the first time after 3 years and can be cultivated for at least 20 years. The

Table 1.18 Approximate yield derived from 1 ton no. 2 yellow corn with 15.5% moisture (International Starch Institute, 2014).

Derived from corn kernel (1 ton)	Derived from starch (0.625 ton)	Dry matter (%)	kg	Market
Starch		88	625	Food, industries
	Dextrose	91.5	714	Food
	HFCS	71	845	Beverage
	Ethanol		(417 l) 329	Mobility
Gluten meal		89	50	Feed
Corn oil		100	27	Food

HFCS, high-fructose corn syrup.

biomass consists primarily of lignocellulose. Miscanthus is grown in Canada yielding up to 44 tons dry mass per hectare. Today, it is utilized thermally but the future potential is seen in second-generation energy carriers and chemicals. It can be cultivated on grassland (globally 3.55 billion hectare (Raschka and Carus, 2012))

Woody biomass from short-rotation trees like poplar yields 10–15 tons per hectare annually within 4–6 years. The trees sprout after harvest again over a period of at least 20–30 years and are harvested every 2 years. Lignocellulosic biomass consists of 42–49% cellulose, 24–30% hemicellulose, and 25–30% lignin. It is today used for generating heat by burning. The material is quite identical with forestry waste (hardwood contains 42–51% cellulose, 27–40% hemicellulose, 24–28% lignin) or oil palm residues which is also basically lignocellulose. Its future potential is in second-generation energy carriers and chemicals. Under European conditions, forests add 12.1 m^3 wood per hectare per year with an average density of 470 (pine) to 690 kg m^{-3} (oak). Global forest area is about 4 billion hectare with 264 million hectare planted (Adams, 2012).

1.3.5 Lignocellulosic Biomass

Straw from cereals represent another lignocellulosic biomass. Today, it is used for livestock bedding and more low-value applications. As it is a by-product of cereal agriculture, thus part of an existing harvesting process, it draws special attention as a future second-generation carbon source. With an average cereal/corn relation of 1 to 1 100 million tons of cereal comes with about 100 million tons of straw. As 20–30% of the straw should be left on the ground for rotting and keeping the soil fertile about 70–80 million tons could be available from 100 million tons of cereal biomass for industrial use.

1.3.6 Algae

Microalgae consist of a single or only a few cells and produce fatty acids, carbohydrates, and some special ingredients by photosynthesis. Algae are cultivated commercially only for special compounds like astaxanthin, which is added to fish and poultry feed giving salmon and egg yolk its typical color. As biodiesel gets growing commercial attention, algae are evaluated as a source of lipids (up to 70% lipids of fresh algae biomass). Microalgae are seen superior to land plants due to high rate of growth, high yield per hectare, high energy content (Table 1.19), lack of lignin, and low cellulose content. They are cultivated in open ponds or specific photoreactors. In both the methods, algae are grown under light and supplied with atmospheric CO_2 or CO_2 emitted by an industrial plant. Therefore, such production systems neither occupy fertile land nor

Table 1.19 Performance of microalgae, corn, and short-rotation trees (Fachagentur nachwachsende Rohstoffe, 2012).

Biomass	Yield (ton dry mass ha^{-1} year^{-1})	Fixed CO_2 (ton ha^{-1} year^{-1})	Energy (MWh ha^{-1})	Energy (MJ kg^{-1})
Microalgae (open pond)	40–60	66–100	166–250	9.0
Microalgae (reactor)	80–120	130–200	333–500	9.2
Corn (whole biomass)	15–20	25–35	75–100	10.8
Short-rotation trees	6–20	10–35	30–100	10.8

compete with food production. However, handling the large volume of water related with algae cultivation (pumping, product isolation) is costly and a significant hurdle in commercialization.

1.3.7 Plant Breeding

Plant performance characteristics and plant compounds are subject to optimization by plant breeding. If features like fruit and biomass yield (ton ha^{-1} year^{-1}) or seed and biomass composition (e.g., starch, lipid, or lignin content) are targeted, genes in so-called output traits are addressed. These are the metabolic pathways controlling growth, biomass composition, and yield. Other breeding programs aim on cultivation parameters like germination rate and resistance against microbial pathogens and insects or draught. These approaches aim on genetic input traits. In any case, plant genes are modified either by *combination breeding*, *smart breeding*, or *genetic engineering*.

Combination breeding is the most commonly used breeding method today. Two parent plants are crossed, and by natural recombination the genomes of both parent plants are rearranged and redistributed according to the laws of heredity (first realized by Gregor Mendel (1822–1884); Austria). Subsequent filial generations are cultivated in green houses and analyzed for the quality of the targeted traits. After selecting improved varieties, the breeding process is repeated. As breeders need to cultivate each generation and analyze the targeted characteristic by checking the whole plant, this method is extremely time consuming. It may take 10 years plus 2–4 years for official registration and seed propagation before entering the market.

In addition to this traditional method, plant breeders increasingly use the tools of molecular genetics. The plant genome is sequenced, and molecular markers tagging the desired quality are introduced by *smart breeding*. The result of genetic recombination becomes much more predictable and by analyzing the genetic markers in the laboratory the desired genetic combination can be identified on cellular level within short time. Plant genome recombination and subsequent selection therefore can be performed much more efficiently.

Genetic engineering also uses selected genome sequences but is able to take advantage of nature's genetic diversity by introducing beneficial genes from other species (plants, bacteria) into the plant genome. In contrast to *smart breeding*, the resulting varieties are classified as genetically modified (GM). Critics point out that foreign genes in a plant may unfold unexpected effects or cross into wild species in an uncontrollable way. Therefore, release of GM plants is strictly regulated. GM corn, soybean, rapeseed, and cotton are cultivated in South and North America, China, and Southeast Asia but Europe acts more restrictive (Tables 1.20 and 1.21).

1.3.8 Basic Transformation Principles

Vegetable biomass and its compounds are subject to various industrial transformation as described here. Depending on the pretreatment of the raw materials, a distinction is drawn between first-, second-, and third-generation processes and product.

1.3.8.1 First Generation

Microorganisms of industrial relevance like *Escherichia coli*, *Corynebacterium glutamicum*, or *Saccharomyces saccharomyces* easily take up and metabolize sucrose

Table 1.20 Leading GM crops (global, 2013) (Compass, 2014).

	Total area (million hectare)	GM plant area (million hectare)	Share GMO (%)
Soy	107	79.0	79
Corn	179	57.4	32
Cotton	34	23.9	70
Rapeseed	34	8.2	24

Table 1.21 Leading areas in GM crop cultivation (million hectare; 2013) (Compass, 2014).

United States	Brazil	Argentina	India	Canada	China
70.1	40.3	24.4	11.0	10.8	4.2

and glucose. Sucrose extracted from sugar crop or glucose (produced from starch crops) can therefore directly be used as carbon source in fermentation. Sugar (and starch) is therefore called first-generation feedstock. This is the state of the art.

1.3.8.2 Second Generation

From lignocellulosic biomass sugar is not directly accessible. Such raw materials need an enzymatica or thermochemical pretreatment to make the enclosed sugar available for the so-called second-generation biotechnological processes. Second-generation ethanol production facilities have been established in recent years in United States, Brazil, Italy, and more are under way.

1.3.8.3 Third Generation

Degrading biomass or other organic materials down to gaseous CO (syngas) and feeding it into gas fermentation is the latest technology. Concerning the original raw material it is most flexible. It even accepts industrial CO emission from a steel mill. Summarized as third generation such processes are running currently on pilot scale in China and the United States.

1.3.9 Industrial Use

Besides traditional applications in food, feed, fiber, paper, and construction materials (wood) biomass is increasingly going to deliver raw materials into the main sectors: energy and chemicals, which are discussed in the following section.

1.3.9.1 Energy

Biomass contributes to all three energy markets: heat, power, and transport. Purpose-grown crops and wood, residues from agriculture and forestry, processing residues from food and wood industries as well as municipal waste are used. Although renewable energy sources (biomass, hydrothermal, wind, geothermal, and solar power) make only 19% (2012) of the global final energy consumption (REN21), biomass

Table 1.22 Renewable energy share of global final energy consumption (2012) (Zervos, 2014).

All energies	(%)	All renewable energies	(%)	Modern renewable energies	(%)
Fossil fuel	78.4				
Renewable	19.0 ≫	Modern renewable	52.6 ≫	Heat: biomass-, geothermal-, solar-, wind-derived	42
Nuclear	2.6	Traditional biomass (e.g., fire wood)	47.4	Hydropower	38
				Power: biomass-, geothermal-, solar-, wind-derived	12
				Bio-fuel	8

Table 1.23 Global growth rate of renewable energy capacity and bio-fuels production (%; end 2008–2013) (Zervos, 2014).

	Power				Heating	Transport	
Geothermal	Hydro	Solar voltaic	Solar heat	Wind	Solar heating	Bio-ethanol	Biodiesel
4.0	4.2	39.0	35.0	12.4	15.7	5.6	11.4

accounts for half of the renewable energy share, that is, 10%. The fact that 60% of that biomass is so-called traditional biomass (fuel wood often collected by hand, crop residues, and animal dung – combusted in open fires or insufficient stoves) illustrates how much optimizing potentials concerning efficiency are in disseminating efficient harvesting, processing, and energy production methods. In contrast, the so-called modern bio-energy uses especially prepared solid, gaseous, and liquid bio-fuels. Examples are wood pellets, biogas, and biodiesel (Table 1.22).

Biomass is clearly a priority in expanding renewable fuel (bio-ethanol, biodiesel) whereas power and heat generation are dominated by solar energy (Table 1.23).

This development demonstrates how the energy sector, which is currently pushed by governmental measures, might trigger renewables in the chemical sector. For example, bio-ethanol can be used not only as fuel but also as a key intermediate in chemical synthesis to ethylene and its follow-up products in the polymer sector. Though today the transport market is driving bio-ethanol, tomorrow chemical markets might build on these investments into bio-ethanol (e.g., feedstock and fermentation capacities). This example illustrates how feedstock and product markets are integrated, compete about the very same feedstock base, and influence each other.

Transport With a total production volume of 116.6 billion liters (2013) bio-based liquid fuel meets about 3% of the global road transport fuel demand. Primarily ethanol and biodiesel are established but aeration fuel is on the horizon.

Biodiesel is produced from vegetable oil by esterification to fatty acid methyl ester (FAME) or by catalytic hydrogenation to hydro-treated vegetable oil (HVO). The largest producer is Europe accounting for 42% of global production (10.5 billion liter FAME and 1.8 billion liter HVO).

Ethanol is made by *Saccharomyces cerevisiae* (baker's yeast) fermentation using sugar as carbon source. First-generation ethanol is made from sugar or starch (plant-derived sugar), whereas second-generation ethanol results from woody raw materials after releasing the lignocellulosic sugars. The United States (50 billion liters) and Brazil (25.5 billion liters) are leading in the production of ethanol.

Recently, *Clostridium* has been tested on pilot scale. These anaerobic bacteria are able to produce ethanol from gaseous CO that comes with industrial off-gases (e.g., from a steel mill) or with synthesis gas (CO, H_2) from gasified biomass or other organic materials.

Power Generation of bio-based power uses wood pellets and chips as well as biogas. In Germany (leading in biogas), 6000 plants produce biogas from specially grown energy crop (corn) and animal dung (e.g., from poultry production). However, the future is seen in biogas from industrial and municipal residues (e.g., food processing). Another option currently under evaluation is gasifying biomass (directly or in the form of municipal waste) and generating power from the resulting synthesis gas. Most of the gas is used to generate power, some goes into heat production, and there is also an option to feed biogas into the natural gas grid as methane is the main component of biogas as well as natural gas.

Wood-based power generation is already an established global market. The high demand in Europe is met by importing feedstock especially from North America (75%), Russia, and Eastern Europe.

Heat Solid biomass is by 90% the most important energy source in bio-based heating (housing, cooking). Europe is leading and burns 15 million tons of wood pellets (2013). North America is also an expanding market. Though most of biogas is used to generate power its role in heating should not be underestimated especially in rural areas with less developed infrastructure.

1.3.9.2 Chemicals

In many industries, bio-based feedstocks are already established, either because of (i) special feedstock suitability (pulp and paper), (ii) missing alternatives (proteins, drugs), (iii) unique characteristics (special polymers, tensides, lubricants), or (iv) customer demand (cosmetics). All the following chapters demonstrates that modern bio-based feedstocks conquer chemical markets according to similar rules of the fossil-based chemicals that emerged 150 years ago: Starting from high-value adding pharmaceutical markets where bio-based materials provide the only alternative, the growing knowledge encourages first to enter fine and bulk chemistry markets and finally to compete directly with fossil-based processes. One of the striking differences to the early chemical industry times is the fact, that science and knowledge are key from the very beginning.

Bio-based raw materials like rubber, wood, starch, and vegetable oils are established in various industrial applications. Figure 1.2 presents an overview about the current material use of biomass.

Natural rubber goes into high-performance tires, for example, for heavy-duty vehicles and special applications like medical products. It is a long-chain polyterpene, which is more uniform than its fossil-based synthetic alternative. About 50% of the global

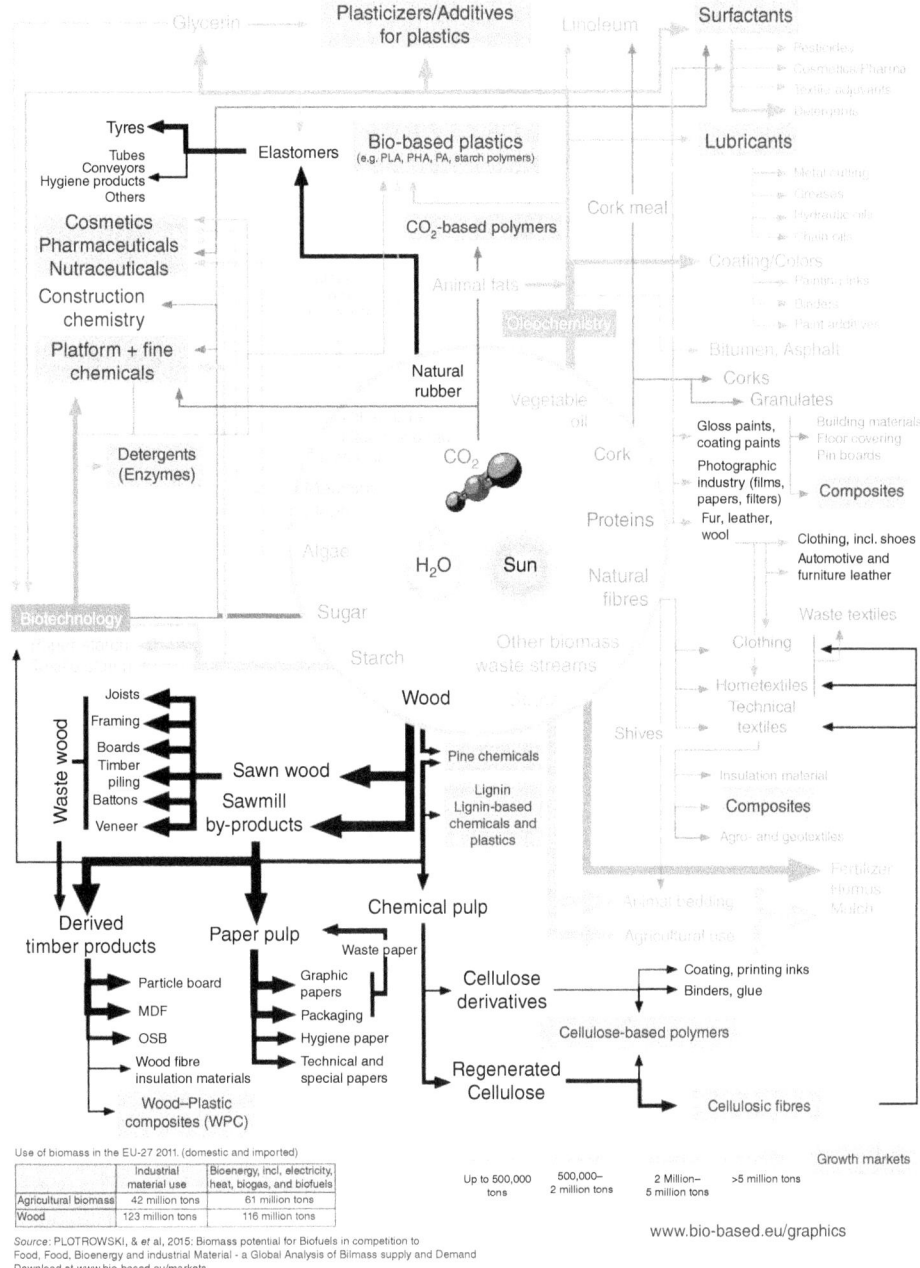

Figure 1.2 Biomass applications and material flow (Germany 2008) (Raschka and Carus, 2012; Anton and Steinicke, 2012).

rubber production of 25 million tons is provided by natural rubber (van Beilen and Poirier, 2008).

Cellulose is used to produce fibers like rayon for the textile industry with a volume of 3.5 million tons (Morris, Welters, and Garthoff, 2011). It is also the basis for paper and cardboard production. More than 500 million m^3 of wood is used to make 160 million tons of pulp. As a side product, 50 million tons of lignin appears, only 2% of that material is used for further chemical products (Morris, Welters, and Garthoff, 2011).

Starch is used to stiffen textiles and increase the mechanical strength of yarns. It goes as filler in very different applications like inks, detergents, and drug tablets. Around 30–40% of the global production of 2.5 billion tons goes into such non-food applications (Morris, Welters, and Garthoff, 2011).

Soap, perfumes, cosmetics as well as paints, wood treatment products, and hydraulic fluids and lubricants are made from or contain vegetable oils.

1.3.9.3 Biocatalysts

In bioprocessing, bio-based feedstock is not only the basic material to be transformed into a product. It is as well the raw material to provide the transforming catalyst itself.

This catalyst can be a whole cell like yeast in ethanol fermentation where yeast cells transform sugar into ethanol. The complex cellular metabolism consisting of enzymatic reaction chains nested in one another is steered by cellular control in a well-balanced way but can be modified to perform defined biochemical transformations and push metabolic pathways to cellular products of industrial relevance. This is the business field of specialized small- and medium-sized enterprises (SMEs).

Purified microbial enzymes are another group of protein biocatalysts, increasing reaction rates by 100 million to 10 billion times faster than normal reactions (Gurung *et al.*, 2013).

Enzymes play a significant role in food and feed processing, medicine, and technical applications (Association of Manufacturers and Formulators of Enzyme Products, kein Datum). Seventy-five percent of technical enzymes catalyze hydrolytic reactions. In food processing, one of the commercially most important enzymes is α-amylase, the enzyme splitting starch into glucose for the beverage industry. In medicine, another commercially relevant application is presented by DNA polymerase. This is the key enzyme in DNA sequencing for genomic research, diagnostics, and forensics. A growing technical application of enzymes is in digesting lignocellulosic biomass as a first process step to use such bio-feedstock industrially. Enzymes are not only processing catalysts but also active ingredients in consumer products. For example, detergents contain proteinases and lipases that remove food stains.

Enzyme production is globally about 100,000 tons making a $3.4 billion business with leading companies like Novozymes (Denmark), Danisco and DuPont (USA), DSM (Netherlands), and BASF (Germany) (Lorenz, 2012). It is no coincidence that big chemical industries complement their synthesis tool set by biological catalysts. Even the very first industrial application of an enzyme in tanning by Otto Röhm in 1908 in Darmstadt (Germany) resulted in a chemical company, which today is part of Evonik Industries (Germany). This example not only shows the close relationship of bio-based and chemical industries but also demonstrates how in the early twentieth century the chemical innovation cycle started its next round into biotechnology. Enzyme science earned its first Nobel Prize in 1947 (James B. Sumner, Cornell University, USA).

1.3.9.4 Pharmaceuticals

The pioneering role of the antibiotic penicillin has already been acknowledged earlier in this chapter. Drugs are still one of the most relevant and profitable applications of fermentative microbial transformation. The regular carbon source is sugar, which is easily taken up and metabolized by the transforming organism. Today 20% of commercial drugs and 50% of drugs under development (Bonnacorso, 2014) are estimated to be produced by biotechnology.

Still relevant are antibiotics of the β-lactam type produced by the fungi *Penicillium* (penicillin) and *Cephalosporium* (cephalosporins). The appearance of resistances initiated the development of derivatives of the natural molecules by adding chemically a wide variety of side chains to the microbial nucleic molecule. The estimated world market is some 10,000 tons (Table 1.24).

A modern group of biopharmaceuticals are monoclonal antibodies (mAbs). The production method of these highly functional proteins is based on the scientific work of the German biologist Georges J.F. Köhler (1946–1995) and the Argentinian chemist César Milstein (1927–2002) at Cambridge University (Great Britain) who both received the Nobel Prize in 1984. They fused antibody-producing B-cells with myeloma cells to the so-called hybridoma cells. By this method, clones of cells producing identical mAbs are propagated. The mAbs are highly active drugs that are used at a daily dose in microgram range. Though produced rather in kilogram than in ton range, the global sales volume is about $40 billion with an annual growth rate of 9% (Table 1.25).

Obviously, there are much more pharmaceuticals, but as drugs are not the main topic of this chapter the examples of antibiotics and mAbs should be sufficient to demonstrate (i) the high value and (ii) the low volume (and feedstock demand) of pharmaceuticals.

Table 1.24 Some semisynthetic antibiotics and their global annual production volume (Franssen, Kircher, and Wohlgemuth, 2010).

Antibiotic	Nucleus	Side chain	ton year^{-1}
Ampillicin	6-Aminopenicillanic acid	D-Phenylglycine	5000
Cephalexin	7-Aminodeacetoxycephalosporanic acid	D-Phenylglycine	4000
Amoxicillin	6-Aminopenicillanic acid	D-Hydroxyphenylglycine	16,000
Cefadroxil	7-Aminodeacetoxycephalosporanic acid	D-Hydroxyphenylglycine	1000

Table 1.25 Indications to be treated by monoclonal antibodies and sales volume (Pohl-Appel, 2011).

Monoclonal antibody	Indication	Company	Sales (billion US$)
Rituximab MabThera®/Rituxan®	Rheumatoid arthritis	Hoffmann-La Roche	4.1 (1998)
Trastuzumab Herceptin®	Cancer	Hoffmann-La Roche	3.6 (2000)
Adalimumab Humira®	Rheumatoid arthritis	Abbott Laboratories	4.5 (2003)
Cetuximab Erbitux®	Cancer	Merck Serono	0.9 (2004)
Bevacizumab Avastin®	Cancer	Hoffmann-La Roche	4.2 (2005)

1.3.9.5 Nutrition

Another commercially important biotechnological product based on sugar is L-lysine. It is one of the essential amino acids not synthesized by humans and animals but by plants. Therefore, humans and animals depend on L-lysine-containing food and feed. However, as the amino acid profile of plant biomass does not meet the demand of livestock like poultry and hogs, it is beneficial to add the limiting amino acids. However, livestock animals eat as long as the demand of the most limiting feed amino acid is satisfied and overflow amino acids are excreted. As amino acids contain nitrogen, consequently there is a significant burden of nitrogen-loaded manure. By adding L-lysine feed transformation efficiency into animal biomass improves and nitrogen excretion is significantly reduced. Mostly the bacterium *C. glutamicum* is cultivated in amino acid fermentation, which uses sugar as carbon source. Production strains are especially optimized to focus the metabolism on L-lysine synthesis and excretion. Another amino acid produced by fermentation from sugar is L-glutamic acid, commercialized as a food condiment especially in Asia. Global production volume is in the range of 1.2 million tons (L-glutamic acid), 0.5 million tons (L-lysine), and only a few hundred tons of special amino acids for medical applications such as L-histidine. The market price parallels the production volume, thus demonstrating the strong impact of *economy of scale* on production cost. Whereas L-glutamic acid, the largest by volume, is commercialized at a price of about \$1 kg^{-1}, L-histidine is marketed at about \$90 kg^{-1} (Lothar Eggeling, 2006).

1.3.9.6 Polymers

Biopolymers reached a share of 1.6 million tons (0.7% of total polymer market) but are expected to grow to more than 6 million tons in 2017 (European Bioplastics, 2014; Statista, 2013). There are two groups of bioplastics: (i) bio-based or partially bio-based nonbiodegradable plastics such as bio-based polyethylene (PE) and polyethyleneterephthalate (PET) and (ii) bio-based and biodegradable plastics such as polylactide (PLA) (Table 1.26).

As bio-PE and bio-PET are identical to their fossil-based counterparts, it is easy to use it in established processes without any modification. Therefore, such compounds are called *drop-in* chemicals. Though on the one hand a perfect *drssop-in* PE demonstrates

Table 1.26 Global bioplastics capacities by material type (1000 tons per year; 2013) (European Bioplastics, 2014).

Nonbiodegradable bioplastics			
Bio-PA	PTT	Bio-PEE	Bio-PET30
80	110	200	600

Biodegradable bioplastics				
PLA	Starch blends	Polyesters	PHA	Cellulose (regenerated)
185	183	175	34	27

PA, polyamide; PTT, polytrimethylene terephthalate; PEE, polyethylethylene; PHA, polyhydroxyalkanoate.

on the other a typical obstacle of many bio-based chemicals. Ethylene (28.05 g mol^{-1}), the monomer to be polymerized to PE can easily be made from bio-ethanol (46.07 g mol^{-1}) by dehydration but the stoichiometric product yield is only 0.609 kg kg^{-1}.

$$C_2H_5OH \rightarrow C_2H_4 + H_2O$$

In contrast, the fossil-based process reaches a carbon yield of 98 kg kg^{-1}. Obviously, the bio-based process is not cost-competitive as long as biomass feedstock does not offer a significant cost advantage.

PLA is based on the monomer lactic acid ($C_3H_4O_2$), a natural intermediate of lactic acid bacteria like *Lactobacillus*, the very same species used in yogurt fermentation. The monomer is produced by fermentation based on sugar (Groot et al., 2011) but an alternative process using lignocellulosic feedstock is under development (Riesmeier, 2013). Subsequent to lactic acid fermentation all further steps to the polymer are performed synthetically. PLA belongs to the polyester plastics and finds applications in packaging, agriculture, automotive, electronics, and textiles industries. The production capacity is expected to grow to close to 2 million tons by 2020 (Grand View Research, 2014).

1,3-Propanediol (PDO) is another sugar-based monomer. Developed since the 1990s by DuPont (USA), it is commercialized in its polymerized form under the trade name Sorona® and finds application especially in replacing fossil-based polytrimethylene terephthalate (PTT), for example, in plastic bottles. What makes this monomer worth to be mentioned here is its biological production system because PDO is a molecule not known to nature. By combining metabolic reaction chains from *S. saccharomyces* and *Klebsiella* in an *E. coli* cell, this host cell was taught to produce a man-made molecule (Demain and Sanchez, 2012). Today, the world capacity exceeds 100,000 tons per year (de Guzman, 2013), and from a financial perspective the market is expected to grow from $157 million (2012) to $560 million in 2019 (Rohan, 2014).

Vegetable oils are also used as precursors in oleochemical synthesis. For example, castor oil is of special industrial value because of the presence of hydroxyl groups on the especially long fatty acid chains of 18 carbons. Long molecular structures give high-performance properties to the resulting polymers to be applied, for example, in sports, aircraft, and medical products. Castor oil is becoming increasingly important in the production of polyurethane plastic, which is an application generally known as natural oil polyols.

In 2004, the US National Renewable Energy Laboratory (NREL) published a study that analyzed bio-based chemicals for their potential in the chemical industry from a technical point of view. It resulted in 12 candidates: 1,4-dicarboxylic acids (succinic, fumaric, and malic), 2,5-furandicarboxylic acid (FDCA), 3-hydroxpropionic acid (3-HPA), aspartic acid, glucaric acid, glutamic acid, itaconic acid, levulinic acid, 3-hydroxybutyrolactone, glycerol, sorbitol, and xylitol/arabinitol. Especially succinic acid draws industrial attention. DSM (Netherlands) and Roquette (France) formed the joint venture Reverdia (France) and Succinity (Germany) has been founded by BASF (Germany) and Corbion Purac (Netherlands).

All these molecules are provided by nature. However, it should be emphasized that the chemical industry to a large extent works with non-natural compounds. Therefore, PDO presents a special future-oriented example because it demonstrates that bio-based processes can provide man-made chemical entities for innovative materials and applications as well. Synthetic biology, emerging since early 2000s, is the principle behind such processes. It provides another example of how basic science is translated into industry.

1.3.10 Expectancy of Resources

As it is the case with fossil feedstock, bio-based raw materials are also not available without any limitation. However, the kind of limitation is different. In contrast to fossil materials, biomass is renewable and regrows continuously. The only limiting factor is the annual growth rate of the biomass source. In principle, vegetable as well as animal biomass is renewable but only plants build biomass through photosynthesis and transform atmospheric CO_2 into carbohydrates according the following chemical equation:

$$6CO_2 + 12H_2O \rightarrow C_6H_{12}O_6 + 6O_2 + 6H_2O$$

Only vegetable biomass is therefore directly involved in the photosynthetic carbon cycle and should be considered as a major industrial feedstock.

Nature is estimated to produce annually 210 billion tons of vegetable biomass. Most of it is lignocellulose (40–55% cellulose, 10–35% hemicellulose, and 18–41% lignin) with an estimated carbon content of about 50%. Therefore, it can be reasonably assumed that photosynthesis fixes globally about 105 billion tons of carbon.

However, because of economical, ecological, and societal reasons not all biomass is available for human purposes. Sustainably available is biomass from agriculture, forestry, and marine sources. The use of agricultural biomass for industrial purposes has already been established. Today, global agriculture produces 14 billion tons of biomass (containing about 7 billion tons of carbon) annually for providing food, feed, fiber, and a little heat, fuel, and chemistry. A significant share is considered waste and utilized not at all, of only little value or under valued: (i) agricultural side streams like straw, rice husks, corn cobs; (ii) silvicultural materials like branches and saw-cut, and (iii) processing residues like milling or food-processing residues (Table 1.31). These resources are going to be realized for industrial purposes by second- and third-generation processing.

In addition to plant breeding, soil management, fertilization, and plant protection, more efficient harvesting and storage methods will still be improved. Plant breeding alone sets since decades the still unbroken trend of annual crop yield improvement of about 2%. On the contrary, a growing world population (9 billion by 2050) will ask for more food, feed, and fiber. Considering currently neglected resources, yield improvement, and food demand a range of 0.5–1.4 billion tons of biomass has been estimated to be sustainably available for industrial purposes by 2030 (Kircher, 2012; Bang, Follér, and Buttazzoni, 2009).

1.3.11 Green House Gas Emission

In the photosynthetic carbon cycle, CO_2 is fixed in biomass and released into the atmosphere again when biomass degrades (or is burned). Theoretically producing and using vegetable biomass should not emit any CO_2. However, plants not only consist of photosynthetic leaves but have roots interacting with the microbial soil flora. Soil management (e.g., tilling) aerates soil thus activating microbial metabolism and the related CO_2 emission further. It is estimated that a CO_2 amount equivalent to about 4% of the harvested biomass is emitted by the microbial soil flora from agricultural areas. Therefore, non-tilling land-management methods have been developed in order to at least reduce this emission. In contrast, if land is not disturbed over longer periods like grassland or forest soil, the microbial soil biomass grows, thus binding carbon up to a volume of about 30% of the grass or wood biomass above ground.

Modern agriculture is based on intensive land management. Farmers use extensive machineries in preparing the soil, seeding, harvesting, processing, storing, and shipping. Today, these machines run mostly on fossil fuel emitting CO_2 equivalent to about 11% of the crop biomass harvest.

Last but not least, the high yields of modern agriculture heavily depend on adding fertilizer to the soil, especially nitrogen. About 2–3% of the nitrogen applied escapes in the form of N_2O into the atmosphere. This small amount appears to be negligible; however, its climate warming impact is 310 times higher than that of CO_2 (Tables 1.27 and 1.28) (Haberl et al., 2012).

Methane (CH_4) emitted from paddy fields and released by livestock husbandry (especially ruminants) gives another biomass-related emission with a GHG (green house gas) effect 21-fold more than that of CO_2 (Table 1.28). In Table 1.29, CO_2 equivalents are given, which consider the specific climate impact of different GHGs.

Biomass production does not come without any GHG emission. It is estimated that about 20–50% of global GHGs originate from agriculture, livestock breeding, and forestry (Table 1.28).

1.3.12 Regional Pillars of Competitiveness

Using renewable raw materials implies a significant impact on global feedstock regions and industrial centers. As discussed before, fossil carbon sources are produced in selected and manageable areas and are easy to ship globally through pipelines, tankers,

Table 1.27 Options for carbon sources from agricultural, forestry, and industrial side streams and carbon content (global; million tons per year) (Kircher, 2012).

	Sugarcane residues	Wheat straw	Palm oil wood	Wood biomass	Sugar processing effluents	Palm oil mill effluents
Volume	530	350	114	900	300	480
Carbon content	265	175	67	450	4	16

Table 1.28 GHG emission associated with biomass production (% CO_2 fixed in harvested biomass) (Haberl et al., 2012).

	CO_2 from soil (%)	CO_2 from fossil fuel (%)	N_2O and CH_4 due to fertilization (%)
Cropland	4	11	12
Grassland	−26	7	20
Forests	−32	3	1

Negative numbers mean that more CO_2 is fixed than emitted.

Introduction to Renewable Biomaterials

Table 1.29 Greenhouse gas sources and climate impact factor as well as share of climate impact weighted for the climate changing potential over the next 100 years (EPA US Environmental Protection Agency, 2014).

GHG	CO_2			Methane	Nitrous oxide (N_2O)	Fluorinated gases
Source	Fossil fuel, power, heat	Other fossil-based (e.g., cement industry)	Biomass degradation, deforestation	Livestock breeding (especially ruminants)	Agriculture (especially fertilizers)	Industry
Share CO_2 (%)	73.8	3.6	22.6			
Share CO_2 equivalent (%)	77.7			14.3	7.9	1.1
Climate impact (rel. to CO_2)	1			21	310	10.000–20.000

Table 1.30 Density, bulk density, and carbon per volume (t m^{-3}) of various materials.

	Fossil oil	Coal	Peat	Wood	Cereal	Straw bale
Density	0.8–0.9					
Bulk density		0.4–0.8	0.3–0.4	0.2–0.6	0.40–0.48	0.05–0.11
C m^{-3}	0.68–0.77	0.32–0.64	0.18–0.24	0.1–0.3	0.2–0.24	0.03–0.05

and railroad to industrial centers. In contrast, biomass and processing residues are distributed over very large areas, in fact, over all global agri- and silvicultural areas. Harvesting from the field or forest is costly, and as biomass is degradable it needs special efforts for storage. In addition, these materials are relatively bulky and not suitable for pipelines. They need to be shipped by truck, railways, and ship. For example, 1 m^3 straw contains 30–50 kg carbon, whereas the same volume of fossil oil gives 800–900 kg carbon. In other words, to get the same amount of carbon from straw requires the transportation of 23-fold volume compared with fossil oil (Table 1.30).

Because harvesting, storage, and shipping are costly and transportation over large distances is difficult, at least preprocessing will be done close to the raw material production site. Preprocessing may include sugar extraction from sugar and starch crops, release of lignocellulosic second-generation sugar, or gasification to syngas (third-generation feedstock). Most of such carbon sources may be shipped directly to industrial centers. Figure 1.3 shows biomass trade routes to be expected by 2020.

Preprocessing of biomass might trigger the integration of more processing steps, thus adding value to more advanced intermediates. At this stage, the very same pillars of competitiveness, which made early fossil-based industrial regions successful become effective. More sophisticated processes need a skilled workforce, ask for accompanying academic research, require an efficient public infrastructure and administration, and last but not least attract investors. Rural areas concerned will turn into industrial feedstock regions. However, these regions will most probably not see bio-refineries of the scale of

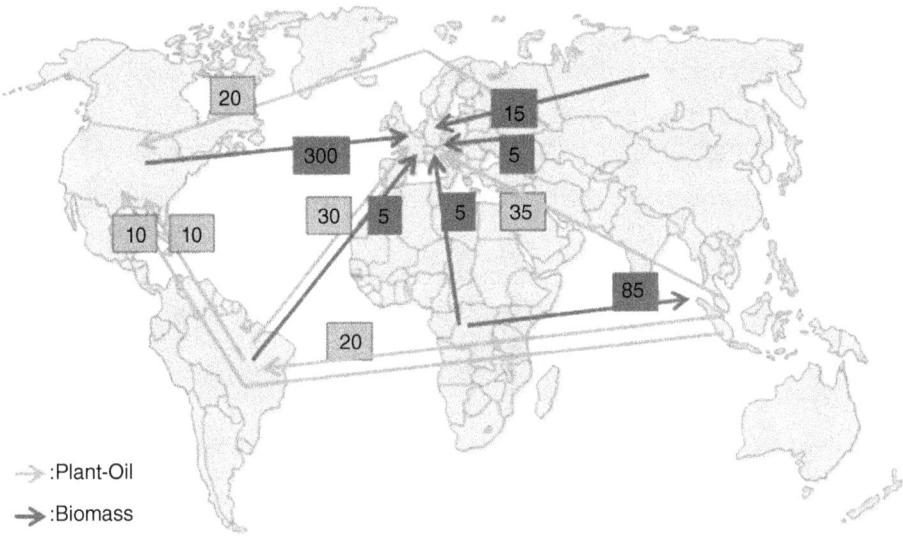

Figure 1.3 Expected biomass trade routes by 2020 (TWh) (King and Hagan, 2010).

oil refineries. Because feedstock transport distances are economically limited, biomass refineries will reach much smaller capacities than big oil refineries. Consequently, the bio-economy is a special opportunity for SMEs.

1.3.13 Questions for Further Consideration

Does bio-feedstock provide an alternative to fossil feedstock? What are the differences? What are limiting economical, ecological, and societal factors in using bio-feedstock? What makes a bio-based production site attractive?

1.4 General Considerations Surrounding Bio-Based Raw Materials

1.4.1 Economical Challenges

Energy carriers and chemicals represent an extremely wide range of value. The more specific the function of a molecule is, the more the added value is. For example, a pharmaceutical active addressing a very specific physiological response such as a monoclonal antibody might be valued millions of dollars per kilogram. Astaxanthin, a feed additive giving salmon its typical red color, earns more than $1000\,kg^{-1}$. However, such molecules reach market volumes of only gram (special pharma actives) or up to 10,000 tons per year and belong to the fine chemicals sector, which makes about 10% of all chemical sales. Specialties are produced in the 100,000 tons range (30% of chemical sales) and bulk products (60% of chemical sales) like monomers for plastics are produced in more than 100,000 and even million tons per year range. In this segment, market prices are in the $1-2\,kg^{-1}$ range or even lower, and therefore only extremely low production costs are tolerated. Besides capital and running cost,

Table 1.31 Share (%) of cost factors in bio-based production of bulk chemicals (Kircher, 2014).

Feedstock	Auxiliary materials	Labor	Maintenance	Energy	Depreciation	Interest
50	15	15	14	4	1	1

especially cost of feedstock are decisive and can make up to 40–50% of total production cost (Table 1.30). Compared to fossil feedstock renewable carbon sources still struggle on a pure cost basis because of expensive (i) harvesting from large areas, (ii) shipping of bulky materials, (iii) storing of degradable biomass, and (iv) preprocessing. Reducing feedstock cost will be key in realizing the bio-economy (Table 1.31).

1.4.2 Feedstock Demand Challenges

Not only cost but also the required volume is a topic to discuss. The global consumption of coal, oil, and gas is about 13 billion tons. When assuming an average carbon content of about 85% in fossil resources, this means that approximately 11 million tons of fossil carbon is consumed annually.

To replace fossil carbon by agriculture alone, this sector would need to expand by a factor of 2.5 from today – 14–36 billion tons of biomass (equivalent to 18 billion tons of carbon). Even when considering future yield increase by plant breeding, cultivation methods, and bringing more areas on top yield level, it seems risky to seek the solution only in more agriculture, as soil erosion, climate change, and fertilizer shortage (e.g., geological phosphate resources are running out) raise more uncertainties.

Obviously relying on agriculture alone is not enough. Industries need to exploit new sustainable carbon sources such as second-generation lignocellulosic raw materials. They present so-called non-food biomass from agriculture as well as forestry. Such options as well as processing side streams and third-generation carbon sources like CO have been discussed earlier. About 0.5–1.4 billion tons of biomass have been estimated to be sustainably available; a volume obviously not sufficient to replace fossil carbon in total.

Such a conflict asks for setting priorities: The modern bio-economy should be focused on products without an alternative to the use of carbon. These are at first organic chemicals and according to the state-of-the-art heavy-duty fuels such as aeration fuel. Today, organic chemicals consume about 7% (280 million tons containing 238 million tons of carbon) of the total fossil oil production. This gives the theoretical minimum volume range to be satisfied by biomass. In fact, the biomass demand will be significantly bigger as the stoichiometric productivity of bio-based processes is generally lower than that of chemical synthesis, but this number gives an orientation. Compared to the biomass volume, which is currently sustainably available (0.5–1.4 billion tons containing 250–750 billion tons of carbon), it appears appropriate to replace fossil carbon in chemical synthesis and in addition produce some bio-fuel. Power generation and short-distance mobility are not really dependent on carbon sources. These markets may run on solar, wind, geo, thermal, hydro, and more alternative power sources and the storage media required.

In summary, the pressing feedstock challenge asks for comprehensive use of carbon-containing materials and prioritization on products depending on carbon.

1.4.3 Ecological Considerations

It is widely accepted that the climate change is caused by anthropogenic GHG emissions. However, the public debate often focuses only on CO_2 from fossil-based activities though agriculture contributes to the whole set of GHG significantly. Therefore, biomass should be produced and used carefully and as energy- and carbon-efficient as possible. Efficiency means in this context to transform biomass into energy or materials with energy and carbon losses as low as possible. This concept analyzes single production steps and complete utilization chains.

For example, sugar might be extracted from sugarcane. The residual biomass (bagasse) might first be used to produce second-generation sugar and subsequently be burned for generating power (what is today the only use). The last step of sugarcane-based ethanol fermentation is an aqueous solution still containing some organic residues plus minerals (vinasse), which is considered today a waste and spread on the field as fertilizer. Before that last utilization step, vinasse could generate biogas, thus yielding energy from the organic fraction but still retaining the minerals. This concept is called cascade-use because biomass energy and carbon are utilized in consecutive steps as completely as possible. The consistent implementation of that concept addresses not only agricultural biomass but also forestry and marine resources, industrial processing residues, and even municipal solid and liquid wastes.

Besides reducing agro-related emissions cascade-use also helps to optimize the use of biomass cultivation areas. This topic is too broad to be discussed here but nevertheless, in view of the growing world population, it is obvious that land use is critical. This topic is not only about food production but also about keeping soils fertile, avoid salinization and erosion, water management, and maintaining biodiversity.

In summary, the modern bio-economy is on the one hand the only alternative to fossil-based processing with all its implications but on the other does not come without any environmental burden. To find the most beneficial biomass, its way of production and later processing in a cascade mode needs careful analysis through life cycle assessment.

1.4.4 Societal Considerations

1.4.4.1 Food Security

Though agriculture today serves at first food and feed markets, about 870 million people, which is more than 10% of the world population, are estimated to suffer from hunger. Mostly regions where agriculture is inefficient because of general under development, political instability, or climate change are affected but, nevertheless, the fact that hunger is still in the world raises the question about food security when agriculture feeds not only people but also industries. This topic has a real global dimension called "indirect land-use change" (ILUC) when increasing bio-based production demands more biomass for industrial purposes in one region leads to land-use change (LUC) in another region.

Governments, NGOs, and private parties agree clearly on the priority of agricultural food production. Industrial feedstock should increasingly be generated from non-food biomass such as agricultural, forestry, and industrial side streams. Exploiting currently neglected resources like municipal waste and industrial emission will help to reduce the biomass demand pressure as well. Together with further disseminating high-yield agricultural methods and technologies it appears feasible to meet the demand of future food and feedstock demand. However, it needs the joint effort of agriculture and industry scientists, farmers, processing engineers, and last but not least economical, societal, and political leaders to make the indispensible agricultural transformation happen.

1.4.4.2 Public Acceptance

The latter paragraph already raised the role of public leadership in gaining acceptance of the bio-economical transformation process. Leadership is the more requested as the bio-economy is not only a technical and economical transformation, it has as well a transformative impact on the society itself. It includes comprehensive recycling of industrial processing streams and after-use consumer products, thus changing our way of using materials and products. Today, the production and consumption chain from a raw material (e.g., fossil oil) up to a consumer product (e.g., the plastic housing for a smart phone) and later to after-use waste is a linear one. There is a beginning with raw material and an end with waste. The bio-economy will form circular production and consumer chains because a product after use becomes the raw material for the next production cycle. To make it a reality, it needs not only technologies but also, for example, (i) the public administration to specify recycled goods in contrast to waste materials, (ii) the manager of an urban waste disposal facility to integrate recycling into operation and find industrial customers for the resulting materials, (iii) the industrial supply manager to accept recycled raw materials, and last but not least (iv) the ordinary consumer to buy products made from recycled materials. In other words, governments, public administration, industries, and the whole economical sector as well as the society as such must pull together. This is what makes the bio-economy and the circular economy transformative.

1.5 Research Advances Made Recently

Academic and industrial scientists work hard on basic bio-economy know-how and industries test methods and processes. Here is a selection of press releases about recent advances. It gives an impression how academia, SMEs, and big industries cooperate internationally to make the bio-economy real.

1.5.1 First-Generation Processes and Products

December 12, 2014 – Algenol Biofuels (USA, founded 2006) was named the recipient of the 2014 Global Energy Award for Industry Leadership in Biofuels, presented by PLATTS Global Energy Awards. Algenol's algae technology platform for production of the four most important fuels (ethanol, gasoline, jet, and diesel fuel) uses algae, sunlight, CO_2, and saltwater for high-yield, low-cost fuel production. The technology recycles CO_2 from industrial sources.

September 19, 2014 – Lufthansa made commercial flight with bio-based jet fuel. It contained 10% farnesane produced by yeast from sugar. The process was developed by Total (France) in cooperation with the spin-off Amyris of UC Berkeley (USA). In addition, Lufthansa intends to test Jatropha oil-based jet fuel.

September 18, 2014 – BASF (Germany), Cargill (USA), and Novozymes (Denmark) announced a process to produce acrylic acid from renewable raw material. Acrylic acid is a bulk chemical and building block for plastics, fiber, coatings, and superabsorbers (absorbing aqueous liquid, e.g., in diapers). The process produces 3-hydroxypropionic acid by fermentation, which is subsequently transformed into acrylic acid. Suitable carbon sources are, for example, sugar and glycerol.

June 26, 2013 – Sugarcane-based low-density polyethylene (LDPE) will be used in all Tetra Pak packages produced in Brazil. Tetra Pak (Sveden) stated that about 13 billion bio-based packages will be produced in Brazil. Bio-ethylene will be produced from sugar-based ethanol by Braskem (Brazil).

June 19, 2013 – Multilayer tube systems with bio-based polyamides from Evonik Industries were tested in a racing car for the first time. The car boasted a number of novel features including a multilayer line for charge-air cooling. The outer layer was made of a polyamide, which is based on castor oil (extracted from the oil crop *Ricinus communis*).

1.5.2 Second-Generation Processes and Products

September 29, 2014 – POET-DSM Advanced Biofuels LLC, a 50:50 joint venture of the bio-ethanol producer POET (USA) and DSM (chemistry; The Netherlands), published the first commercial-scale cellulosic ethanol facility in Iowa (USA). It will daily convert 770 tons of biomass into 75 million liters of ethanol.

November 18, 2013 – M&G Chemicals (Italy) announced to construct a second-generation bio-refinery in China for the conversion of 1 million tons of straw biomass into bio-ethanol and bio-glycols. The lignin resulting as a by-product from the bio-refinery will feed a 45 MW cogeneration plant, which will be constructed at the same time as the bio-refinery in the same site.

October 6, 2014 – Energochimica SE (Slovakia) signed an agreement with Biochemtex and Beta Renewables for the construction of a second-generation bio-ethanol plant and an annexed energy block to deliver 55,000 metric tons per year of cost-competitive cellulosic ethanol using non-food biomass as its feedstock.

1.5.3 Third-Generation Processes and Products

November 11, 2014 – The bio-news agency Biofuel Digest voted LanzaTech (USA) the hottest company in bio-energy. This company developed a gas fermentation process to bio-ethanol. Carbon source is gaseous CO directly fed into the fermentation broth to cultivate the bacteria *Clostridium*. CO might come with industrial flue gas or with synthesis gas from gasified biomass. The process has been successfully tested on pilot scale in a Chinese steel mill. The academic basis has been laid especially at British and German universities.

December 9, 2013 – Evonik Industries (Germany) and LanzaTech (USA) have signed a 3-year research cooperation agreement, which will see Evonik combining its existing biotechnology platforms with LanzaTech's synthetic biology and gas fermentation expertise for the development of a route to bio-processed precursors for specialty plastics from waste-derived synthesis gas. In this route, microorganisms placed in fermenters are used to turn synthesis gas into chemical products. Synthesis gases comprise mainly of either CO or CO_2 and H_2 and can come from a variety of gasified biomass waste streams including forestry and agricultural residues and gasified municipal solid waste.

1.6 Prominent Scientists Working in this Arena

Joško Bobanović (physicist) is an investor and helps to finance bio-economical innovation. He joined the Sofinnova Green Seed Fund (France) dedicated to seed activities in green chemistry and bio-energy in 2010. He is currently a director on the boards of Metgen, Synthace, and Cellucomp. Joško holds a BSc. in physics from University of Zagreb, PhD in physical oceanography from Dalhousie University, and an MBA in finance and marketing from McGill University.

Karl-Erich Jaeger (biologist) is professor for molecular biology at the Heinrich-Heine-University Duesseldorf (Germany). His research interests focus on bacterial enzymes, their production, and application in biotechnological processing. Erich Jaeger is vice chairman of the German bio-economy cluster CLIB2021 and co-founder of evocatal (Duesseldorf), a start-up specializing on industrial enzymatic catalysis and biomass transformation.

Jay D. Keasling (biologist) is professor for chemical and bio-engineering at UC Berkeley as well as director for biophysics at the Lawrence Berkeley National. He is one of the leading scientists in synthetic biology, systems biology, and environmental biology. One of his prominent projects is the biotechnological production of the anti-malarial drug artemisinin in bacteria. Other projects target on first and second-generation bio-fuel and chemicals. Jay Keasling co-founded Amyris Biotechnology in California in 2003.

Ray Miller (chemical engineer) is Chief Business Officer at Verdezyne, Inc. (Carlsbad, California), a start-up focusing on the development and production of bio-based chemicals. Before he took that position in 2012, he has been Program Director of DuPont's bio-based polymer platform since 2002. This program developed bio-based PDO, which became one of the first commercially successful biopolymers. Ray Miller was part of the joint DuPont/Genencor team that received the 2003 EPA Presidential Green Chemistry Challenge Award.

Marc van Montagu (biologist) is chairman of the Institute of Plant Biotechnology for Developing Countries (IPBO) of the Ghent University (Belgium), and the president of the European Federation of Biotechnology. Jozef Schell (scientist; 2003) has been director at the Max Planck Institute for Plant Breeding in Cologne, professor at the University of Cologne and Collège de France (Paris). Both being molecular biologists they realized the potential of the soil bacterium *Agrobacterium tumefaciens* to work as gene vector into plants. In their laboratories, plants have been genetically engineered first in 1983. Only 11 year later in 1996 transgenic plants started their commercial success. The contribution of both scientists to modern agriculture has been awarded with many prizes, among other in 1998 with the World Food Prize.

Christian Patermann (lawyer) is the initiator of the European concept of the "Knowledge-based Bio-economy" published in Cologne (Germany) in 2004. Since 2004 he was Programme Director for Biotechnology, Agriculture & Food Research. He also served for 4 years as co-chair in the EC-US Task Force Life Sciences and Biotechnology Research. Earlier he held positions in International Science Organisations like ESA, ESO and EMBL as well as with the German Federal Ministry of Research, Education and Science. Christian Patermann studied law, economics, and languages in Germany, Switzerland, and Spain.

Sean Simpson is the co-founder and Chief Scientific Officer of LanzaTech (USA) and leads the development and commercialization of LanzaTech's core technology. Simpson's leadership has encouraged collaboration between biologists, fermentation specialists, process and design engineers, and business development teams to develop the technology and the company to become a global leader in gas fermentation. Simpson holds a PhD from York University, UK, and a Masters degree from Nottingham University, UK.

1.7 Summary

About 150 years ago the feedstock change from natural materials to fossil feedstock triggered the industrial revolution. Coal, natural gas, and mineral oil offered raw materials with high carbon and energy content. Especially the use of oil started the ongoing innovation cycle in chemistry, the growth of energy and chemical markets and the development of a global supply chain infrastructure. However, these resources are limited and in addition their use is leading to CO_2 emission and consequently climate change.

Biomass is the only alternative. Above all, it is renewable and therefore an apparently endless resource. However, energy and chemicals are not the only markets to be satisfied; the growing world population asks for more food, feed, and fiber as well. Therefore, priorities have to be set: Nutrition is first, and in industrial use biomass needs to be focused on products that depend on carbon, that is, chemicals and heavy-duty fuel. In addition, also biomass production does not come without any emission. Industrial processing should therefore target on high carbon efficiency to keep the raw material consumption as low as possible.

First-, second-, and third-generation processes and products are under development and continuously enter the market. What we see today is again the launch of an innovation cycle triggered by a feedstock change. The time is right for scientists, engineers, and entrepreneurs to deepen the academic insights into biomass fundamentals, develop and implement transformation technologies, and explore the commercial potential of the bio-economy!

1.8 Study Problems

Is biomass going to replace fossil resources in energy and chemical production?
Can cascade-use of biomass disburden the feedstock demand?
How can the nutrition versus industrial biomass use conflict be solved?
Which unconventional resources to be explored?
What makes the bio-economy transformative?

1.9 Key References

Hacker (2012)
Höfer (2009b)
Ulber, Sell, and Hirth (2011a)
Wim Soetaert (2010)

References

Adams, E. I. (2012). *Eco-Economy Indicators.* Abgerufen am 25. 12 2014 von www.earth-polica.org/indicators/C56/forests_2012

Anton, C. and Steinicke, H. (2012). *Bioenergy – Chances and Limits.* German National Academy of Sciences Leopoldina. Halle: Deutsche Akademie der Naturforscher Leopoldina.

Association of Manufacturers and Formulators of Enzyme Products. (kein Datum). Abgerufen am 13. 12 2014 von www.amfeb.org/content/list-enzymes

Bang, J. K., Follér, A., and Buttazzoni, M. (2009). *Biotechnology – More Than a Green Fuel in a Dirty Economy?* World Wildlife Fund.

van Beilen, J. B. and Poirier, Y. (2008). Production of renewable polymers from crop plants. *Plant J*, **54**, S. 684–701.

Benzing, T. (2013). Rohstoffeinsatz in der chemischen Industrie. *2.ibi Fachsymposion.* Halle/Saale.

Birol, F. (2008). *International Energy Agency.* Abgerufen am 23. 12 2014 von www.worldenergyoutlook.org/media/weowebsite/2008-1994/weo2008.pdf

Bonnacorso, M. (2014). Italian Bioeconomy. *NanaotechITALY2014; Venice, 26.-28.11.2014.*

Compass, G. (2014). *Genetically Modified Plants: Global Cultivation on 174 Million Hectares.* Abgerufen am 25. 12 2014 von www.gmo-compass.org/eng/agri_biotechnology/gmo_planting/257.global_gm_planting_2013.html

Davis, S. (2011). *Chemical Economics Handbook Product Review: Petrochemical Industry Overview.* S. Consulting (Hrsg.).

Demain, A. L. and Sanchez, S. (2012). Microbial synthesis of primary metabolites: Current trends and future prospects. In E. M. T. El-Mansi, C. F. A. Bryce, A. L. Demain, and A. R. Allman (Hrsg.), *Fermentation Microbiology and Biotechnology* (S. 78–99). CRC Press.

Diamond, J. (2005). *Collapse; How Societies Choose to Fail or Succeed.* New York: Penguin Group.

EKT Interactive Oil and Gas Training. (2014). Abgerufen am 23. 12 2014 von www.ektinteractive.com/introduction-oil-gas/what-is-downstream/

EPA US Environmental Protection Agency. (2014). *National Greenhouse Gas Emissions Data.* (E. U. Agency, Hrsg.) Abgerufen am 28. 12 2014 von www.epa.gov/climatechange/ghgemissions/usinventoryreport.html

European Bioplastics. (2014). Abgerufen am 14. 12 2014 von en.european-bioplastics.org/market: www.european-bioplastics.org

Fachagentur nachwachsende Rohstoffe. (2012). Abgerufen am 12. 12 2014 http://energiepflanzen.fnr.de/pflanzen/algen

FAO. (2014). Abgerufen am 25. 12 2014 von faostat.fao.org/site567/DesktopDefault.asp#ancor

Franssen, M. C. R., Kircher, M., and Wohlgemuth, R. (2010). Industrial biotechnology in the chemical and pharmaceutical industries. In E. J. Wim Soetaert (Hrsg.), *Industrial Biotechnology* (S. 323–350). Wiley-VCH.

Grand View Research. (2014). www.grandviewresearch.com/industry-analysis/lactic-and-poly-lactic-acid-market. Abgerufen am 14. 12 2014 von www.grandviewresearch.com

Groot, W., van Krieken, J., Sliekersl, O., and de Vos, S. (2011). Production and purification of lactic acid and lactide. In R. Auras, L.-T. Lim, S. E. M. Selke, and H. Tsuji (Hrsg.), *Poly(lactic acid): Synthesis, Structures, Properties, Processing, and Applications* (S. 3–24). John Wiley & Sons.

Gurung, N., Ray, S., Bose, S., and Rai, V. (2013). A broader view: Industrial enzymes and their relevance in industry, medicine and beyond. *Biomed Res Ind*, 2013, S. 329121.

de Guzman, D. (2013). www.greenchemicalsblog.com. Abgerufen am 14. 12 2014 von greenchemicalsblog.com/2013/03/18/bio-pdo-market-update/

Haberl, H., Körner, C., Lauk, C., et al. (2012). *Bioenergy – Chances and Limits.* German National Academy of Sciences Leopoldina. Berlin: Deutsche Akademie der Naturforscher Leopoldina.

Hacker, J. (2012). *Bioenergy – Chances and Limits.* German National Academy of Sciences Leopoldina. Halle (Saale): German Academy of Sciences Leopoldina.

Harald Andruleit, H. G. (2011). *Energiestudie 2012; Reserven, Ressourcen und Verfügbarkeit von Energierohstoffen.* Abgerufen am 23. 12 2014 von www.bgr.bund.de/DE/Themen/Energie/energie_node.html

Herrmann, M. and Weber, J. (2011). *Öfen und Kamine: Raumheizungen fachgerecht planen und bauen.* Beuth Verlag.

Höfer, R. (2009a). History of the sustainability concept – Renaissance of renewable resources. In R. Höfer (Hrsg.), *Sustainable Solutions for Modern Economies* (S. 1–11). RSC Publishing.

Höfer, R. (Hrsg.). (2009b). *Sustainable Solutions for Modern Economies.* RSC Publishing.

International Starch Institute. (2014). Abgerufen am 25. 12 2014 von www.starch.dk/isi/stat/img/RawMaterialPic.gif

King, S. and Hagan, A. (2010). Abgerufen am 27. 12 2014 von www3.weforum.org/docs/WEF_FutureIndustrialBiorefineries_Report_2010.pdf

Kircher, M. (2012). The transition to a bio-economy: Emerging from the oil age. *Biofuels, Bioproducts, Biorefineries,* **6**, S. 369–375.

Kircher, M. (2014). Personal communication.

Landesverband der Kartoffelkaufleute Rheinland-Westfalen. (2013). Abgerufen am 25. 12 2014 von Die Kartoffel: www.die-kartoffel.de/im-boden/anbaugebiete-weltweit/

Lorenz, P. (2012). *Industrielle Herstellung von Enzymen für Lebensmittel.* Abgerufen am 13. 12 2014 von www.git-labor/news/aus-der-wissenschaft/erfolgsfaktor-enzyme?page=1

Lothar Eggeling, W. P. (2006). Amino acids. In B. K. Colin Ratledge (Hrsg.). Cambridge University Press.

Marland, G., Boden, T. A., and Andres, R. J. (2007). Global, regional, and national CO_2 emissions. In Carbon Dioxide Information Analysis Center and Oak Ridge National Laboratory (Hrsg.), *Trends: A Compendium of Data on Global Change.* US Department of Energy.

Michelsen, E. R. (1941). *Beitrag zur Chemie des Torfes, dessen Schwelung und Extraktion.* Dissertation, Eidgenössische Technische Hochschule in Zürich.

Morris, P. C., Welters, P., and Garthoff, B. (2011). Plants as bioreactors: Production and use of plant-derived secondary metabolites, enzymes and pharmaceutical proteins. In R. Ulber, D. Sell, and T. Hirth (Hrsg.), *Renewable Raw Materials* (Bd. 2001, S. 7–32). Wiley.

Olivier, J. G., Jansens-Maenhout, G., Muntean, M., and Peters, J. (2014) *Trends in Global CO_2 Emissions: 2013 Report.* PBL Publishers.

Pohl-Appel, G. (2011). Monoklonale Antikörper. BIOspectrum, **17**, S. 116–117.

Raschka, A. and Carus, M. (2012). *Stoffliche Nutzung von Biomasse; Basisdaten für Deutschland, Europa und die Welt.* nova-Institut GmbH.

Riesmeier, J. (2013). www.direvo.com/fileadmin/user_upload/DIREVO-PressRelease_No1_2013.pdf. Abgerufen am 14. 12 2014 von www.direvo.com

Rohan, M. (2014). www.marketsandmarkets.com/PressReleases/1-3-propanediol-pdo.asp. Abgerufen am 14. 12 2014 von www.marketsandmarkets.com

Statista. (2013). www.statista.com. Abgerufen am 14. 12 2014 von www.statista.com/statistics/281126/global-plastics-production-share-of-various-countries-and-regions/

Statista. (2014). *Erntemenge der wichtigsten Anbauländer von Mais weltweit im Jahr 2012.* Abgerufen am 2. 12 20154 von de.statista.com/statistik/daten/studie/156066/umfrage/groesste-maisproduzenten-der-welt/

Ulber, R., Sell, D., and Hirth, T. (Hrsg.). (2011a). *Renewable Raw Materials – New Feedstock for the Chemical Industry.* Wiley-VCH.

Ulber, R., Muffler, K., Tippkötter, N., Hirth, T., and Sell, D. (2011b). Introduction to renewable resources in the chemical industry. In R. Ulber, D. Sell, and T. Hirth (Hrsg.), *Renewable Raw Materials* (S. 1–7). Wiley VCH.

US Energy Information Administration. (2013). *International Energy Outlook 2013.* Abgerufen am 23. 12 2014 von www.eia.gov/forecasts/ieo/pdf/0484(2013)pdf

USDA. (2009). *Oilseeds: World Market and Trade.* FOP1-09.

USDA. (2013). usda01.library.cornell.edu/usda/fas/sugar//2010s/2013/sugar-05-23-2013.pdf.

Wicke, L., Schellnhuber, H. J., and Klingefeld, D. (2012). *The 2 max Climate Strategy – A Memorandum.* Abgerufen am 23. 12 2014 von www.pik-potsdam.de/research/publications/pikreports/.files/english_short_pr116

Wim Soetaert, E. J. (Hrsg.). (2010). *Industrial Biotechnology – Sustainable Growth and Econoic Success.* Wiley-VCH.

Zervos, A. (2014). *Renewables 2014 Global Status Report.* Abgerufen am 27. 12 2014 von www.ren21.net/Portals/0/documents/Resources/GSR2014/GSR2014_full%20report_low%20res.pdf

2

Fundamental Science and Applications for Biomaterials

Ali S. Ayoub[1,2] and Lucian A. Lucia[2]

[1] *Archer Daniels Midland Company, ADM Research, Chicago, IL, USA*
[2] *North Carolina State University, Department of Forest Biomaterials, Raleigh, NC, USA*

> *The fuel of the future is going to come from fruit like that sumac out by the road, or from apples, weeds, sawdust - almost anything. There is fuel in every bit of vegetable matter that can be fermented. There's enough alcohol in one year's yield of an acre of potatoes to drive the machinery necessary to cultivate the fields for a hundred years.*
> Henry Ford, 1925

2.1 Introduction

The basic construct of biopolymer matrices remains a virtually insurmountable obstacle to the "best laid plans of mice and men" of providing products to compete with petro-based chemicals and associated commodity items. A more robust and precise understanding of the factors that limit a widespread use of lignocellulosic substrates in society is perhaps the most pressing challenge that the emergent bio-economy faces. The goal, therefore, of this chapter is to elucidate the fundamental physico-chemistry of the biomaterials, emphasize their value proposition for supplanting petrochemicals, tackle the challenges of conversion, and ultimately provide a milieu of possibilities for the biomaterials. The reader will be conversant and knowledgeable of the critical issues that surround the field of lignocellulosic intransigence, possible successful strategies to cope with their inertness, and potential pathways for the successful use of lignocellulosics and starch in the new bio-economy.

2.2 What are the Biopolymers that Encompass the Structure and Function of Lignocellulosics?

In the history of energy usage, wood has occupied a particularly noteworthy and prominent role for most of humankind. It was not until about 100 years ago (the early part of the twentieth century) that its hegemony and utility came into serious question principally due to the discovery of a cheap, seemingly inexhaustible, and

Introduction to Renewable Biomaterials: First Principles and Concepts, First Edition.
Edited by Ali S. Ayoub and Lucian A. Lucia.
© 2018 John Wiley & Sons Ltd. Published 2018 by John Wiley & Sons Ltd.

easily implementable product known as petroleum (literally, "rock oil"). At that time, a famous entrepreneur and industrialist, Mr. Henry T. Ford, opined that it behooved mankind to foster the exploitation and use of natural materials (such as wood) for their use as fuels; for example, he planned to power his Model Ts with ethanol while early diesel engines were run with peanut oil. The ability to use wood, plants, and their by-products for energy and other valuable products was important at that time, and it continues to be even more paramount now with the specter of the scarcity of petroleum reserves looming and the concomitant ever-increasing price of gasoline. However, efficient use of such raw materials demands a keen and in-depth understanding of their constituents and the processes in place to unlock their fuel and material value. A good understanding of these materials is key to their future use.

Wood is a raw material that has served humankind very well over its history. Locked within its macrostructure are three significant polymers (biopolymers) whose utility can rival that of petroleum and its by-products. The principal building blocks of wood (and hence nearly all lignocellulosics) are cellulose, heteropolysaccharides (or "hemicelluloses"), and lignin. These raw polymers are among the most abundant materials in the biosphere. They are only rivaled by chitin, another polysaccharide like cellulose that is found in the exoskeletons of marine and terrestrial life forms, as to their dominance in the material world. The symmetry found in the natural world is extraordinary especially when comparing the exquisite twofold screw structure of cellulose, the most dominant biomaterial on land, to its analogue, chitin, and the most dominant biomaterial in the sea. Figure 2.1 demonstrates this awesome symmetry.

2.2.1 Cellulose

Cellulose, the major constituent of lignocellulosics, occupies up to 50% or more of the overall composition of the matter [1]. In general, it is considered to be the most abundant renewable material on the planet, with annual cellulosic biomass production in the order of approximately 2 teratons (or 2×10^{12} tons). Although it has the same structural motif in each material that it comprises (i.e., the same corkscrew twofold symmetry), its degree of polymerization (DP or "n" shown in Figure 2.1) and crystallinity (degree of packing order as evidenced by X-ray crystallography) can vary widely. Estimates of the DP found in the literature show that it can vary from about $n = 300$ (in wood) to upward of 10,000 (cotton and bacterial cellulose, BC). It can also show wide trends in the overall crystalline morphology; for example, it can show a relatively low crystallinity in wood,

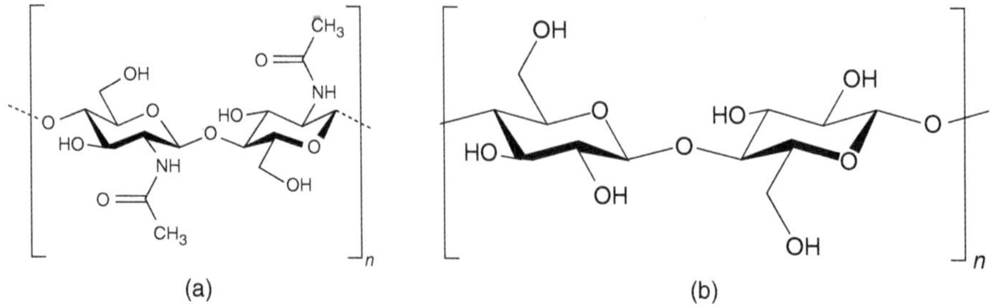

Figure 2.1 The archetypal structures of the most abundant biomaterials on the planet.
(a) The repeating unit (N-acetylglucosamine) of the biopolymer chitin. (b) The repeating unit (glucose) of the biopolymer cellulose.

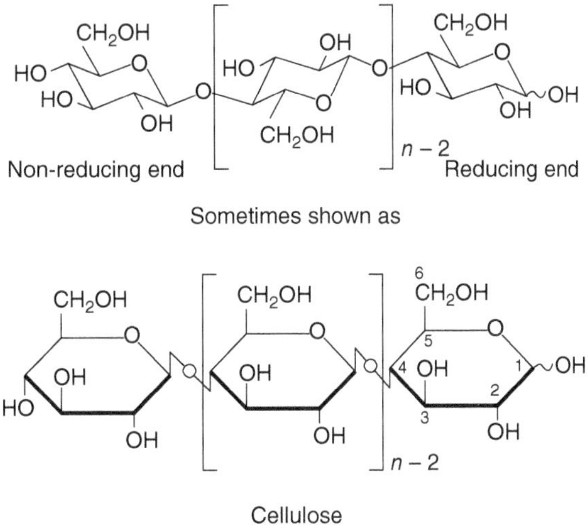

Figure 2.2 A simplified representation of the stereochemical asymmetry present in cellulose: the existence of a non-reducing end group (NREG) versus an opposite reducing end group (REG) give cellulose different terminal chemistries.

whereas it can be highly crystalline in *Valonia* [2]. Nevertheless, cellulose tends to be rather nonreactive to chemicals other than cellulose enzymes. It is a linear polymer that is made up of β-1,4-linked D-glucopyranose units.

One of the glucose units within the chain can be described as having a rotation of 180° relative to its neighbor within the chain. The chemical description of a "dimer" or two units is cellobiose unit, which technically could be used to describe the polymeric structural pattern. Classically, a cellulose chain, in as much as its glucose monomer makeup, presents two different terminal hydroxyl groups: the non-reducing end group (NREG) and the reducing end group (REG). The NREG consists of a 4-hydroxyl group (at the 4 carbon), whereas the REG is categorized in organic chemistry as a hemiacetal linkage (or aldehyde hydrate group). Figure 2.2 shows a classical representation of the NREG and REG.

Each glucose unit presents three reactive hydroxyl groups (C_2, C_3, and C_6), which expedite the formation of both intra- and interchain hydrogen bonding. The stiffness and compactness of the chain can almost exclusively be attributed to interchain interactions. However, the structural uniformity of cellulose in any plant is lacking; typically, there exist both highly ordered (crystalline) domains and amorphous (low degree of order) domains, which as a rule depend on the raw material and the treatments it has been subjected to. The relative robustness of cellulose is critically dependent on these features.

Cellulose generally provides the main mechanical properties of any lignocellulosics owing to its packing and hydrogen-bonded structure; in general, it provides the overall load-bearing capacity of wood and plants but does not typically complex with the other biopolymers in the lignocellulosic matrix.

Cellulose is synthesized in vascular plants[1] in the plasma membrane of the rosette terminal complexes (RTCs), a class of macromolecular protein structures approximately 25 nm in diameter. These structures contain the cellulose synthase enzymes (at least

---

1 Vascular plants (also known as tracheophytes or higher order plants) possess the so-called lignified tissues – a distinguishing characteristic from avascular plants – that are primarily responsible for circulating water, nutrients, and photosynthetic by-products within the plant.

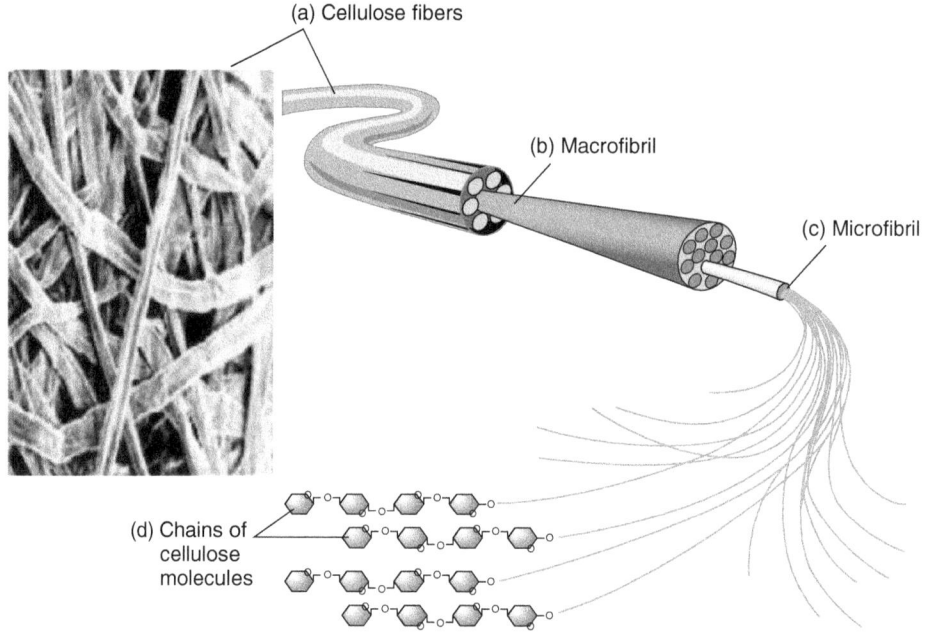

Scheme 2.1 A simplified pictorial summary of the contextual development of cellulose fibers: the evolution of cellulose chains into microfibrils, an elementary unit in cellulose, which contribute to the higher order elementary structures (e.g., the macrofibril is composed of a bundle of microfibrils). Reprinted with permission from http://alevelnotes.com/Carbohydrate-polymers/65.

three different sythases are involved that are encoded by $CesA$[2] genes) that are responsible for synthesizing the individual cellulose chains. The RTCs are able to spin a bundle of cellulose chains known as a microfibril into the cell wall. Scheme 2.1 shows a simplified representation of the production of microfibrils leading to cellulose fibers.

Native cellulose whose production is demonstrated in Scheme 2.1 is known from crystallographic data to have two distinct crystalline phases: Iα and Iβ [3]. Recently, exact representations of the principal phases of cellulose (Iα and Iβ) were obtained using atomic-resolution synchrotron and neutron diffraction data. The resulting structure of Iα was a one-chain triclinic unit cell (shown in Figure 2.3 with the principal axes: a, b, and c). The conformation of the glucosidic linkages and hydroxymethyl groups is identical, which is not the case for Iβ. In Iβ, there are two conformationally distinct chains in the monoclinic unit cell (corner and center chains) that requires adjacent glucosyl residues (in the same chain) to be the same. Both triclinic and monoclinic crystal systems derive from the seven crystal systems that are available to describe highly ordered systems.

The triclinic system can be best described as a crystal whose vector descriptors possess unequal length AND are not mutually orthogonal perpendicular to each other as in a classic three-dimensional Cartesian coordinate system. Similarly, a monoclinic system

---

[2] $CesA$ genes are likely responsible for encoding the catalytic subunit of cellulose synthase.

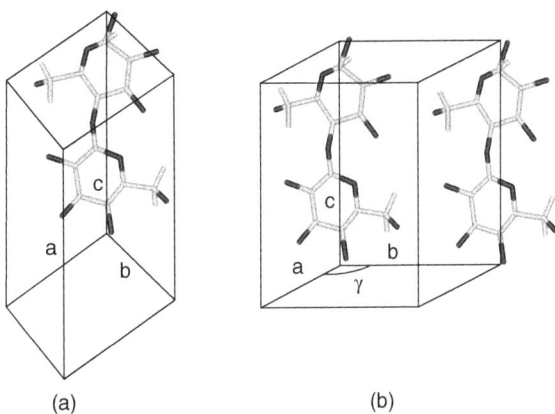

Figure 2.3 A representation of the unit cell mode of chain packing for cellulose. (a) The triclinic unit cell (Iα). (b) The monoclinic unit cell (Iβ).

has vectors of unequal length, but they form a rectangular prism with a parallelogram – a base indicating that two vectors are orthogonal, but the third angle, γ, $>90°$.[3]

2.2.2 Heteropolysaccharides

Akin to cellulose in form, but not in function, are the heteropolysaccharides (or "hemicelluloses," a more colloquial, but inexact term for this class of biomaterials) [4, 5]. This class of biomaterials is likely the second most dominant terrestrial material available after cellulose. In general, the amount of hemicelluloses is typically 20–30% of the dry weight of wood. They constitute a variegated class of materials without the chemical precision and homogeneity of cellulose, yet they are naturally produced annually in the order of 60 billion tons. This class of biomaterials is almost mandatorily associated with cellulose in any cellulosic matrix. They act in general to maintain the physical integrity of the cellulosic microfibrils and likely engage in a covalent complex (the lignin–carbohydrate complex, or LCC) with the lignin polymer of the plant cell [6]. They adopt a multiplicity of structural motifs, the most popular form being "xylan," which is likely the third most abundant biomaterial on the planet [7]. Figure 2.4 shows several representations of xylan, a typical and globally dominant heteropolysaccharide.

As shown in Figure 2.4, heteropolysaccharides are very much unlike cellulose in a structural sense. Cellulose is crystalline, that is, well organized or packed, of a high molecular weight, and possesses a low polydispersity.[4] However, heteropolysaccharides tend to be rather amorphous in their structure (often they are branched as shown in Figure 2.4), of a low molecular weight, and display a low PDI. In fact, the heteropolysaccharides typically do not have just a single repeating unit (such as xylan) but can often contain a multiplicity of monomers. For example, the dominant heteropolysaccharide is xylan, which tends to be fairly localized in angiosperms (hardwoods), whereas the dominant heteropolysaccharide in gymnosperms (softwoods) is glucomannans.

---

3 For more information on the theory and technical aspects of X-ray crystallography and crystal lattices, the reader is encouraged to obtain and inspect a free copy of "Elements of X Ray Diffraction" by B.D. Cullity (1956) from http://archive.org/details/elementsofxraydi030864mbp.

4 The polydispersity or polydispersity index (PDI) is a measure of the distribution of molecular masses in a polymer sample. It is calculated by obtaining the quotient of the weight average molecular weight (M_W) divided by the number average molecular weight nM_n). It is a reliable metric for the "purity" or uniformity of the chain length of the polymer in the sample.

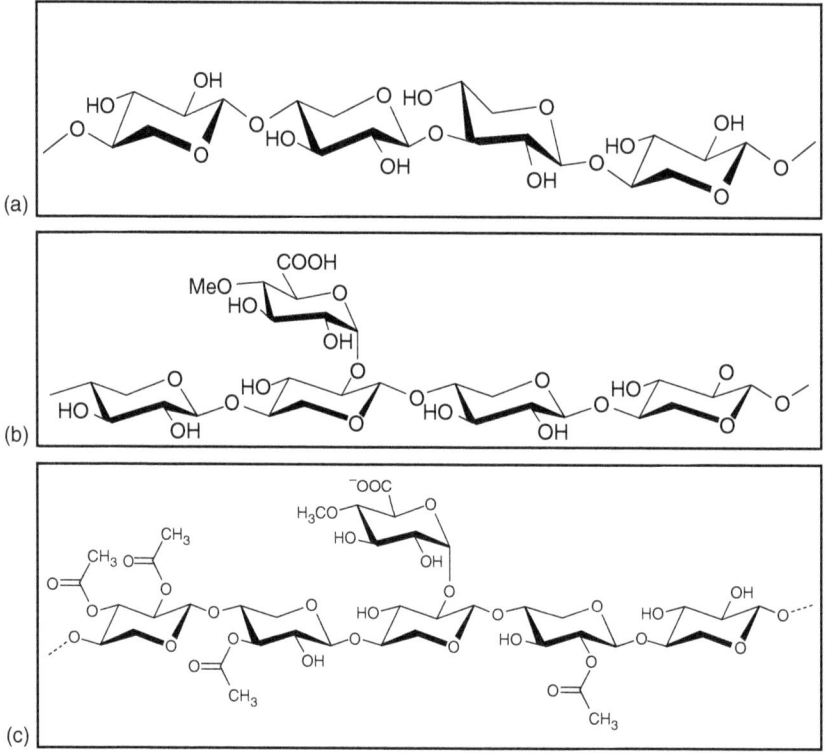

Figure 2.4 Various structural representations of xylan. (a) Simple xylan backbone composed of non-functionalized xylose monomers. (b) Xylan backbone with a pendant 4-O-methylhexenuronic acid residue at the 2-carbon. (c) Representation of a xylan backbone decorated with pendant acetyl groups.

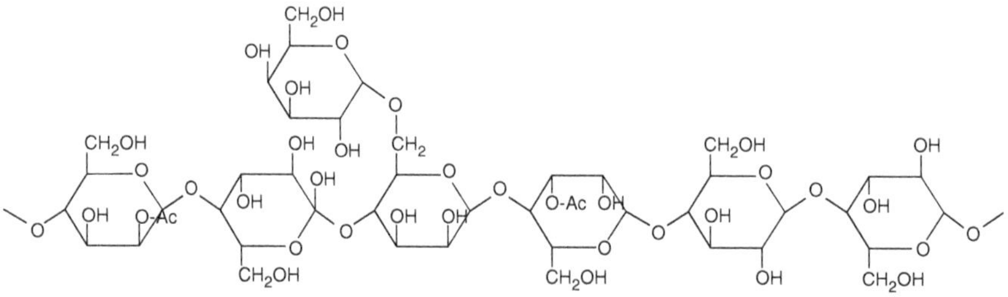

Figure 2.5 Representative structure of the galactoglucomannan heteropolysaccharide. Note that the glucomannan backbone has several Ac (acetyl) groups decorating it, while it possesses a pendant galactose residue on the 6 carbon of a mannose residue.

In addition, there are galactoglucomannans, arabinoxylans, and so on. The list is nearly endless in the natural diversity of heteropolysaccharides that are found. Figure 2.5 shows the structure, for example, of galactoglucomannan, a typical branched heteropolysaccharide.

Given the various physical characteristics stated here for heteropolysaccharides, it is no surprise that they tend to be more susceptible to degradation from both chemical and enzymatic attack. In addition, they can sometimes be more amenable to chemical derivatization and reaction than their cellulose analogue.

2.2.3 Lignin

This last polymer, albeit neglected for many years because of its molecular heterogeneity and seeming lack of tractability, is one of the most important polymers on the planet. It forms the last element of the lignocellulosics that primarily comprise the biomass on this planet. The term "lignin" is in fact taken from the Latin word *lignum*, which is translated to "wood." It is found virtually in all biomass especially vascularized plants including herbs and grasses, but it is chiefly located (on a per capita basis) in the cell wall of woody tree species [8, 9]. On a mass basis, up to approximately 30% of all carbon in the biosphere can be attributed to lignin. A representation of its diverse native structure is provided in Figure 2.6 [10]. The structure seemingly lacks the signature monomer associated with polymer structures, but in wood chemistry circles, it is assumed that the monomer is based on a C_9 phenylpropanoid residue that will be further elaborated within this section.

The structure elucidated here is speculative at best in nature and represents a best guess attempt because not only is its X-ray structural characterization impossible, but it is extremely divergent in its structure between species and even within the same species. In terms of its role among the natural polymers in woody and plant species, lignin maintains unique and powerful functionality within the bio-system. It is essentially a natural "glue" that helps to keep the polysaccharide bundle intact and whole. The current theory for its particular niche in lignocellulosics is that it covalently complexes with the heteropolysaccharides in a so-called LCC. In addition to its chemical attachment to polysaccharides, it is believed that lignin also acts as a plant's second line of defense (after the bark/extractives) against microbial attack/infestation.

Lignin as shown in Figure 2.6 is a complex aromatic network polymer (but it is three-dimensional) that is polydisperse and contains a number of divergent functionalities and branching points. Although PDI may not be objectively applicable in the case of lignin, lignin nevertheless is a branched network polymer that has its origins in well-characterized lignol subunits. Figure 2.7 shows the principal monomeric units that constitute the biosynthesized lignin.

Each of the structures shown in Figure 2.7 is available in differing amounts in different classes of woods or plants. For example, monocotyledonous grasses tend to have almost exclusively the *p*-hydroxyphenyl monomer unit, whereas gymnosperms are almost exclusively constructed out of the guaiacyl units. Angiosperms, on the other hand, tend to have some combination of syringyl (principal component) and guaiacyl units.

The lignin extant in nature generally has a support role for the polysaccharide matrix. In woody tissue, it tends to surround the wood cell helping to ensure not only integrity but to maintain proper surface energetics (hydrophobicity) to allow for the uninterrupted circulation of water and nutrients. Figure 2.8 demonstrates the respective localization of lignin within a representative wood cell.

The cell wall is the region that contains the most lignin polymer, but it is highly concentrated between cells, in essence, to ensure good mechanical integrity among

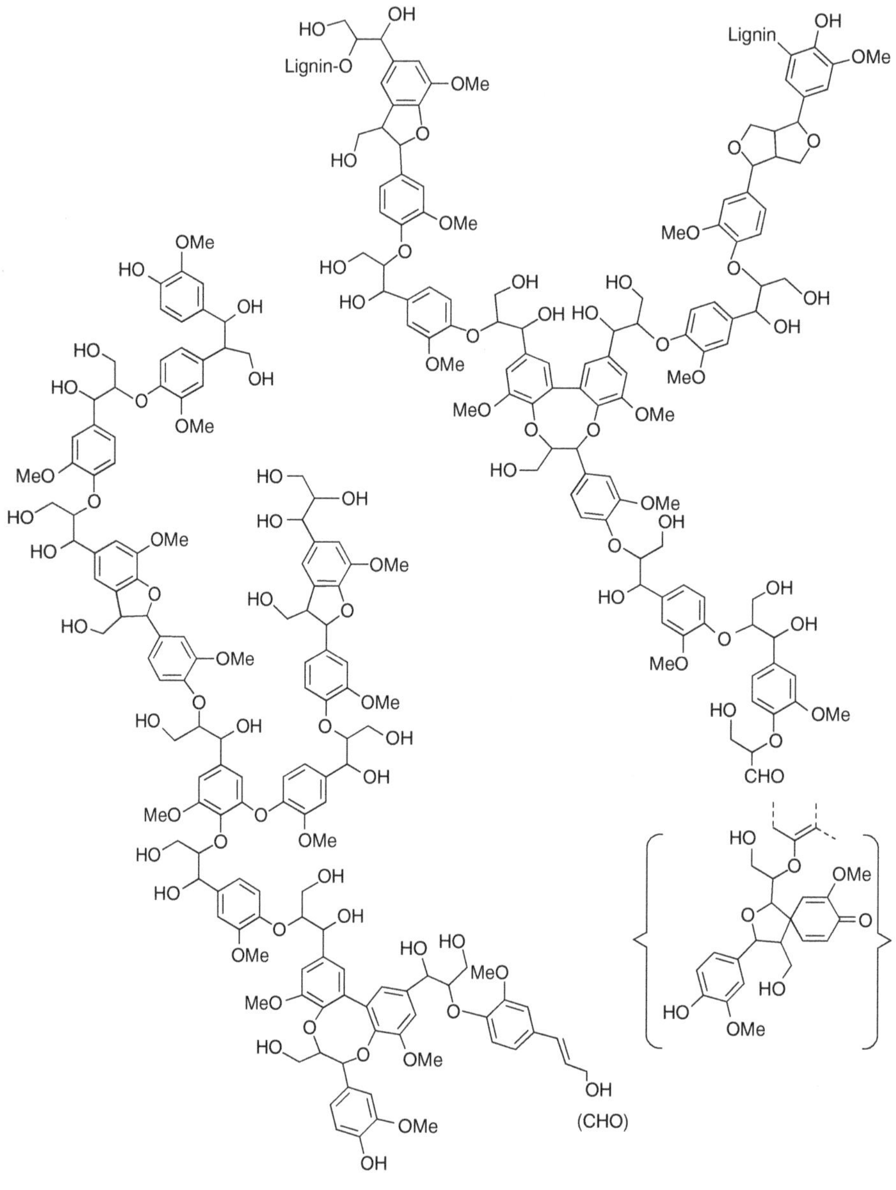

Figure 2.6 A generic representation of the lignin that is believed to be localized within angiosperm wood cells.

the macrofiber bundle [11]. This material has for roughly 130 years been the targeted polymer for deconstructing cellulosic fibers for papermaking. A huge global paper pulp industry has emerged since 1879 when Dahl, a German chemist, used sodium sulfate as a makeup chemical for soda pulping to regenerate sodium hydroxide. Serendipitously, sodium sulfide was formed and unexpectedly gave far better pulping results in terms of rates and pulp quality. The process was termed *kraft* from the German for "strong" [12].

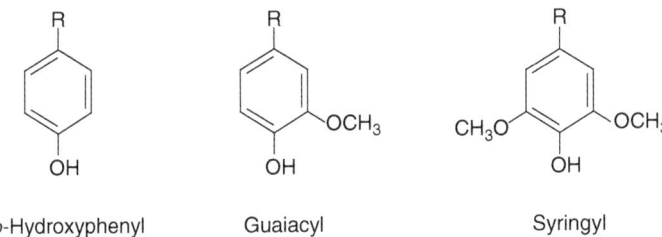

p-Hydroxyphenyl   Guaiacyl   Syringyl

Figure 2.7 The generic lignin monomeric structural motifs that constitute the totality of all of the lignin polymers that are extant in nature.

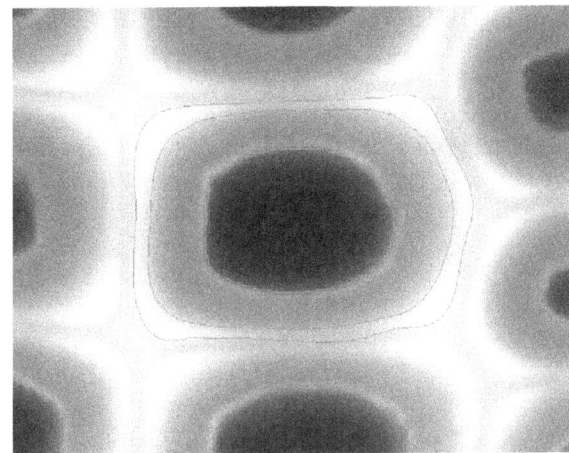

Figure 2.8 A fluorescence micrograph obtained from a cross-section of wood cells. Shown in the micrograph are the major elements of the wood cell starting from the inner lumen (dark, no fluorescence), normal lignified cell wall (second part extending from the inner dark sphere), to the outer middle lamella.

2.2.4 The Discovery of Cellulose and Lignin

Anselme Payen (1795–1891) was born in Paris to Marc and Jean Payen, a family in which there was great respect for science, law, and chemistry. In fact, young Anselme (at the tender age of 13) began a lifelong love of studying science first with his father, a man whose entrepreneurial spirit led him to establish chemical factories, and then Anselme went off to study chemistry, physics, and mathematics at the École Polytechnique under the tutelage of Louis Nicolas Vauquelin and Michel Eugène Chevreul. Interestingly, after this quasi-internship, he returned to work for his father as the superintendent of a borax refining plant when he was only 23 years old (1818). What was so unique about this situation was that Payen's father had devised a better way of producing borax from boric acid that allowed him better market opportunities. After Payen's father died in 1820, he held sole custody of the family estate. Payen turned his interest to one of his father's factories that refined sugar from beets. Interestingly, he employed activated charcoal to decolorize the sugar, a method that has been in vogue ever since that impressive discovery. In addition, this work helped to expedite the transition, on a world stage, from obtaining sugar from cane to beets. One of his most remarkable efforts and perhaps what he may be best known for is the discovery of diastase (from the Greek for "separate"), an enzyme that converts starch to glucose. He was able to isolate this substance from malt extract in 1833; interestingly, it was the first isolated enzyme, an organic system that demonstrates catalysis, that is, enhancing the rate of reaction without being consumed.

His choice of terminology, *diastase*, led to the tradition of using the suffix *-ase* in biochemistry for the naming of enzymes. In 1834, Payen began the systematic study of wood from which he discovered a substance from the plant cell walls that could be hydrolyzed (broken down with water) to glucose units (similar to starch). He called the substance "cellulose," which began another tradition of using "*-ose*" as the suffix to name carbohydrates. In 1835, he departed from industrial work in favor of a professorship of industrial and agricultural chemistry at the École Centrale des Arts et Manufactures. In 1839, he then accepted a joint appointment at the nearby Conservatoire des Arts, which he held in conjunction with his original appointment until his death in 1871. He was a prolific researcher, publishing over 200 papers and 10 books on topics such as dextrans, sugars, lignin, cellulose, and starch. His seminal paper on cellulose and lignin isolation was published in 1838. These experiments in 1838 revealed that in addition to cellulose, wood contained an oxidizable crusty substance that was later designated as "lignin" in 1857 by Schulze. Payen treated the wood with a concentrated nitric acid solution and later washed the residue with an alkaline solution (sodium hydroxide) to dissolve the crusty substance. Payen noted significant differences between the wood and the crusty substance; he is also credited with the first Klason lignin isolation method whereby he used concentrated sulfuric acid to remove the polysaccharides, which then allowed for the isolation of the "lignin" fraction. Today, the American Chemical Society (ACS) honors Anselme Payen's memory by awarding each year a prize in his name to the scientist who in the opinion of the Cellulose and Renewable Materials Division (CELL) of the ACS has contributed the most to the science, engineering, and technology of cellulose and renewable materials (more information can be found at the ACS site: http://cell.sites.acs.org/anselmepayenaward.htm).

2.3 Chemical Reactivity of Cellulose, Heteropolysaccharides, and Lignin

A proper understanding of the chemical, biological, and mechanical reactivity of the biopolymers in biomass is crucial to determining the best approaches for their strategic utilization. Throughout history, a basic understanding of lignocellulosics has lagged behind their usage on a societal level. This is not surprising because pragmatism dictates that this should be the case; people will always be practical in approaching the usage of materials before a fundamental survey of the properties of the materials is accomplished. However, advances in bio-fuels and biomaterials cannot be made from a pragmatic perspective because there are too many technical hurdles to overcome and, additionally, the economic barriers to implementation are severe. Therefore, this section attempts to survey the individual reactivities of the biopolymers under scrutiny. The survey examines several different stressors or reactants used on the biopolymers and their responses.

2.3.1 Cellulose Reactivity

One of the complicating factors in the proper understanding of the reactivity of cellulose is its accessibility in addition to its chemical makeup. It is organized elegantly within a lignocellulosic matrix that consists of its interplay with the heteropolysaccharides and the lignin. Today, the reactivity of cellulose is a topic of grave importance because of its ability to supply an ever burgeoning bio-fuels and biomaterials community [13, 14]. As demonstrated earlier, the reactivity of cellulose is a function of accessibility, which is severely hampered by its compact structure. This compact structure is a function of the presence of a very strong hydrogen bonding network that gives rise to highly ordered region.

Critical factors that have been suggested by a number of experts include the number and size of pores in the cellulose structure, the molecular size and type of chemical that is added to the cellulose, the internal surface as controlled by the size of fibrils/aggregates, and the morphology of the cellulose macromolecules. These critical factors can only be addressed by (i) ensuring that the cellulosic pores are sufficiently open to accommodate the chemicals/reagents added, (ii) deconstructing fibrillar aggregates, and (iii) deconstructing the ordered regions to no longer adopt a stiff, compact network structure, that is, the hydrogen bonding network must be disrupted. An optimal chemical interaction, therefore, must consider these latter three criteria; recently, chemical, mechanical, and biological treatments have been tested, with a great push toward the environmentally benign biological treatments.

Chemical treatments of cellulose [15] are employed to enhance the swelling of the cellulose fibers. This swelling not only facilitates the passage of chemical agents, but its primary purpose is to disrupt the hydrogen bonds because of the high osmotic pressure induced by the swelling phenomenon. Thus, the hydrogen-bonded network becomes disrupted, and as a result the compact structure is no longer available leading to a more accessible structure. A pictorial representation of this phenomenon is shown in Figure 2.9.

The scheme illustrates the inclusion of water molecules that interfere with the hydrogen bonding between the C_6, and the C_3 and C_2 hydroxyls of chain neighbors and act to therefore swell the cellulosic substrate. As a result of this inclusion of water molecules, the swelled structure loses its ordered nature and in some cases, there is a complete loss of crystallinity, which consequently leads to an increase in the active surface area or exposed hydroxyl groups (which were heretofore buried within the tightly packed cellulosic crystallite). There have been a number of studies within this

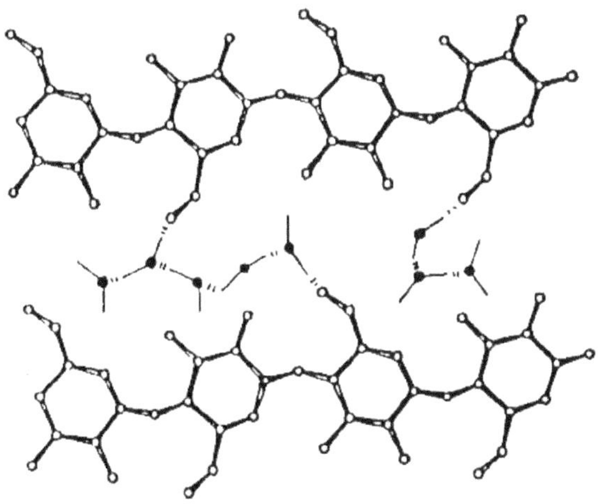

Figure 2.9 Shown is a simple cartoon that illustrates the effect of introducing water molecules within the H-bonded network structure of cellulose. Reprinted with permission from Westermark, S. "Use of Mercury Porosimetry and Nitrogen Adsorption in Characterisation of the Pore Structure of Mannitol and Microcrystalline Cellulose Powders, Granules and Tablets." Academic Dissertation, November 2000, University of Helsinki, Helsinki, Finland.

particular area that have shown such accessibility to solvents including among others the use of sodium hydroxide and its combination with urea. Urea is conjectured to insert within the interchain cellulosic matrix and disrupt the hydrogen bonding paradigm. The purpose of such treatments is to facilitate accessibility for a particular

In addition to chemical treatments, cellulosics can be treated with a mechanical process [16]. Mechanical treatment of the cellulose fibers is used in the pulp and paper industry because of its capacity to enhance fiber–fiber bonding, to cut or make the fibers stronger, and to produce changes on the cellulose structure. For instance, strong bonds among fibers give the printing paper strong and smooth properties. When the pulp is subjected to mechanical treatment, the interfibrillar bonds, which are mainly located in the primary wall and in the outer lamella of the S1 layer of the cell wall, are disrupted. This effect leads to an increase in the reactive surface area of the fibers, improving the accessibility of the cellulose. In several studies, mechanical treatment has been used in combination with other treatments.

Moreover, enzymatic treatments can be used as well [17]. Enzymes have broad industrial applications. They have been used in the detergent, food, and pharmaceutical sectors. Enzymes have also been studied in the pulp and paper industry, and they are currently used for several applications, including deinking and as bleaching agents. The effect of enzymatic treatments on cellulose reactivity has also been investigated. It has been reported that enzymatic treatments, especially that of cellulases on dissolving pulps, hold a great potential for increasing cellulose reactivity.

One of the enzymes is the cellulases: monocomponent endoglucanases. Cellulases are enzymes that hydrolyze the 1,4-β-D-glucosidic bonds of the cellulose chain. There are three major groups of cellulases: endoglucanases, cellobiohydrolases or exoglucanases, and glucosidases. These enzymes can act alone or together on the cellulose chain or together. When they act together, a synergistic phenomenon is often generated, resulting in an efficient degradation of the cellulose structure.

Endoglucanases are enzymes that randomly cleave the amorphous sites of the cellulose creating shorter chains (oligosaccharides) and, therefore, new chain ends. Cellobiohydrolases or exoglucanases attack the reducing and non-reducing ends of the cellulose chains, generating mainly glucose or cellobiose units. This type of cellulose can also act on microcrystalline cellulose by a peeling mechanism. Glucosidases act on cellobiose generating glucose units. It has been suggested that there are three primary parameters affecting the degree of enzymatic hydrolysis: the crystallinity, the specific surface area, and the degree of polymerization of the cellulose.

Most cellulases consist of two domains. The first is a catalytic domain, which is responsible for the hydrolysis of the cellulose chain. The catalytic domain of endoglucanases is "cleft shaped." Exoglucanase, on the other hand, has a "tunnel-shaped" catalytic domain structure. The second is a cellulose-binding domain (CBD), which helps the enzyme to bind to the cellulose chain bringing the catalytic domain close to the substrate. An interdomain linker serves as a connection between the two domains.

2.3.1.1 Reactivity Measurements

Several methods have been developed to measure cellulose reactivity, including iodine sorption water retention value of pulps, swelling water coefficient, and viscose filter value [18–20].

2.3.1.2 Dissolving-Grade Pulps

For regenerated cellulose manufacturing, it is generally known that the raw material is required to have a high cellulose content (over 90%) and low levels of hemicellulose, lignin, extractives, and minerals [21–23]. Today, the raw materials used for the production of regenerated cellulose are dissolving-grade pulps and, to a lesser extent, cotton linters. Dissolving-grade pulps are produced mainly by two different processes: the sulfite process and the prehydrolysis kraft process. Other pulping processes have been investigated for the production of these pulps, including organosolv pulping. This process is based on the use of organic solvents; however, the expense of solvent recovery is the biggest drawback of this process. Dissolving-grade pulps are costlier than kraft pulps. This can be attributed to several factors, such as wood costs (the production of these pulps has a lower yield since hemicelluloses are dissolved and washed away); capital costs (because the yield is low, more equipments may be needed to have a high production); chemical costs; production rates (lower than for paper-grade pulps); and the inventories and storage space (the pulps are produced for specific customers with certain requirements, which implies a high control of the inventory). As a consequence, the viability of converting paper-grade pulps into dissolving-grade pulps arises.

2.3.1.3 Converting Paper-Grade Pulps into Dissolving-Grade Pulps

In recent years, several studies have used different methods to examine the feasibility of modifying paper-grade pulps for further use as dissolving-grade pulps [24, 25]. These studies have focused mainly on the optimal removal of hemicelluloses because in the production of viscose, hemicelluloses can affect the viscose filterability, the xanthation of cellulose, and the strength of the end product. Several methods have been reported for the removal of hemicelluloses, including treatments with alkaline extraction, nitren and cuen extraction, and a combination of pretreatments using xylanases and alkaline extraction. However, little attention has been paid to the effect of changes in the accessibility and reactivity of the cellulose after these treatments.

2.3.2 Hemicellulose Reactivity

Several methods have been used to extract hemicellulose from woody tissues [26–28]. Those methods include dilute-acid pretreatments, alkaline extraction, alkaline peroxide extraction, liquid hot-water extraction, steam treatment, microwave treatment, and ionic liquid extraction.

Dilute-acid pretreatment, often involving about 0.5–1% sulfuric acid, is a useful procedure for hemicellulose isolation. By using this method, the majority of the original amount of hemicellulose from the poplar tree (hardwood) can be recovered as dissolved sugar. However, in the course of such treatment, a high amount of the hemicellulose monomers are degraded, leading to the generation of by-products. On the other hand, extraction of hemicellulose from sugarcane bagasse by using hot-water extraction at a temperature of 150–170 °C recovers almost 90% of the hemicellulose as dissolved sugar, but with relatively less monomer degradation than that of the dilute-acid method. Because hot-water treatment cleaves some of the acetate groups from hemicellulose, the pH decreases. The reduced pH results in additional generation of acetic acid, leading to a phenomenon called "autohydrolysis." It has been shown that autohydrolysis can be promoted by irradiating wood with microwaves in water.

Due to severe treatments caused by acidic and partly applied heated extraction processes, hemicellulose degrades by losing its chain length, and consequently, exhibits high polydispersity. Steam and microwave treatments are combined with chemicals to dissolve the hemicellulose; however, due to the complexity of those methods, they have been applied mainly on a small scale for hemicellulose extraction. In steam treatment, ester bonds in the hemicelluloses will be cleaved and then the wood is treated with steam, resulting in the formation of acetic acid. This will lower the pH and, thus, induce autohydrolysis of the glycosidic bonds in the hemicelluloses. Such a process will generate a low-molecular-weight, water-soluble hemicellulose. From this method, one can get a low percentage yield of hemicellulose with significant contamination of dissolved cellulose and lignin and degradation products of the hemicellulose.

An ionic liquid/cosolvent extraction system has been used by Froschauer and coworkers to separate hemicellulose and cellulose from wood pulp. In this study, hemicellulose-rich birch kraft pulp was selectively separated, with high levels of purity, into pure cellulose and hemicellulose fractions by using mixtures of cosolvents (water, ethanol, or acetone) and the cellulose-dissolving ionic liquid 1-ethyl-3-methylimidazolium acetate (EMIM OAc) with reaction conditions of 60 °C for 3 h under stirring. This process was used to generate dissolving pulp that met the manufacture of revived cellulose products and cellulose derivatives because of its extraordinary cellulose content, which was assumed to be more than 90%, as well as the product's high brightness and uniform molecular-weight distribution.

Alkaline extraction is well studied and is known as a strong and efficient method for hemicellulose extraction. Indeed, alkaline treatment can cleave the ester linkage between ferulic acid of lignin and the glucan and arabinan residues of hemicellulose in the cell wall, thus releasing oligo- and polymeric hemicellulose rather than sugars, as obtained from dilute-acid extraction or hot-water extraction. The extraction of hemicelluloses from various biomasses is expected to be closely related to the amount of LCCs. The LCCs entail covalent linkages between lignin and carbohydrates, mainly hemicelluloses. The major types of LCCs include phenyl glycoside, benzyl ether, and benzyl ester types of linkages. Different biomasses have different frequencies of LCC, for example, the amounts of ether and ester LCC linkages in pine and aspen cellulolytic enzyme lignin (CEL) have been reported to be about 2.2–2.5 and 0.3–0.6 per 100 monomeric lignin units, respectively.

The decision on which method to use for hemicellulose extraction is highly dependent on the final application of the recovered hemicellulose. For example, if the extracted hemicellulose is targeted for the production of bio-ethanol, then the monomeric form of sugars is required, and thus, it may be preferred to use dilute acid or hot water. Some other applications require a high-molecular-weight polymer (blended plastics, hydrogels, and others). For those applications, it may be extremely important to use the alkaline method for the isolation of hemicellulose from the biomass.

2.3.2.1 Structural Characterization of Hemicellulose

Great interest in converting lignocellulosic biomass into valuable, green fuels and chemicals has challenged researchers to develop methods for determining the structure, accurate chemical composition, quantity, and potential uses of hemicellulose in the lignocellulosic biomass [26]. Several analytical methods have been used to characterize hemicellulose: high-performance liquid chromatography (HPLC), gas chromatography (GC), size-exclusion chromatography (SEC), gas chromatography mass spectrometry (GC-MS), high-performance anion-exchange pulse amperometric detection (HPAE-PAD) chromatography, 1D and 2D NMR spectrometry, and matrix-assisted laser desorption/ionization time-of-flight (MALDI-TOF) MS.

2.3.3 Lignin Reactivity

Many new techniques were developed to isolate lignin formulations to study [29–33]. Generally, three approaches were taken toward investigating lignin structure: degradation reactions, biosynthetic work, and spectroscopy studies. In the most common approach, lignin formulations were subjected to various degradation reactions yielding identifiable products that gave useful structural information. Many new analytical techniques were developed during these times that were highly useful for the lignin chemists. In the final approach, model compounds were synthesized from postulated lignin precursors to produce new products for study. This was a time in which much work was applied to developing new methods to isolate lignin. In 1936, Bailey at the University of Washington showed by microdissection that the middle lamella was composed mainly (72%) of lignin rather than pectin. In 1935, Van Beckum and Ritter at the US Forest Products Laboratory (USFPL) removed lignin from plant tissues with hypochlorite followed by NaOH. The material that remained termed holocellulose consisted of the total carbohydrate mass present in the plant tissues. In 1939, Brauns reported that neutral solvent extraction of woody tissue and subsequent purification created a few percent of what he named native lignin or Braun's lignin. Presently, many investigators view this material as a mixture of lower molecular weight lignins and/or lignans. This work reflected the continuing search by some classical organic chemists for a lignin that could be extracted simply by use of solvents without chemical reaction. In 1947, Richie and Purves at McGill University oxidized wood at pH 4 with aqueous 5% sodium periodate. The periodate lignin preparation thus obtained was 86–96% Klason lignin, insoluble in organic solvents even at boiling temperature but completely soluble under conditions of sulfite pulping. A periodate lignin from spruce closely duplicated the behavior of spruce lignin in situ toward many degradative procedures.

2.4 Composite as a Unique Application for Renewable Materials

Plastic composite processors worldwide are becoming increasingly aware that environmentally sustainable products have become mainstream, and it can no longer be considered only a niche market that can be ignored [34–37]. Moreover, in the light of the recent Paris climate agreement in 2015, development of environmentally sustainable new technologies and materials is of growing importance; state and local governments are mandating it; and now, even the largest retailers are building it into the foundation of their marketing strategies. The development of renewable/sustainable materials is perceived by the industry as a hedge against the prospect that traditional plastics will be much more costly in the future due to dramatically higher petroleum prices. The sustainability movement is further seen as a positive development for plastic processors since it will drive further innovation and a new generation of materials with properties more comparable to commodity plastics. For instance, packaging and containers constitute a nearly $500 billion global market. Plastic container sales alone account for $130 billion worldwide.

The combination of lignocellulose and starch would mean a further step ahead in the utilization of bio-based materials for challenging applications such as Styrofoam-like foams, plastics, and packaging made from petroleum resources.

After an extensive literature review on this topic, we have concluded that there is a great need for systematic and accurate mapping of product structure characteristics.

2.4.1 Rationale and Significance

Plastics industries manufacture has a wide variety of materials, which plastics are produced or derived from major non renewable energy sources such as crude oil, natural gas and coal. Nearly 6% of the world's crude oil production is used for making approximately 245 million metric tons of plastics globally on an annual basis. These are used to meet both the requirements of cheap mass production and of highly specific applications. The worldwide economy dependence on petroleum-based plastics is not sustainable, based on the extremely volatile oil and energy situation, coupled with major changes in supply and demand patterns. The fluctuation in costs is a challenge but in addition, the limited feedstock availability is tightening and impacting supply and demand worldwide, and putting the industry under tremendous pressure.

Moreover, plastics now make up a significant part of a typical municipal solid waste (MSW) stream, and represent the fastest growing component. A huge 44 billion pounds of plastics enter United States' MSW stream each year, equivalent to $\frac{1}{2}$ pound per day per person. In the United States, plastics on average account for 10% by weight of MSW, more than metals (8%) and glass (6%). The cost of waste management is now a matter of great public concern. There is more carbohydrate on earth than all other organic material combined. Polysaccharides are the most abundant type of carbohydrate and make up approximately 75% of all organic matter. The use of biodegradable, starch- and wood-based products as proposed here would be a significant improvement to the economy, environment, and society. Furthermore, starch-cellulosic fibers are viable feedstocks for enzymatic conversion to ethanol or are high heat value material suitable for combustion, alternatives to landfilling.

Many industry announcements regarding new and innovative plastic products occur on an ever more frequent basis. *Coca-Cola* recently announced that they will begin utilizing polyethylene terephthalate (PET) bottles containing 30% renewable content from sugarcane-derived ethylene glycol. They also announced plans to convert all their plastic packaging to the new material by 2020. *Heinz* will use the same material to make 120 million bottles for their ketchup products this year. *PepsiCo* claims to have developed the world's first totally bio-based PET bottle. It is made from biomass including switchgrass, pine bark, and corn husks. Pilot-scale production began in 2012. Other interesting new materials entering the market include a new family of resins (Panacea) containing 10–40% finely ground soy-based protein and an injection mold-grade cellulose-based resin. The cellulose-based resin is being used to make the first biodegradable tubes for toothpaste.

Much of the focus on renewable and sustainable plastics involves the use of starch either as a feedstock or as a component for industrial products. Industrial products in the United States that utilize starch have grown from 13 million metric tons (MMT) in 1975 to over 160 MMT today. Starch is inexpensive, widely available, and one of the most abundant biomass products in nature. It is produced in many different plant organs including roots, leaves, seeds, and stems. However, the hydrophilic nature of starch and its tendency to embrittle with age do not make it suitable as a replacement for plastics – hence, the importance of adding lignocellulose to the starch matrix system.

2.4.2 Starch-Based Materials

The development and production of biodegradable starch-based materials have been spurred by oil shortages and the growing interest in easing the environmental burden of petrochemically derived polymers. Starch is one of the most studied and promising raw materials for the production of biodegradable plastics, which is a natural renewable carbohydrate polymer obtained from a great variety of crops (Figure 2.10). Starch is low cost material in comparison to most synthetic plastics and is readily available. Starch has been investigated widely for the potential manufacture of products such as water-soluble pouches for detergents and insecticides, flushable liners and bags, and medical delivery systems and devices. Native starch commonly exists in a granular structure, which can be processed into thermoplastic starch (TPS) under the action of high temperature and shear by melt extrusion. Unfortunately, the properties of TPSs are not satisfactory for some applications such as packaging materials and composites engineering products.

One of the unique characteristics of starch-based polymers is their processing properties, which are much more complex than conventional polymers. The processing of starch-based polymers involves multiple reactions, for example, water diffusion, granule expansion, gelatinization, decomposition, melting, and crystallization (Figure 2.11).

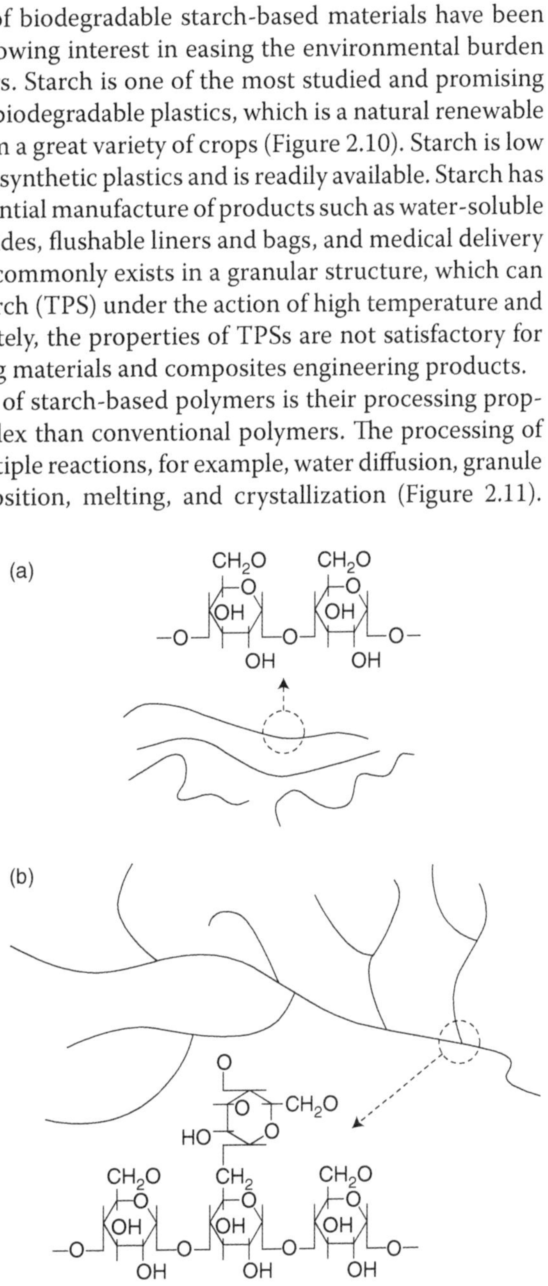

Figure 2.10 Chemical structures and physical schematic representation of (a) amylose starch and (b) amylopectin starch.

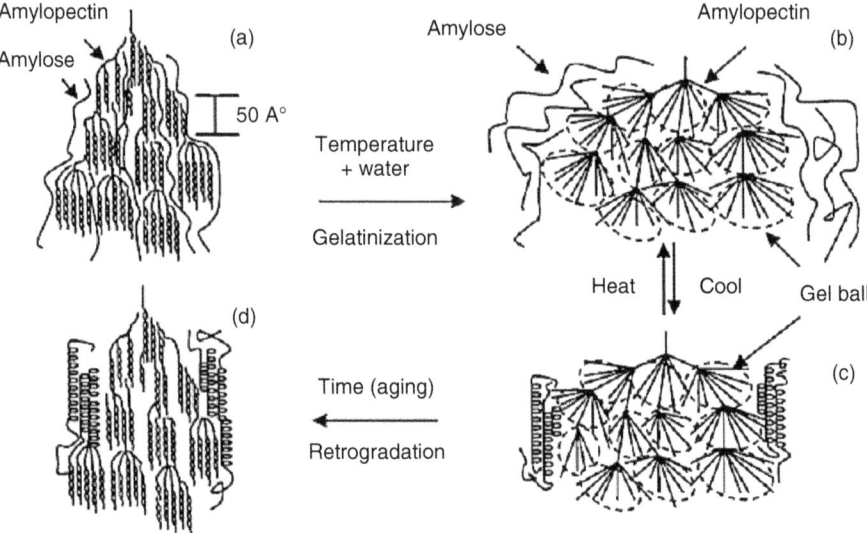

Figure 2.11 Schematic representation of the phase transitions of starch during thermal processing and aging.

Among the various phase transitions, gelatinization is particularly important because it is closely related to others, and it is the basis for the conversion of starch to a thermoplastic. Furthermore, the decomposition temperature of starch is higher than its melting temperature before gelatinization. Without physical force (shear stress), the process of gelatinization depends mainly on water content and temperature conditions. A previous study has shown that shear stress can result in the fragmentation of starch granules during extrusion. Indeed, both the mechanical and thermal energies are transferred to starch dough during extrusion in molten medium. The main objectives of most starch-processing techniques are melting and mixing, which are adjusted to minimize chain degradation.

Starch use in papermaking dates back to the invention of paper 2000 years ago, when it was applied to obtain a stronger and smoother writing surface. Starch contributes to paper manufacturing because it serves as a binding agent that can enhance the mechanical properties of paper and improve paper manufacturing by increasing paper pulp retention on the paper machine. Starch was also chemically modified into cationic forms to further improve interconnections between fibers, increasing thus the paper strength. Non-covalent binding of starch and cellulosic fiber in biocomposites were created using a polymer suspension either by, for example, drying or hot-pressing.

2.4.3 Starch-Based Plastics

Plastic ranks as the second most used packaging material in the United States. In contrast to paper, only 7% of plastic generated as waste is recycled. This explains why more plastics ultimately end up in landfills than paper or any other packaging material. China, one of the world's largest plastic producers, noted that the largest source of its marine pollution was from discharging wastewater to sea. In Europe alone, an estimated 2–3 million tons of plastics is used each year in agricultural applications. Polyethylene films are used extensively to increase yields, extend growing seasons, reduce the usage of pesticides and herbicides, and help conserve water. Films made of starch blends were some of the first films containing renewable content tested as agricultural mulch.

Packaging and containers make up the largest sector (29.5%) of plastic waste in MSW. Plastic packaging has become an integral part of the global marketplace. Today, packaging design has begun integrating sustainability as never before, in part, because sustainability itself has become a marketing angle. Retailers are now realizing that customers respond positively to products marketed in more sustainable or green packaging.

The interest in utilizing starch as a replacement for plastics started in the 1970s and intensified in the 1980s right along with the dramatic growth in the use of plastics worldwide and the concerns about the effects of plastics on the environment (Figure 2.12)

But starch by itself is a poor substitute for petroleum plastics due to its moisture sensitivity and inferior mechanical properties. However, numerous strategies have been tested resulting in commercialized technologies for incorporating starch in plastics. A survey of the more common approaches has been included in this section as follows:

2.4.3.1 Novamont

One of the early innovators in developing TPS blends for commercial production is the Italian company, Novamont. Founded in 1989, Novamont has developed four different classes of materials all based on blends of TPS and synthetic polymers. The film-grade product is based on blends of TPS and polycaprolactone. This grade degrades in a composting environment in about 1 month. The injection molding grade is a blend of TPS and cellulose derivatives. It degrades in about 4 months and is a rigid material that may replace polystyrene (PS). A foaming grade is also available that contains more than 85% TPS. The foam product is used as a replacement for PS foam for loose-fill packaging. The current production at Novamont has exceeded 60,000 MT.

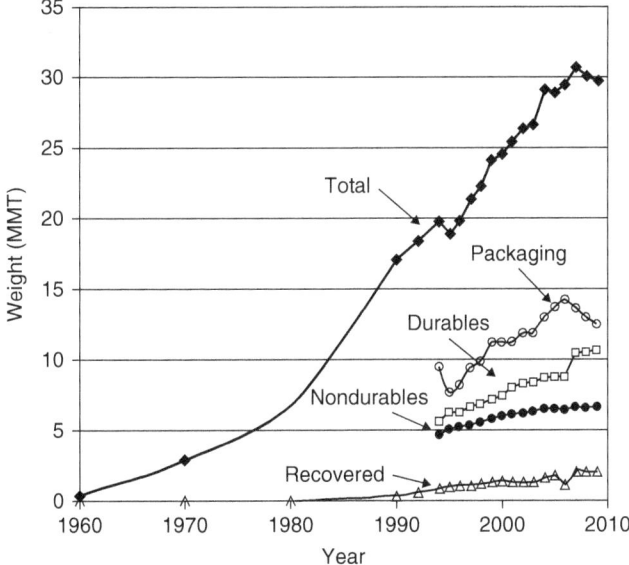

Figure 2.12 Growth of the plastics worldwide.

Novamont continues to develop their technology. In 1997, Novamont purchased the Warner-Lambert patents for TPS/polymer blends. They later acquired (2004) the technology from Eastman Chemical to produce the polyester (Estar Bio). This polyester is now being produced using oil from a non-food crop and is being blended with TPS. The bio-based polyester has enabled Novamont to increase the renewable polymer content of their resins to about 50%. They also intend to commercialize "nanostarch" particles for use in film grades of Mater-Bi. The nanostarch allows for higher renewable content in films while maintaining good strength and clarity.

2.4.3.2 Cereplast

Cereplast is located in El Segundo, CA. It was founded in 1996 and began marketing the Novamont product in North America. In 2000, Cereplast began developing their own TPS technology and started marketing Cereplast products in 2001. Cereplast recently opened a 36,000 MT facility in Indiana.

2.4.3.3 Ecobras

BASF, the world's leading chemical company, has entered the starch blend market. BASF produces the biodegradable polyester, Ecoflex (polybutyrate adipate terephthalate, PBAT). In 2007, BASF aligned itself with Corn Products International and began selling a starch/PBAT blend for the Latin American market. The blend contains about 50% corn starch and is designed for making films although it can be injection molded as well.

2.4.3.4 Biotec

Biotec GmbH in Germany has a capacity of about 11,000 MT. Its products include pure TPS and various blends of starch and copolyesters. It has six commercial formulations for injection molding, rigid and flexible extrusion, and foams. Many of the finished products include biodegradable carrier bags, bin liners, and refuse bags. Most sales are in Europe.

This results in a large number of uses and applications for compostable packaging, short-term consumable articles and special products. Examples from the various production fields are as follows:

Blown films	Flat films	Injection molding
Sack, bags	Trays for food and non-food articles	Disposable cutlery
Trash bags	Flower pots	Cans, containers
Mulsh foils	Freezer products and packaging	Performed pieces
Hygiene products	Cups	CD trays
Diaper films	Pharmaceutical packaging	Gold Trees
Air bubble films		Cemetery articles
Multiple layer films		Golf tees
Protective clothing		Toys
Gloves		
Double rib bags		
Labels		
Barrier ribbons		

2.4.3.5 Plantic

Plantic Technologies was incorporated in Victoria, Australia, but is located today in Melbourne, Australia. Plantic acquired technology in 2001 for making high-amylose corn starch TPS plastic sheets and trays. They found that TPS sheets could be thermoformed into trays and used to package fatty foods or products with a water activity of 35–70%. The business started making trays for a candy company. Plantic entered the global market in 2004 and recently developed multilayered polymer films with a starch film core that has improved moisture resistance, gas barrier properties, and physical properties. Plantic has announced joint ventures with several companies in recent years including DuPont for making cosmetic and food packaging and Bemis Co., Inc., Neenah, WI, to develop blown film for dry-goods packaging.

2.4.3.6 Biolice

Biolice was developed by Limagrain, a leader in the European agricultural sector. Biolice is a TPS made from cereal flour that is blended with biodegradable polyesters. Biolice is a rigid packaging material that can be thermoformed into single-use items like drink trays and cups. Films can also be made from the resin for agricultural mulch and carrier bags. The product is being marketed in France and is completely biodegradable.

2.4.3.7 KTM Industries

KTM Industries is a company located in Lansing, MI; now it is called Green Cell Foam. The company uses an extrusion process similar to that used to make PS foam sheets. The process involves extruding TPS through an annular die to form a foam tube. The tube is sliced and opened flat to form sheets of starch foam that can then be used for packaging operations. The foam sheets can be cut and glued to form padding for specific packaging applications. The company also makes colored loose-fill products for children craft projects. Other companies using TPS for making starch-based loose-fill products include StarchTech, Inc. and National Starch with its Eco-Foam product.

2.4.3.8 Cerestech, Inc.

Cerestech, Inc. was incorporated in 2001 in Montreal, Canada. The company produces blends of TPS and commodity thermoplastics in a one-step extrusion process. The process involves preparing starch/glycerol/water blends of approximately 48%, 32%, and 20%, respectively. The starch preparation is fed into a twin-screw extruder to form a TPS melt. A second single-screw extruder is attached to the twin-screw extruder in a perpendicular position to allow a thermoplastic polyolefin such as high-density polyethylene to be melted and injected directly into the TPS melt. The melt blend is compounded further using high shear to form a blend of the two incompatible resins. Although the polyolefin and TPS form an incompatible blend, the domains of the respective polymers range from several micrometers to less than 1 µm. Blends containing up to 50% starch have been produced with excellent mechanical properties and moisture resistance. The carbon footprint of these blends is significantly reduced compared to the neat polymer due to the starch content. A family of blends (Cereloy) based on starch and various polyolefin resins is being developed. Cerestech has granted a worldwide license to Teknor Apex to produce the blends. The blends are being sold at a similar or lower price than the neat polymer.

2.4.3.9 Teknor Apex

Teknor Apex is also marketing blends of starch and recycled PP and PE to further improve the environmental profile of its products.

We can conclude from the examples cited here that starch as biomaterials is poised to establish an even stronger role in the manufacture of sustainable plastics and other bioproducts largely because it is abundant, renewable, and inexpensive. The cost and availability of starch may improve even further in the future if lignocellulose materials get involved in the process. The proposal herein will boost the strategies for improving the properties of starch-based plastics such as blending starch with other polymers as cellulose materials or others, using starch in composite materials. The prospects for biomaterials in the packaging sector continue to become brighter as the market for sustainable plastics drives further innovation and development in our era.

2.5 Question for Further Consideration

1. What are the challenges for sustainable bio-fuel production?
2. Could the lignocellulose replace the potential applications of starch new business opportunities?
3. What could be the right technology to transform the biomaterials to potential products?

References

1 I. Siro and D. Plackett. Microfibrillated cellulose and new nanocomposite materials: a review. *Cellulose*, **17**, 3, 459–494 (2010).
2 J. Blackwell, P. D. Vasko, and J. L. Koenig. Infrared and Raman Spectra of the cellulose from the cell wall of *Valonia ventricosa*. *Journal of Applied Physics*, **41**, 11, 4375–4379 (1970).
3 A. French. Idealized powder diffraction patterns for cellulose polymorphs. *Cellulose*, **21**, 2, 885–896 (2014).
4 A. Esker, U. Becker, S. Jamin, S. Beppu, S. Renneckar, and W. Glasser. Self-assembly behavior of some co- and heteropolysaccharides related to hemicelluloses. In: *Hemicelluloses: Science and Technology*, Chap 14, pp 198–219, ACS Symposium Series, **864** (2003).
5 D. M. Alonso, S. G. Wettstein, M. A. Mellmer, E. I. Gurbuz, and J. A. Dumesic. Integrated conversion of hemicellulose and cellulose form lignocellulosic biomass. *Energy & Environmental Science*, **6**, 76–80 (2013).
6 M. Lawoko, G. Henriksson, and G. Gellerstedt. Structural differences between the lignin–carbohydrate complexes present in wood and in chemical pulps. *Biomacromolecules*, **6**, 6, 3467–3473 (2005).
7 A. Ebringerova and T. Heinze. Naturally occurring xylans structures, isolation procedures and properties. *Macromolecular Rapid Communication*, **21**, 542–556 (2000).
8 A. Duval and M. Lawoko. A review on lignin-based polymeric, micro- and nano-structured materials. *Reactive and Functional Polymers*, **85**, 78–96 (2014).

9 M. Norgren and H. Edlund. Lignin: Recent advances and emerging applications. *Current Opinion in Colloid and Interface Science*, **19**, 5, 409–416 (2014).
10 S. Laurichesse and L. Averous. Chemical modification of lignins: Towards biobased polymers. *Progress in Polymer Science*, **39**, 7, 1266–1290 (2014).
11 H. P. S. A. Khalil, M. S. Alwani, and A. K. M. Omar. Chemical composition, anatomy, lignin distribution, and cell wall structure of Malaysian plant waste fiber. *BioResources*, **1**, 2, 220–232 (2006).
12 F. S. Chakar and A. J. Ragauskas. Review of current and future softwood kraft lignin process chemistry. *Industrial Crops and Products*, **20**, 2, 131–141 (2004).
13 E. Quintana, C. Valls, A. G. Barneto, T. Vidal, J. Ariza, and M. B. Roncero. Studying the effects of laccase treatment in a softwood dissolving pulp: Cellulose reactivity and crystallinity. *Carbohydrate Polymers*, **119**, 2015, 53–61 (2015).
14 D. Klemm, B. Heublein, and A. Bohn. Cellulose: Fascinating biopolymer and sustainable raw material. *Angewandte Chemie*, **44**, 22, 3358–3393 (2005).
15 M. M. Kabir, H. Wang, K. T. Lau, and F. Cardona. Chemical treatments on plant-based natural fiber reinforced polymer composites: An overview. *Composites Part B: Engineering*, **43**, 7, 2883–2892 (2012).
16 C. Tian, L. Zheng, Q. Miao, C. Cao, and Y. Ni. Improving the reactivity of kraft-based dissolving pulp for viscose rayon production by mechanical treatments. *Cellulose*, **21**, 5, 3647–3654 (2014).
17 F. M. Gama, J. A. Teixeira, and M. Mota. Cellulose morphology and enzymatic reactivity: A modified solute exclusion technique. *Biotechnology and Bioengineering*, **43**, 5, 381–387 (1994).
18 A.-C. Engstrom, M. Ek, and G. Henriksson. Improved accessibility and reactivity of dissolving pulp for the viscose process: Pretreatment with monocomponent endoglucanase. *Biomacromolecules*, **7**, 6, 2027–2031 (2006).
19 E. S. Welf, R. A. Venditti, M. A. Hubbe, and J. Pawlak. The effects of heating without water removal and drying on the swelling as measured by water retention value and degradation as measured by intrinsic viscosity of cellulose papermaking fibers. *Progress in Paper Recycling*, **14**, 3, 1–9 (2005).
20 U. Weise, T. Maloney, and H. Paulapuro. Quantification of water in different states of interaction with wood pulp fibers. *Cellulose*, **3**, 1, 189–202 (1996).
21 H. Sixta, M. Iakovlev, L. Testova, A. Roselli, M. Hummel, M. Borrega, A. van-Heiningen, C. Froschauer, and H. Schottenberger. Novel concepts of dissolving pulp production. *Cellulose*, **20**, 4, 1547–1561 (2013).
22 H. Wang, B. Pang, K. Wu, F. Kong, B. Li, and X. Mu. Two stages of treatments for upgrading bleached softwood paper grade pulp to dissolving pulp for viscose production. *Biochemical Engineering Journal*, **82**, 183–187 (2014).
23 L. Testova, M. Borrega, L. K. Tolonen, P. A. Penttila, R. Serimaa, P. T. Larsson, and H. Sixta. Dissolving-grade birch pulps produced under various prehydrolysis intensities: quality, structure and applications. *Cellulose*, **21**, 3, 2007–2021 (2014).
24 D. Ibarra, V. Kopcke, P. T. Larsson, A.-S. Jaaskelainen, and M. Ek. Combination of alkaline and enzymatic treatments as a process for upgrading sisal paper-grade pulp to dissolving-grade pulp. *Bioresource Technology*, **101**, 19, 7416–7423 (2010).
25 J Shen, Z. Song, X. Qian, and W. Liu. Modification of papermaking grade fillers: A brief review. *BioResources*, **4**, 3, 1190–1209 (2009).

26 W. Farhat, R. Venditti, M. Hubbe, M. Taha, F. Becquart, and A. Ayoub. A review of water resistant hemicellulose based materials: processing and applications. *ChemSusChem*, **10**, 2, 305–323 (2017).

27 M. Marinova, E. Mateos-Espejel, N. Jemaa, and J. Paris. Addressing the increased energy demand of a kraft mill biorefinery: The hemicellulose extraction case. *Chemical Engineering Research and Design*, **87**, 9, 1269–1275 (2009).

28 H. Y. Celebioglu, D. Cekmecelioglu, M. Dervisoglu, and T. Kahyaoglu. Effect of extraction conditions on hemicellulose yields and optimization for industrial processes. *International Journal of Food Science and Technology*, **47**, 12, 2597–2605 (2012).

29 A. J. Bailey, Lignin in Douglas fir: composition of the middle lamella. *Industrial & Engineering Chemistry Analytical Edition*, **8**, 1, 52–55 (1936).

30 W. G. Van Beckum and G. J. Titter. *Paper Trade Journal*, **105**, 18, 127 (1937).

31 C. Laaksometsa, E. Axelsson, T. Berntsson, and A. Lundstrom. Energy savings combined with lignin extraction for production increase: case study at a eucalyptus mill in Portugal. *Clean Technologies and Environmental Policy*, **11**, 77–82 (2009)

32 X.-F. Sun, R. Cang, P. Fowler, and M. S. Baird. Extraction and characterization of original lignin and hemicelluloses from wheat straw. *Journal of Agricultural and Food Chemistry*, **53**, 4, 860–870 (2005)

33 M. Schwanninger and B. Hinterstoisser. Klason lignin: Modifications to improve the precision of the standardized determination. *Holzforschung*, **56**, 2, 161–166 (2005).

34 A. Dufresne, D. Dupeyre, and M. R. Vignon. Cellulose microfibrils from potato tuber cells: processing and characterization of starch–cellulose microfibril composites. *Journal of Applied Polymer Science*, **76**, 14, 2080–2092 (2000).

35 L. Averous, C. Fringant, and L. Moro. Plasticized starch–cellulose interactions in polysaccharide composites. *Polymer*, **42**, 15, 6565–6572 (2001).

36 A. A. S. Curvelo, A. J. F. de Carvalho, and J. A. M. Agnelli. Thermoplastics starch-cellulosic fibers composites: preliminary results. *Carbohydrate Polymers*, **45**, 2, 183–188 (2001).

37 X. Ma, J. Yu, and J. F. Kennedy. Studies on the properties of natural fibers-reinforced thermoplastic starch composites. *Carbohydrate Polymers*, **62**, 1, 19–24 (2005).

3

Conversion Technologies

Maurycy Daroch

Peking University, School of Environment and Energy, Shenzhen, China

> *(…) a new type of thinking is essential if mankind is to survive and move toward higher levels*
> Albert Einstein, May 25, 1946

3.1 Learning Objectives

By the end of this chapter, students should be familiar with and cognisant of the following concepts:

1. Major energy resources and their alternatives in the context of energy return on investment and environmental impact.
2. Biomass and fossil fuels as forms of solar energy that has been captured and stored in a form of chemical molecules.
3. Types of biomolecules present in biomass and their potential for bio-fuels and biomaterials.
4. Biomass conversion platforms and energy perspective on conversion of biomass-derived molecules.
5. Themochemical and biochemical methods of biomass conversion.
6. Policy aspects of bioenergy and biomaterials development.

3.2 Energy Scenario at Global Level

3.2.1 Why Our Energy is so Important?

It may sound like a trivial question at the first glance and an immediate answer 'of course it is important, we need it for everything from cooking the food to moving our cars' may sound like an obvious answer to most of us. Since for at least three generations we are living in the world of cheap and abundant energy, we sometimes forget the effects

Introduction to Renewable Biomaterials: First Principles and Concepts, First Edition.
Edited by Ali S. Ayoub and Lucian A. Lucia.
© 2018 John Wiley & Sons Ltd. Published 2018 by John Wiley & Sons Ltd.

of the abundant energy supply on how our societies function in broader context. Our modern life is fundamentally connected to the energy use. Availability of abundant energy that can be used on demand is a primary factor shaping our modern world. Since Industrial Revolution, the development of fossil fuel conversion technologies allowed unprecedented growth of world economy, population, and life quality that even aristocracy of pre-industrial times could not have dreamed of. Energy is not only required to increase the comfort of our life: cook our food, keep us warm in winter or provide entertainment. Most importantly, energy provides cheap physical work what translates to unparalleled productivity of mechanised processes over those performed by humans or animals. Mechanisation not only resulted in huge increases in productivity, what is equally important is that it made physical work of humans expensive, inefficient, and ultimately unnecessary. To put this into numerical context, 1 l of diesel oil is equivalent to about 38 MJ of energy. A worker consuming 3000 kcal of food (12.6 MJ) is capable of producing about 3.5 MJ of useful energy per day [1]. This simple calculation shows that every litre of diesel that the economies use is in energy sense equivalent to the work of additional ten workers for one day. Based on the recently collected data of energy use [2] an average US citizen uses 7.08 tons of oil equivalents (toe) per year (additional 232 workers per day), EU citizen 3.29 toe (additional 108 workers per day) and Chinese citizen 2.06 (additional 68 workers for per day). To achieve current levels of productivity, the technology has moved from low efficiency work provided by humans or animals to mechanised work powered by chemical energy of fossil fuels. This shift had huge impact on social landscape of the world. Abundance of cheap energy from fossil fuels made social systems based on slavery or serfdom obsolete, and machine work was much more effective in providing useful work than any of the servants ever would. Consequently, these social systems quickly disappeared in industrialised parts of the world. Additionally, industrialisation made most of time-consuming activities easier, relieving people from physical work. As a result, entire masses of societies were given chance to raise their social status and/or develop in science, arts or philosophy, the disciplines that have been previously reserved for the elite at the very top of the social ladder. Since the beginning of the industrial age, world economy fuelled by fossil resources has experienced unprecedented exponential growth, which in turn exponentially increased the consumption of fossil resources.

3.2.2 Black Treasure Chest

Considering the importance of energy supply for both economic and social aspects of our modern civilisation, the existence of fossil fuels can be paralleled to a treasure chest. The treasure chest of fossil fuels contains a finite content of different riches (coal, crude oil and natural gas) that have has been hidden underground for millions of years. This chest has been opened with the beginning of Industrial Revolution, and ever since the content of the chest is being depleted.

All fossil fuel resources such as crude oil, natural gas and coal were produced over millions of years and as such considered non-renewable. These compounds were formed through partial decomposition of organic matter at high pressure and temperature and low availability of oxygen. Under these conditions, a part of carbon from the carbon cycle has been removed and stored underground for millions of years. With the start of the Industrial Revolution, humans started to utilise these long-stored compounds – initially

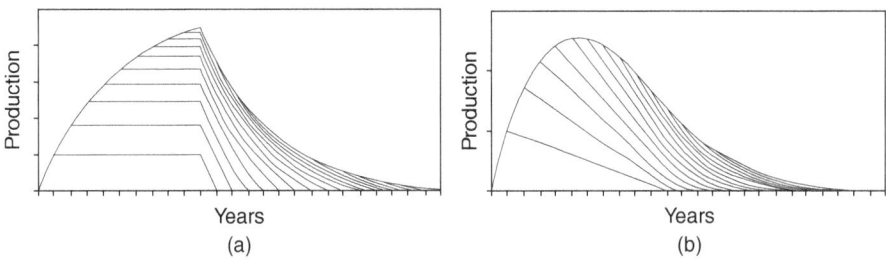

Figure 3.1 Simple theoretical models of the combined output from a group of gas or oil fields: both graphs assume that fields are found 1 year apart, larger fields are found earlier. (a) Field production follows a trapezoidal profile, possibly typical of gas fields. (b) Field production follows a profile typical of oil fields, that is, a rapid build up, followed by a slow decline. Bentley 2002 [3, 4]. Reproduced with permission of Elsevier.

to release the energy from their chemical bonds and transform it into heat and work and later to transform these molecules into useful materials such as plastics. Continued utilisation of non-renewable resources on a finite planet must result in eventual resource depletion, it is only the matter of question when. For every non-renewable deposit, the productivity achieved from the extraction of this deposit resembles a bell-shaped curve (Figure 3.1) called Hubbert curve. Once new deposit is found, the production quickly rises until it reaches the peak, called Hubbert peak and then it starts dropping due to resource depletion [3, 4].

At this point, the production needs to be replaced by another resource to maintain the output. For individual non-renewable resources, these bell-shaped cures have been observed not only at the levels of individual oilfields but also at national and regional levels. It is just a matter of time when peak production is achieved on global level, and many would argue that the peak for conventional oil has been already achieved [5]. Once peak production is reached the inevitable shift towards alternative resource needs to take place as the original resource will eventually become depleted. The reality of year-by-year growth of supply will be replaced by a reality of year-by-year decline most likely with a 'bumpy plateau' at the top due to buffering effects of natural gas and unconventional oil [5]. Recent reports suggest that new additions to the fuel sector are mainly from natural gas and non-conventional oil [6]. This may suggest that we have already entered the 'bumpy plateau' at the top of Hubbert curve. The length and shape of this plateau will depend on the efficiency of the introduced replacements to conventional oil such as: non-conventional oil, natural gas or bio-fuels. Overall the issue is not really when we exhaust our supplies of fossil resources but when our demand for them will exceed the capacity to provide these resources or their replacements. Once this inflexion point is reached, the economies and citizens will face a very different landscape from the one we currently know and are used to. In terms of economic impact, the ratio between the production rate and its increase or decrease and the consumption rate and its increase or decrease is most important. To put current situation into larger perspective, most of our current fossil resources come from deposits discovered in the 1960s and 1970s in the Middle East, Western Siberia, Alaska and North Sea. Many of these deposits are still rich in resources but they will inevitably achieve their peaks in the near future. Since then, despite huge advances in investment and technologies, only relatively small oil fields were found. Over past few decades, the ratio of production to new discoveries has gradually fallen and is currently estimated to about three to one. For every discovered barrel of oil we consume three. It is therefore evident that resources at our disposal are shrinking fast. At the same time, more and more regions of the world are seeking high-quality lifestyles that are resource intensive. Until relatively recently (about 30 years ago), high consumption of energy was reserved for the developed economies of the 'West'. Since then rapid development of other countries such as China, India or Brazil resulted in huge increase for energy sources worldwide. The entire population of OECD countries is estimated to be about 1.25 billion people and their primary energy use as 4.37 toe per capita [2]. When China, India and Brazil, altogether about 2.75 billion people, approach even conservative 'European' levels of fossil resources usage (3.29 toe per capita) the additional supply exceeding the current use of all OECD countries will be required. It is difficult to envisage how this demand could be met with non-renewable resources in the medium to long term.

3.2.3 Conventional Fossil Resources and their Alternatives

Currently, our economies rely on three predominant sources of energy and carbon all of which are fossils.

3.2.3.1 Light Crude Oil (Conventional Oil)

Light fractions of crude, also known as conventional oil are the most important of fossil resources. Conventional oil is a liquid of relatively low density that flows freely at room temperature. It is composed of a mixture of light hydrocarbons that have large utility as fuels (gasoline and diesel) and platform chemicals for petrochemical industry. Conventional oil is deposited in geological formations trapped by impermeable rocks. These deposits are usually under enough pressure to allow the flow of the oil to the surface once a well is drilled.

3.2.3.2 Coal

Coal is the first fossil resource that initiated Industrial Revolution. It is a combustible sedimentary rock rich in organic carbon compounds (mainly aromatic). Coal is used primarily for power generation and refining metals. The resource is extracted from underground deposits by either shaft or strip mining.

3.2.3.3 Natural Gas

Natural gas is arguably the cleanest fossil fuel available. It is essentially pure methane which upon combustion yields only CO_2 and H_2O. It is also utilised in a series of important chemical reactions that yield hydrogen (steam reforming) and subsequently ammonia (Haber–Bosch process), which as such becomes not only hugely important for our energy supply but also agriculture. Due to depletion of the resources of light crude oil and cleaner combustion characteristics liquefied natural gas is increasingly used as transportation fuel.

The reserves of each of these resources have been estimated to be sufficient for 55, 132 and 71 years of utilisation at 2011 consumption levels, respectively [6]. These numbers however do not show the full picture. First, the growth of the economy will require additional resources to be allocated to fuel this growth. Second, as we have seen in the earlier section, it is the ratio of the resource supply to demand that really matters more than absolute quantity of the resource available. Third, most of the highest quality resources have been already depleted, and the remaining ones are increasingly difficult to extract. Combination of these reasons has prompted the search for alternative sources of carbon among both renewable and fossil resources. At least in principle there are alternatives to traditional light oil among other fossil resources. These resources are usually called unconventional oil and might be used to supplant and possibly replace currently used crude oil and provide carbon and energy to the economy once current fossil fuel deposits are exhausted. The most widely considered alternatives with their short characteristics are presented in the following sections.

3.2.3.4 Shale Oil (Tight Oil)

Shale oil is similar in structure to light oil. The difference between the two is their location and method of extraction. Conventional oil deposits can be continuously retrieved from the source until it is depleted. Shale oil on the other hand is much more complicated to extract. Oil is entrapped in porous rock formations that need to be fractured to release it. The fracturing is performed by drilling multiple vertical–horizontal wells and injecting hectolitres of water with numerous chemicals often not disclosed by the drilling companies. Created fissures make the oil flow into the surface where it can be collected. A number of refractures needs to be performed to maintain the flow of shale oil from the rock to surface.

3.2.3.5 Oil Sands, Bitumen Extra Heavy Oil

This group of non-conventional oil contains oxidised hydrocarbon mixtures in which lighter fluid fractions have evaporated. These heavier hydrocarbons contained in bituminous sands cannot be extracted as easily as conventional oil. First, they cannot be pumped, hydrocarbons mixed with sands are strip mined and extracted with hot water. Second, the extracted hydrocarbons require advanced upgrading to be transformed into fuels. Combination of these two factors makes this technology expensive and limited to regions with abundant water resources and cheap energy, both of which are increasingly scarce. Additionally, the extraction process produces a lot of wastewater contaminated with hydrocarbons that are usually stored in ponds that create environmental and health hazards.

3.2.3.6 Shale Gas

In recent years, the advances in the extraction of the so-called shale gas gave the oil and gas industry large hopes for significantly extending the life span of gas technologies. From the chemical point of view, there is no difference between shale gas and 'conventional' natural gas; they are both predominantly composed of methane. What differs however is the method of extraction, which is very similar to the one used for shale oil. Shale gas is entrapped in porous rock formations that need to be fractured before the gas is released. The fracturing (fracking) is performed by drilling multiple vertical–horizontal wells and injecting water with chemicals. Fractured rocks release the gas into the water and ultimately to surface; secondary fracking needs to be performed to maintain the flow of gas.

3.2.3.7 Methane (Gas) Hydrates

Gas hydrates are ice-like solids formed from water and natural gas. Natural gas is entrapped inside the cages made of frozen water particles. These cages are stable at the bottom of the ocean at high pressures and low temperatures. When either of these components is disrupted the structure of gas hydrates breaks into its individual components. Generally, the lower the water temperature the lower the pressure required to maintain the stability of methane hydrates. Methane hydrates pose very different risks of extraction compared with other fossil fuels. The latter are extracted from rather stable and well-defined geological structures; methane hydrates on the other hand are expected to be extracted from deep oceans where their stability is maintained by pressure and low temperature. Proposed extraction methods are very likely to rely on local disruptions of this to release the gas. If any of the parameters of this process is not set up correctly, there is a catastrophic risk of destabilising the entire deposit and releasing toxic methane to the surface and subsequently atmosphere. Ignition of destabilised hydrate is likely to cause an explosion that might result in a destructive tsunami.

3.2.3.8 EROI – How Much Fuel in Fuel?

The history of the human development to date has been a history of replacing low-quality fuels and working with higher quality ones. Civilisations expanded with their move towards better resources: from human muscle power to draft animals to water and wind mills to coal and ultimately to petroleum [5]. Now, for the first time, civilisations need to deal with the opposite process: how to replace high-efficiency energy means with lower efficiency ones. The main reason is the depletion of good-quality fossil resources. There is nothing particularly surprising in this finding; it is the fundamental characteristic of humans and a component of how our economy works [5]. Whatever the resource, humans start its utilisation from the best available and move towards lesser available ones until they prove to be useless or too expensive in monetary or work input terms. The economic reality dictates that an individual needs to use best resources at the lowest cost possible first otherwise the competitors will gain an advantage and price an individual out of the market.

An indicator that is used to measure the efficiency of a particular source of energy is called energy return on investment (EROI) [5]. EROI is a ratio describing the content of energy that one obtains from an activity compared to the energy it took to generate that activity; in other words, it describes how much usable energy is left in the resource after it is obtained. Investment of a joule of energy in a resource of high EROI produces many of joules of energy that can be spent on another activities [5]. Good energy sources have high EROI factors, whereas poor energy sources are closer to 1. EROI equal to 1 is the ultimate physical barrier, which determines usefulness of a resource for energy production; it means that the process generates no net energy and is therefore pointless. The relationship between useful energy extracted from the resource and its EROI is exponential. As long as the EROI of an energy source is above 10, its EROI makes little difference as over 90% of the energy remains useful. Lower values of EROI, however, fall into the region called 'net energy cliff' where even slight differences in EROI of a resource make huge impacts onto how much useful energy can be actually obtained from this resource [7] (Figure 3.2).

First oil fields that were exploited in the beginning of twentieth century provided EROI in excess of 100 [9]. Once these fields became depleted, the less accessible resources were used and EROI of 50–30 was characteristic for the oil extracted in the 1970s [5]. Nowadays, conventional oil fields offer EROI in the range of 18–11 [5, 9] and are being

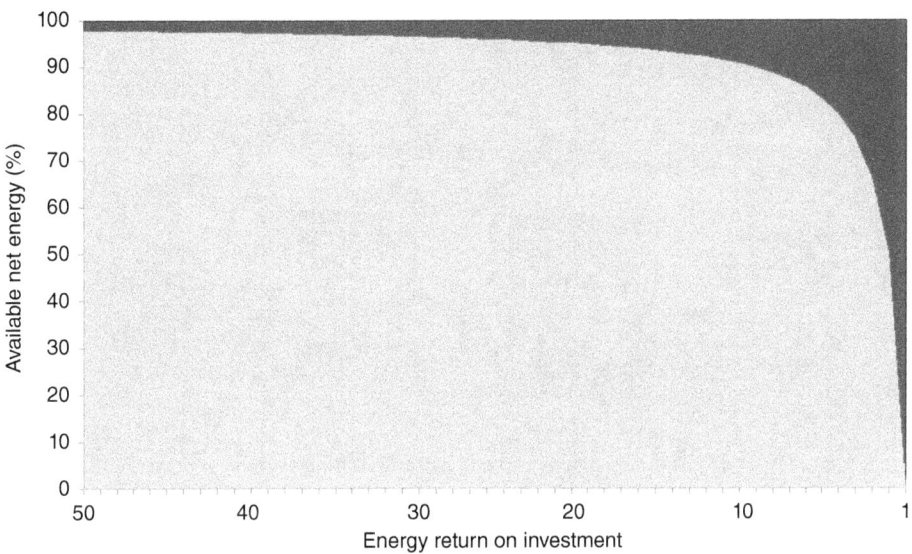

Figure 3.2 Relationship between EROI of an energy resource and percentage of available energy that can be used per unit of this resource. Light grey area represents fraction of energy of the resource that can be delivered to the economy, dark grey area represents fraction of energy of the resource that is spent to extract/produce resource itself. Please note the net energy cliff at EROI lower than 10. X-axis in reverse order. Adapted from Mearns 2008 [8].

depleted, forcing the industry to move for resources deposited in more difficult sites (deep water, extreme environments such as Arctic) and those more difficult to extract and convert like non-conventional oil. Consequently, economies are inevitably moving towards resources of even lower EROI to fulfil their carbon requirements. These resources, however, require much more investment and result in much lower energy outputs and often dangerously approach the net energy cliff. Another important finding from EROI factor is that low EROI resources require much greater volumes of energy to be extracted to provide equal amount of useful energy. This results in the expansion of financial and infrastructure requirements for the extraction of resources from both conventional but not easily accessible and non-conventional sources. This creates a feedback loop when replacement of easily accessible resources with high EROI by low EROI alternatives such as remote deposits or non-conventional resources creates a need for additional energy and financial investment to expand infrastructure required to extract these low EROI resources. The further the resource on net energy cliff the larger the required infrastructure and investment to utilise this resource.

3.2.3.9 Environmental Effects of Fossil Resource Utilisation

Utilisation of fossil resources results in the gradual but certain deterioration of the environment and adverse conditions to human health and most importantly to the climate as a whole. Combustion of fossil fuels releases the stored chemical energy and oxidises various forms of organic carbon into carbon dioxide. Since the beginning of Industrial Revolution, approximately 1200 billion metric tons of previously stored CO_2 has been released to the atmosphere [10]. It took millions of years to achieve carbon balance in the

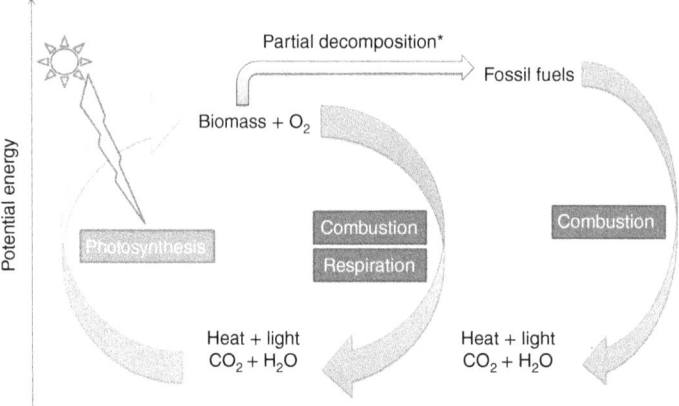

Figure 3.3 Biomass and fossil fuels – origins and energy content. Biomass is an important element of carbon cycle. In the process of photosynthesis, carbon dioxide and water are converted into carbohydrates and other structures called biomass. Biomass can be then digested or combusted to recover stored chemical energy and release oxidised compounds: carbon dioxide and water. Fossil fuels are biomass that underwent fossilisation process, that is, slow partial decomposition in the absence of oxygen powered by heat and pressure from geological sources. In a process of fossilisation, biomass lost significant content of oxygen and became composed of hydrocarbons having higher energy content than original biomass. Bentley 2002 [4]. Reproduced with permission of Elsevier.

atmosphere and mere 250 years to disrupt it. In principle, carbon fixation by photosynthetic organisms could deal with the excessive CO_2 concentration due to the inherent features of carbon cycle (Figure 3.3).

In reality, however, the ability of photosynthetic organisms to mitigate the effects of anthropogenic release of CO_2 is limited. It is estimated that no more than about a half of current anthropogenic emissions (15 of 33 billion tons CO_2 annually [11]) can be mitigated by photosynthetic activity; clearly not sufficient to stop CO_2 accumulation in the atmosphere and prevent anthropogenic climate change. This problem is aggravated by other anthropogenic activities such as deforestation, soil erosion and other processes that decrease primary productivity. Another major carbon sink is ocean. It is estimated that about a third of all anthropogenic emissions has been dissolved in the oceans. It may help to alleviate the effects of global warming but results in the acidification of the oceans. Since the beginning of industrial revolution the pH level dropped by 0.1 unit. It is expected that continued release of CO_2 to atmosphere will lower the pH by another 0.2–0.3 units having profound effects on marine organisms such as corals and plankton [12]. The excessive accumulation of dissolved CO_2 in the oceans will also shift the future absorption pattern towards the atmosphere, and it is expected that the capacity of the oceans to absorb more CO_2 will be about 60% lower in the year 2100 than it is today, aggravating the effects of climate change [12].

It remains to be seen how non-conventional resources impact the environment but the initial findings are not promising. Non-conventional fossil resources have lower EROI than traditional resources and as such require much larger investment, infrastructure and volume of production to deliver an equal content of usable energy. Since all these values change exponentially so does the environmental impact of these processes. Emissions of carbon dioxide from non-conventional resources are larger than those of the corresponding conventional ones due to more complex methods of extraction. Currently known technologies of extracting non-conventional resources rely on utilisation of significant inputs of energy and water and come with risks of significant methane emissions. Especially water is a matter of concern; water is becoming a scarce resource in numerous locations and its expanded use of non-conventional fossil resources may additionally worsen current strain on the freshwater resources and result in the complete destruction of unique ecosystems of Canadian tundra, Orinoko River Delta and many others. Additionally, many of non-conventional technologies rely on aggressive methods that can seriously impact ecosystems like strip mining.

To summarise, there are fossil-based alternatives to renewable sources of energy and carbon, but their large exploitation is likely to aggravate climate change issues more than those of conventional resources; therefore a sensible path to developing low carbon alternatives for energy and chemicals is of primary importance. Among many proposed solutions, biomass is seen as one of the alternatives that could help to supplement the energy market and most importantly provide renewable platform chemicals and materials to support economies in the future.

3.3 Biomass

3.3.1 Renewable Energy and Renewable Carbon

In the simplest terms, biomass could be defined as a biological material derived from living or recently living organisms. These organisms are mainly plants and other photosynthetic organisms. In principle, all the organisms that utilise chemical energy from plant biomass indirectly by indigestion or decomposition of plant matter through the food web can be considered as biomass. Throughout this chapter, however, we focus predominantly on plants as the most important contributors to biomass production on earth, and unless stated otherwise biomass will be synonymous with plant matter.

Plants are photosynthetic organisms that produce their cellular structures through photosynthesis, an unique process that converts the energy of the sun and two simple inorganic molecules – water and carbon dioxide – into chemical energy stored in plant biomass (Figure 3.3). Biomass is therefore a form of solar energy that has been captured and stored in the form of large chemical molecules like carbohydrates, lipids, aromatic compounds, proteins and others. Once collected, this chemical energy can be stored for prolonged periods of time and released on demand to yield other forms of usable energy like heat, work or other useful chemical compounds. There are many compounds that can be produced from biomass, for example, fuels – transportable form of chemical energy, but also platform chemicals, biopolymers and other compounds of high utility to the economies. This places biomass conversion technologies at the forefront of most important technologies needed for the transition from fossil-based economy to the low-carbon economy of the future. For the realisation of low-carbon economy, two aspects are absolutely essential: renewable energy and renewable carbon (renewable materials). Whereas there exist numerous alternatives to produce renewable energy such as hydro, wind, solar, geothermal, wave and tidal technologies, there are only limited array of options to replace fossil-derived platform chemicals such as olefins (ethylene or propylene), aromatics (benzene, toluene, xylene) or butadiene. These platform chemicals are currently processed into an array of consumer products that we use on everyday basis such as plastics, synthetic rubber and textiles, cosmetics, fragrances, adhesives, dyes, cleaning agents and so on. With the depletion of fossil resources and inevitable transition to low-carbon economy, most of these products will need to be replaced with renewable alternatives to maintain the standards of living. Because of its chemical composition and ability of various microorganisms to ferment carbohydrates into array of different products, utilisation of biomass is the most suitable way for the replacement of fossil-derived chemicals. The renewable chemicals and renewable materials are equally important for the development of carbon free economy as are renewable energy and renewable fuels.

Plant biomass is composed of many different chemical compounds synthesised at various stages of plant growth and development. Both presence and relative abundance of these compounds vary significantly between different plant structures like seeds, fruits, stems or leaves. From the chemical point of view, the structures synthesised in plant cells can be divided into carbohydrates, lipids and phenolics, whereas taking into consideration their function for the plant they could be divided into two groups: structural and storage compounds (Figure 3.4). Biomass fraction composed of carbohydrates include cellulose, starch, sucrose and heteropolysaccharides (hemicelluloses and pectins); the phenolic fraction of biomass consists of lignin, whereas the lipid fraction comprises triglycerides. The knowledge of basic chemistry of each of these groups will help to understand the implications that the chemistry of these compounds has for potential routes of conversion; therefore we strongly advise readers to familiarise with the

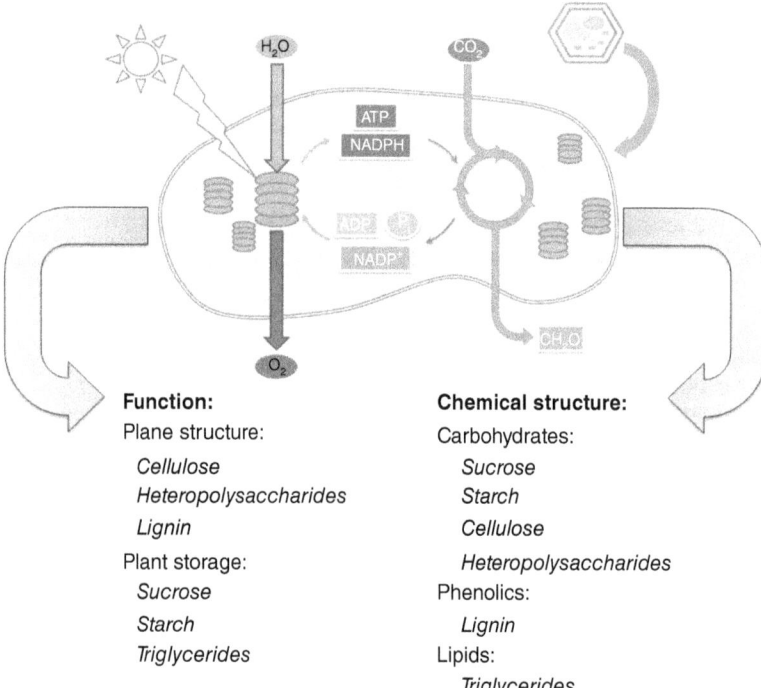

Figure 3.4 Ultimate biological compounds produced via CO_2 fixation in chloroplasts during the process of photosynthesis divided according to their function and chemical structures.

content of others chapters before proceeding to the detailed explanation of conversion technologies.

3.3.2 Why Different Types of Biomass have the Properties they Have?

The energy captured from the sun by the process of photosynthesis can be diverted into a number of routes called fluxes. In this section, we focus on two fluxes that can yield renewable carbon for the conversion of biomass into bioenergy and biomaterials. These fluxes are structural compounds and storage compounds. When a plant is cultivated in abundance of resources such as sunlight, water, micro and macronutrients, plant cells undergo rapid division at the specialised growth regions called meristems [13]. These new cells are relatively unspecialised and their thin primary cell walls are predominantly composed of cellulose and heteropolysaccharides (hemicelluloses and pectins) [14]. As new plant cells are being produced at the meristems the older cells gradually undergo specialisation, and their primary cell walls are reinforced by secondary cell walls composed of lignocellulose [14]. Secondary cell walls are much thicker than the primary ones and are composed of coalesced polymers of cellulose and lignin (hence lignocellulose) with the addition of heteropolysaccharides (hemicelluloses) that act as a 'molecular glue' and help to cross link other two components of lignocellulose to maintain its structural integrity.

The synthesis of this complex biomaterial was one of the key evolutionary traits that allowed terrestrialisation of life and development of complex life forms as we know them. Lignification and formation of lignocellulose structure of secondary cell walls evolved to provide protection to the plant from physical, chemical and microbial damage and to maintain the structural integrity of the plant under conditions of high gravity and oxidative stress [15]. These features of terrestrial organisms can be easily observed when comparing the structure of their cell walls with those of their ancestors – vascular algae. Aquatic environments are much milder in terms of environmental stresses such as oxidative damage, temperature changes and gravity. This translates into significant differences in the chemical composition of terrestrial plants and algae [16]. Algal cell walls are composed of cellulose and various non-crystalline, often sulphated heteropolysaccharides yielding structure of limited mechanical strength. Plant cell walls on the other hand are composed of robust lignocellulose that allows many plants to create large structures that can withstand severe weather conditions and reach heights of 100 m. This recalcitrance of lignocellulose to environmental conditions made it the main structural component of plants and the most abundant polymer on earth. The same features that made lignocellulose such a successful biomaterial in evolutionary sense made it a difficult feedstock for the conversion to other materials such as fuels or chemicals.

In addition to lignocellulose, plants also synthesise other compounds that are more accessible for conversion, these compounds are storage compounds like sucrose, starch and triglycerides. Depending on the function of these compounds within the organism they could be characterised as either short- or long-term storage compounds. Short-term storage compounds, such as sucrose and starch, function as a storage tank of chemical energy; when plants cannot acquire it through photosynthesis; they become primary source of energy during the night. These pools of energy are very easily accessible and are synthesised and broken down on a daily basis in each plant cell. Although in most crops sucrose is used only a short-term storage compound and converted into other forms of storage as more reserves accumulate, in the so-called sugar crops sucrose is the ultimate form and is accumulated in storage parenchyma throughout the vegetative season of the plant [17]. This makes sucrose a very accessible pool of carbohydrates for extraction and conversion. Most crops however store majority of their easily accessible forms of chemical energy in seeds. When plants mature and are ready for reproduction, new tissues are synthesised to promote the spread of genetic material to new territories. The process and methods or reproduction vary between plant species but ultimately after successful pollination plants produce seeds that propagate into a new territory. The seeds contain long-term storage compounds: starch and triglycerides. From the chemical point of view, these are the same substances as the ones use for the short-term storage; these reserves however are much more abundant. The biological function of long-term reserves is to fuel the entire germination process of the plant from the embryo until the development of photosynthetic tissues is completed. Such process requires a significant amount of chemical energy; therefore the pool of energy available in seeds is significantly larger than that accumulated in short-term storage compounds. When plants germinate from their seeds, these storage compounds are broken down into their respective building blocks by the action of specific enzymes and are conveyed to the embryo to fuel the germination process [13]. Energy reserves stored in seeds are also accessible for other uses. During thousands of years of agriculture development, people aimed to maximise the content of these energy reserves, collect them and use for food and feed before they reach the germination stage. The pursuit of grain yield has resulted in modern high yielding food varieties of wheat, corn, soybean, rapeseed and so on where these pools of carbohydrates and lipids are much larger than they were in the ancestral species. Because of their original function as energy reserves storage compounds such as starch and triglycerides are relatively easy to break down into their respective building blocks, which c an be assimilated and converted to other molecules by other organisms like bacteria or yeast.

Depending on the predominant chemistry present in biomass, feedstocks can be divided into the following groups:

Sugar feedstocks – This group includes plants that store their reserves in soluble form as sucrose, such as sugarcane, sweet sorghum and sugar beet. Sucrose is easy to extract and can be either used directly or hydrolysed to simple sugars with enzymatic treatment.

Starchy feedstocks – These feedstocks store their reserves as starch granules mainly in seeds and also fruits or tubers. These include the most important cereal crops such as corn, wheat, rice and so on as well as tuberous crops: potato or cassava. Starch can be easily isolated from seeds and tubers and converted into simple sugars with enzymatic treatment.

Oily feedstocks – These feedstocks store their reserves as triglycerides in oil bodies situated within the seeds. These include the most important oilseeds such as soybean, rapeseed, sunflower and oil palm; oils can be extracted from oilseeds with an array of methods used by edible oil industry and used directly for conversion process into fuels or chemicals.

Lignocellulose feedstocks – They are composed of coalesced polymers of cellulose, lignin and hemicelluloses packed together to create a very robust structural biomaterial.

They include both dedicated lignocellulose crops that can be grown for biomass production such as *Miscanthus*, switchgrass, willow or poplar as well as lignocellulosic parts of other crops like corn, wheat, rapeseed, sorghum and forestry residues such as woodchips or sawdust. Although the properties of lignocellulose differ from source to source they can all be considered as difficult for conversion into other materials as well as their individual components.

Aquatic feedstocks – These feedstocks include micro- and macroalgae. This group contains a wide variety of chemicals within their structural and storage compounds; storage materials of aquatic organisms can contain both carbohydrates (starch, glycogen, laminaran and others) and triglycerides. Algal cell walls do not pose similar recalcitrance as lignocellulose do, but their composition makes the conversion challenging as many of the compounds present in algal cell walls cannot be easily fermented by microorganisms currently used in the industry.

In summary, biomass is composed of three major pools of chemicals: carbohydrates, phenolics and triglycerides. These chemicals are distributed between different cellular components: lignocellulose is a structural material composed of carbohydrates and phenolics. Lignocellulose is robust and resistant to conversion. Plant storage compounds contain both carbohydrates like sucrose and starch and triglycerides. These storage compounds are more accessible for different conversion routes than lignocellulose. The properties of these groups will determine their applicability for various processes of biomass conversion.

3.4 Biomass Conversion Methods

Biomass conversion platform is usually understood as a set of related technologies that could convert a feedstock (biomass of a certain characteristics) into products (array of outputs from this process). Biomass conversion platforms are usually divided into thermochemical and biochemical. The former is a thermal decomposition of biomass into chemicals and energy, the latter utilisation of microorganisms to transform biomass components into desired chemical compounds.

3.4.1 Conversion of Biochemical Energy Perspective

Various types of fuels have different functionalities and are therefore useful for various applications. Solid fuels like coal or wood can be stored relatively cheaply and safely for prolonged periods of time but their combustion produces significant amounts of particulate matter and ash that need to be appropriately disposed. Liquid fuels like petroleum, diesel and biodiesel have high volumetric energy content and can be easily transported through variety of routes such as pipelines or tankers. Gaseous fuels like methane, propane or hydrogen have very high energy content by mass unit, but their volumetric energy content largely depends on the pressure. Additionally, gaseous fuels burn with much cleaner characteristics than solid or liquid fuels do but tend to be explosive and need to be tightly controlled for any possible leakage. Each of these molecules has its particular application either as a fuel or as a chemical feedstock, and its handling depends on their particular properties. From the energy release point of view, what matters most is the chemical composition and bond energies that are present in these molecules. The energy content of fuels or in other words chemical energy available per unit mass is described by heating values, which are corresponding to net enthalpy released during the reaction of a particular fuel with oxygen under isothermal conditions. Heating value is usually expressed in MJ kg^{-1} or related units of energy per unit of mass. The chemical energy stored in these fuels can be released through a process of oxidation. Some typical bond energies are summarised in Table 3.1, and changes of energy related to the combustion of typical fuels are presented in Figure 3.5.

Release of the energy from fuels is divided into two phases: an endothermic step (energy investment) and exothermic step (energy pay off). First, the supply of energy in form of heat is required to break the bonds of the fuel molecule (endothermic reaction), and then in the second phase of the reaction the formation of new bonds results in the release of energy (exothermic reaction). The products of this reaction have lower energy than substrates of the reaction had, and this difference can be collected and utilised

Table 3.1 Typical bond energies present in energy carriers and products.

Bond	Bond energy (kJ mol^{-1})	Bond	Bond energy (kJ mol^{-1})
H=H	432	O=O	494
C=C	347	C=O	799
C=C	611	O=H	460
C=C (aromatic)	519	C=O	360
C=H	410	N=O	623

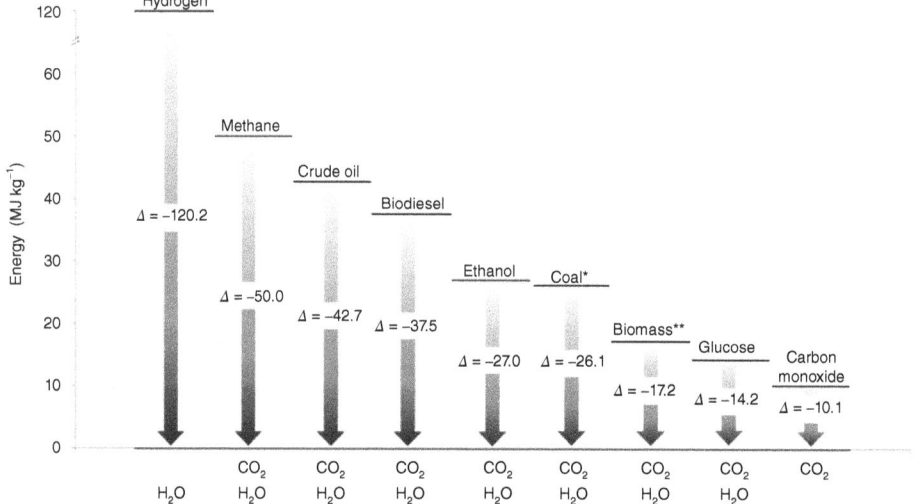

Figure 3.5 Energy released (ΔE) from complete combustion of typical fuels and their major combustion products. Note: discontinuous y axis; * bituminous coal; ** herbaceous biomass. Values from GREET, The Greenhouse Gases, Regulated Emissions, and Energy Use In Transportation Model. US DOE.

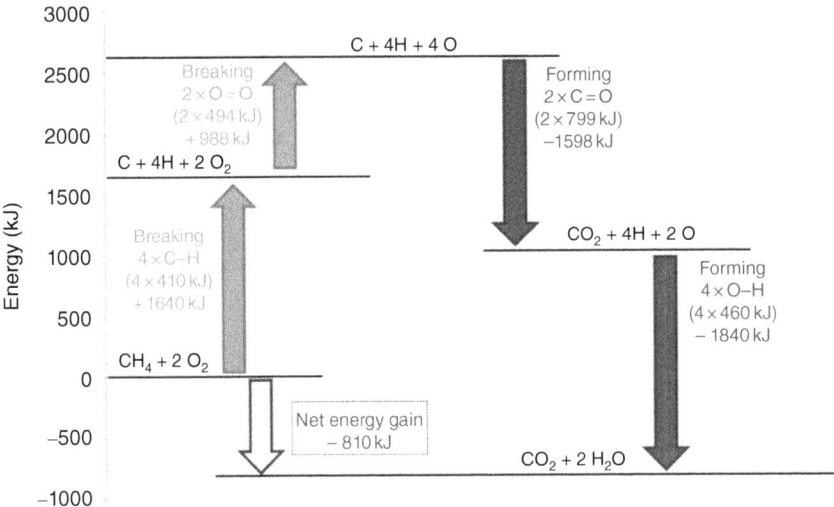

Figure 3.6 Complete combustion of methane, overview of bond energy changes. Energy investment phase in marked with upward arrows, energy payoff phase with downward arrows. The net energy gain is the difference in energy between reactants and products.

as heat or converted into work and power. The energy change during combustion of simplest hydrocarbon, methane, is presented in Figure 3.6.

There are many factors that contribute to the quantity of energy released during oxidation process and the resultant products, but the most important one is the availability of oxidant, in most cases oxygen. Based on this criterion, two types of processes can be distinguished.

Complete oxidation takes place in the abundance of oxygen. During oxidation, carbon molecules are oxidised to carbon dioxide and hydrogen molecules to water. Complete oxidation releases maximal amount of energy during the conversion process.

Incomplete oxidation takes place when supply of the oxygen is limited. It produces less energy and results in the formation of numerous products. The range of these products is largely dependent on the conditions of the conversion, that is, oxidant availability, temperature, pressure and so on.

Another important factor is a type of fuel and more specifically its elemental content and types of chemical bonds that are broken during combustion. In general, the lower the oxygen content of a particular fuel, the higher the energy content. During the process of combustion, oxygen is the terminal electron acceptor and cannot be oxidised; consequently oxygen atoms in the fuel produce no energy during conversion process. The relative content of hydrogen and carbon is another important parameter. Higher ratio of these two atoms is indicative of higher level of carbon–carbon bond saturation and consequently higher energy content of such fuel and an indication of lower molar CO_2 emissions per energy unit.

Biomass and biomass-derived fuels have lower energy content than fossil fuels. First, biomass contains significant amount of bound water (moisture). Moisture content of biomass has a significant impact on the heating values and possibilities of biomass conversion. Thermochemical processes such as combustion, gasification or pyrolysis require low content of moisture to yield high energy conversion rates. Biochemical processes on the other hand can utilise biomass with high moisture. Enzymatic reactions that are essential to this type of conversion are performed in aqueous environment and therefore additional content of moisture in the biomass does not impact their efficiency. Second reason for biomass and biomass-derived fuels having lower energy content is their elemental composition. Biomass and bio-fuels contain more oxygen than fossil fuels. The elemental composition of typical biomass corresponds to the following formula $CH_{1.4}O_{0.66}$ [18] compared to the typical formula of coal that can be approximated to $C_{135}H_{96}O_9NS$ [19]. Comparing these two formulae indicates that the ratio of energy yielding elements, that is, carbon and hydrogen versus non-energy-yielding oxygen is much higher in fossil fuels than biomass. This is due to the fossilisation process in which fossil fuels were created, that is, partial decomposition of biomass in the absence of air and under high pressure and temperature. This process resulted in the deoxygenation of these substances increasing their calorific value.

3.4.2 Overview of Biomass Conversion Technologies

As it was introduced earlier, biomass conversion methods can be broadly divided into thermochemical and biochemical. Thermochemical conversion utilises temperature, pressure and catalysts to transform biomass into various chemical compounds of high utility such as fuels or chemicals. Biochemical conversion on the other hand utilises microorganisms and their ability to transform carbohydrates, lipids and other biomass components into an array of compounds including fuels and chemicals. Although at the first glance these processes are very different and utilise completely different approaches, closer look at them will reveal numerous parallels and similarities especially at the energy level. It has been stated in the earlier that biomass has rather complex chemical composition and various types, and even different fragments of biomass have different chemical compositions. To address this complexity, a number of different thermo- and biochemical routes of biomass conversion have been developed. The choice of particular process of biomass conversion will largely depend on biomass availability and the type of the product that is desired: heat, electricity, fuels, chemicals, food or feed components and so on. Routes that allow conversion of various types of biomass into these respective products have been developed and are summarised in Figure 3.7.

3.4.3 Thermochemical Conversion of Biomass

Thermochemical conversion of any fuel including biomass is composed of four common steps, and all major technologies of biomass conversion take their names from the most important step of a particular conversion process. The four thermochemical transformations are as follows:

Drying – Drying is the first step of any thermochemical processes of biomass conversion. During this endothermic process, water (moisture) is removed from biomass particles. The process takes place above 100 °C and results in the increase of the heating value of biomass at the expense of energy. Although potentially any form of biomass may be dried, it is only practicable if the initial content of moisture is less than 50%, preferably in range of 10%. Otherwise, energy used for drying may significantly outweigh the energy gain from the conversion itself. In case of feedstocks with high moisture content, biochemical routes are more applicable than the thermochemical ones.

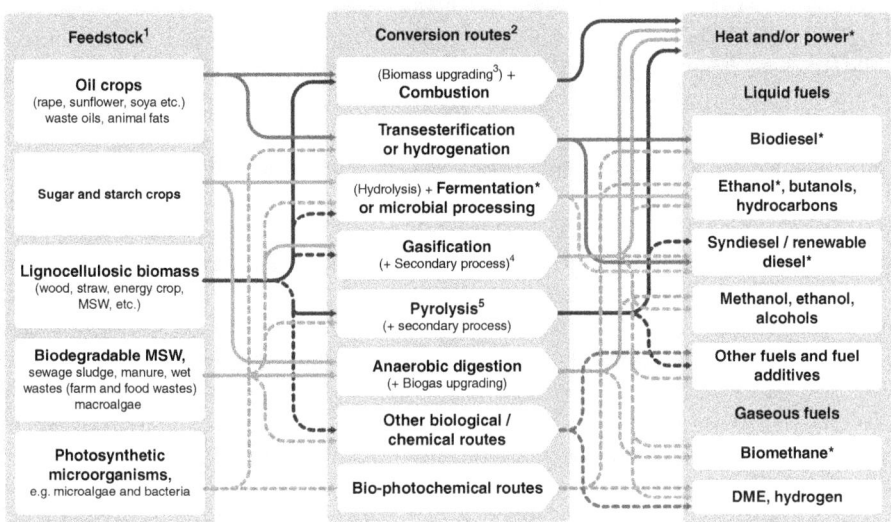

Figure 3.7 Schematic view of the variety of commercial (solid lines) and developing bioenergy routes (dotted lines) from biomass feedstocks through thermochemical, chemical, biochemical and biological conversion routes to heat, power, CHP and liquid or gaseous fuels. Note: commercial products are marked with an asterisk; [1] Parts of feedstock can be used in other routes; [2] Routes can yield co-products; [3] Biomass upgrading can include densification; [4] Anaerobic digestion also produces minor gasses; [5] Including related thermochemical routes like liquefaction. Reproduced with permission from IPCC 2012: IPCC Special Report on Renewable Energy Sources and Climate Change Mitigation. Prepared by Working Group III of the Intergovernmental Panel on Climate Change, Figure TS.2.3. Cambridge University Press.

Pyrolysis – Pyrolysis as a process may be described as thermal decomposition of a material under anoxic or limited oxygen conditions. The products of pyrolysis depend on actual conditions of the transformation and are composed predominantly of volatiles, gasses and to lesser extent high-molecular-weight hydrocarbons and char (carbon enriched biomass) [18]. Pyrolysis is a platform for another two processes: combustion and gasification.

Combustion – Combustion is process that releases energy from various products that were obtained in the pyrolysis reaction. In excess of oxygen all these products (volatile gasses and char) are oxidised to carbon dioxide (CO_2) and water (H_2O) releasing significant content of energy that can be captured. This highly exothermic process compensates for the energy used in the processes of drying and pyrolysis and provides net energy gain.

Gasification – Gasification is a process that transforms molecules through partial oxidation, as opposed to combustion in which the transformation is performed in excess of oxygen. Partial oxidation results in the production of less energy but higher quantity of intermediate compounds that could be used as chemical feedstocks.

All these transformations take place in typical routes of thermochemical conversion of biomass. These changes can be either endothermic or exothermic and produce various intermediates that can become useful products when thermochemical conversion is designed to maximise their production. In the first step of the conversion process, drying, biomass loses moisture, that is, increases its heating value at the expense of energy coming from external source. In the second step, pyrolysis, biomass is converted into higher energy intermediates such as tars, other gaseous products and solid char. Although these transformations are endothermic in nature, the gaseous products of these transformations such as hydrogen (H_2), carbon monoxide (CO) or methane (CH_4) can be combusted to recover the energy to power this process. Oxidation is the next step of thermochemical transformation. Depending on the availability of oxygen and temperature of oxidation, the transformation can either proceed to completion or not. Complete oxidation, that is, combustion occurs when all the gaseous products from previous steps are rapidly oxidised to release energy in an exothermic process. In case of oxidant shortage, both exo- and endothermic transformations take place in the process of gasification. These transformations between gaseous products and solid chars as well as gaseous products themselves result in the production of syngas. In all thermochemical processes, each of these reactions takes place to some extent and the names of predominant conversion technologies are named according to the most important step of the conversion process, that is, combustion, pyrolysis and gasification. Figure 3.8 summarises major steps and transformations between various compounds.

3.4.4 Biomass Combustion

Direct combustion of biomass is both the easiest and the most developed technology of converting biomass into energy. From the historical perspective, the controlled usage of

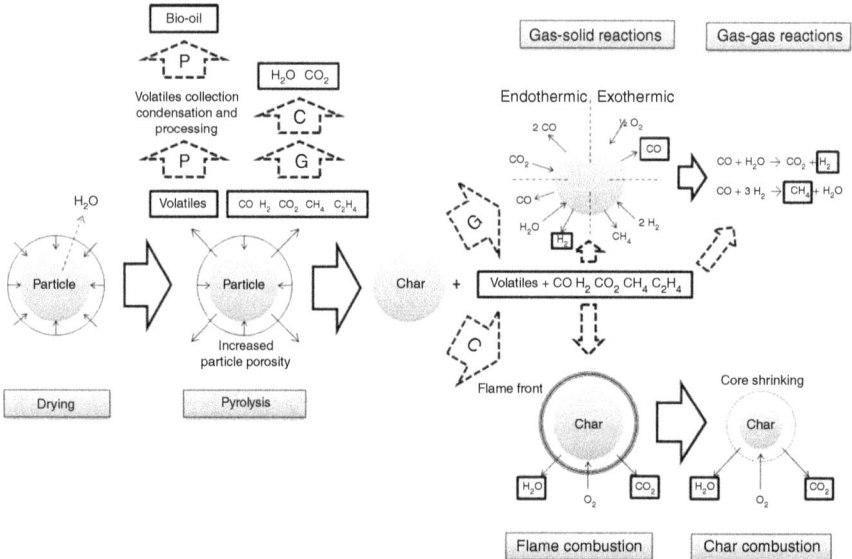

Figure 3.8 Summary of thermochemical transformations in thermochemical conversion processes. Four major transformation steps are annotated in grey boxes; major products from each transformation are outlined in solid lines; common processes are in solid arrows, specific routes are in dashed arrows, abbreviations: C, combustion; G, gasification; P, pyrolysis.

fire to burn biomass was probably one of the first milestones that allowed the development of human civilisation. Early humans burned biomass to generate heat, cook food and provide protection from wild animals. Fast forward to present day and combustion of various fuels still generates about 95% of global primary energy supply worldwide [20]. Vast majority of economies from highly developed to those of low development still rely on combustion to provide heat, power and electricity. The success of combustion process over other methods of generating usable energy can be accredited to its relative simplicity, high portability and energy return per unit mass of fuel.

The process of combustion comprises several phases that release energy from a fuel, fossil or renewable alike. Combustion process starts with the supply of heat from the external source to initiate endothermic processes of drying and pyrolysis. Depending on the type of the fuel, these processes may take significant amount of time and energy or be very brief and will not require significant energy inputs. In general, solid fuels require more time and energy for these initial endothermic steps than liquid fuels do and much more than gaseous fuels, which are already similar to the products of pyrolytic transformations and as such can be easily combusted. If we take solid biomass as our fuel of choice, the combustion process starts with drying, that is, removal of water molecules bound in the biomass. When heat is applied to biomass its temperature increases to about 100 °C, then water changes its state from liquid to gas and evaporates from the biomass making it accessible for subsequent thermochemical transformations. Removal of this moisture is a prerequisite for further steps of conversion and therefore the lower initial content of moisture of biomass the less energy is needed in the drying step and the higher energy return from biomass can be achieved. Once the moisture is removed pyrolysis of biomass takes place. Biomass decomposes into volatile products such as gasses and tars and solid particles of char. The ratio of these products is dependent on the conditions of combustion such as temperature, oxygen availability, heating rate, particle size and effects of catalysts [21]. At the end of pyrolysis stage, two of the products enter oxidation phase and combust to yield energy: volatile gasses (flame combustion) and char (char combustion). Gaseous products of pyrolysis: hydrogen, methane and lower hydrocarbons burn rapidly in bright yellow flame releasing energy, CO_2 and H_2O as products. Simultaneously, carbon contained in the solid product of pyrolysis, char, combusts to yield energy and carbon dioxide in a process. The combustion takes place in the surface of char particles and results in gradual decrease of the particle volume [21]. These processes are summarised in the combustion section of Figure 3.8. For the efficient conversion of the fuel to energy through the process of combustion, the proper adjustment of three 'Ts' is essential, which include temperature, turbulence and time [21]. It is critical to maintain these parameters optimal to prevent the formation of incompletely oxidised products such as carbon monoxide, polycyclic hydrocarbons and soot [21]. Appropriate adjustments of these parameters will limit incomplete combustion and release maximal amount of energy stored in the fuel.

3.4.5 Gasification

Gasification is a process that converts fuel into gasses through partial oxidation. The difference in oxidant content is the major difference between combustion and gasification. Partial oxidation of biomass results in the formation of higher energy compounds such as hydrogen (H_2), carbon monoxide (CO), methane (CH_4) and to lesser extent carbon dioxide (CO_2) and water (H_2O), which are typical of combustion. Due to differences in oxygen content, the products of gasification process have higher bond energy and therefore higher utility than final products of combustion. Consequently, due to partial oxidation gasification releases less energy than combustion does. These higher energy compounds produced in a gasification reaction can be therefore used as both fuel and chemical feedstocks for the production of other materials. The second product of gasification is a solid called char, which is made of mineral matter and unreacted carbon. Char has numerous potential applications including solid fuel, activated carbon or soil conditioner. The ratio of solid and gaseous products depends on the reaction conditions of a particular gasifier. Two initial endothermic transformations, that is, drying and pyrolysis are analogous to combustion process. Pyrolytic products such as hydrogen, methane, volatile lower hydrocarbons, tars and solid char enter partial oxidation in gasification reaction. Depending on the conditions of operation of a gasifier, a series of chemical transformations between gasses themselves and between gasses and char take place. These reactions are both endothermic and exothermic and their sequence and the composition of final product (syngas or producer gas) depend on the content of oxygen, steam, residence time, temperature and design of a gasifier. Because gasification is the process in which both endo- and exothermic reactions take place it is possible to manipulate the conditions of the process so the exothermic reactions provide energy for the endothermic changes and formation of desired products. Although each process is specific to the conditions of operation, in general, low temperatures of operation favour production of methane, whereas high temperatures yield hydrogen formation [21]. Another parameter that has major effect on gas composition is gasification medium, which can include air, oxygen, steam, nitrogen or hydrogen among others [22]. The process of gasification and chemical transformations between gaseous products and solids and gasses alike are presented in the gasification section of Figure 3.8.

Syngas produced in the process of gasification is composed of high-energy compounds such as hydrogen, carbon monoxide or methane. It is therefore possible to further utilise these compounds as fuels or feedstocks for the production of materials. The most straightforward application of this producer gas is its utilisation as low-heating-value fuel. Syngas can be combusted directly to generate steam for heat and power generation via a steam turbine. This application is mostly suitable for low calorific value syngas obtained though gasification of biomass in the presence of air. High content of inert nitrogen present in the air significantly dilutes the producer gas and limits its applications for the synthesis of higher value compounds and direct applications in gas engines [22, 23]. Other designs of gasification reactor and utilisation of other gasification medium such as oxygen, steam or hydrogen results in higher quality producer gas of higher calorific value. Provided that this syngas is cleaned from residual tars and dust particles, it can become a very valuable fuel for gas engines or gas turbines that are more efficient generators of power then steam turbines are. Most importantly, however, such high-quality syngas is applicable for the synthesis of chemicals. An array of compounds that are formed in gasification process opens numerous possibilities for the synthesis of platform chemicals and subsequently high value compounds from these intermediates. The most typical reactions that can provide these platform chemicals from syngas intermediates include the following:

Hydrogen enrichment – Hydrogen is one of the products of biomass gasification. Due to high utility of hydrogen for chemical synthesis, gasification process can be tweaked to maximise its production [23]. Two reactions between syngas components can maximise its formation: steam reforming and water-gas shift reaction. To maximise the relative content of hydrogen in gaseous products, CO_2 is removed by adsorption on CaO [23].

Steam reforming

$$CH_4 + CO_2 \leftrightarrow 2H_2 + 2CO$$

Water-gas shift reaction

$$CO + H_2O \leftrightarrow H_2 + CO_2$$

Fischer–Tropsch process – Fischer–Tropsh synthesis (FTS) is one of the most known applications of syngas. The technology was developed in Germany in the 1920s and is successfully utilised for large-scale synthesis of fuels and chemicals from coal in South Africa. FTS allows the synthesis of numerous hydrocarbons from two major constituents of syngas, hydrogen and carbon monoxide, in the presence of a catalyst, usually iron or cobalt. The products from the FTS vary depending on the catalyst formulation and process conditions [24]. Fundamental reactions of FTS are as follows [24]:

Synthesis of paraffins

$$nCO + (2n + 1)H_2 \leftrightarrow C_nH_{2n+2} + nH_2O$$

Synthesis of olefins

$$nCO + 2nH_2 \leftrightarrow C_nH_{2n} + nH_2O$$

Synthesis of alcohols

$$nCO + 2nH_2 \leftrightarrow C_nH_{2n+1}OH + (n - 1)H_2O$$

The flexibility of FTS in the production of various hydrocarbons makes it an excellent platform for the production of renewable fuels and chemicals from biomass. Numerous compounds such as gasoline, diesel oil, kerosene, olefins, alcohols, waxes and many others can be synthesised using this route [24]. There are, however, two major drawbacks that currently hinder the widespread utilisation of FTS for the synthesis of renewable fuels and chemicals. First, efficient production of chemicals through FTS requires high-quality producer gas, free of tars and dust, and gas cleaning still remains one of the bottlenecks of gasification process. Second, for the successful execution of FTS an appropriate ratio of two gaseous substrates is required. The ratio of H_2 to CO should be close to 2:1 for the successful synthesis. Although the exact composition of syngas depends on a particular gasifier, biomass gasification usually produces gas with different composition mostly due to high content of oxygen in the gasified biomass. In order to successfully utilise FTS, the gas composition needs to be adjusted, which includes removal of free methane and adjusting the content of hydrogen. Two reactions particularly useful for these purposes [23] are steam reforming and water-gas shift reaction. Once these technical hurdles are resolved and costs of installations are brought down, FTS can become one of the mainstream technologies in the synthesis of renewable fuels and chemicals.

Synthesis of methanol and dimethyl ether (DME) – Both methanol and dimethyl ether (DME) are promising chemicals with application in fuels and beyond. Hydrogenation of gasification products results in the synthesis of methanol. Both CO and CO_2 can be hydrogenated in the presence of various catalysts yielding methanol as an end product [23]. Condensation reaction between the two molecules of methanol results in the production of DME.

$$CO + 2H_2 \leftrightarrow CH_3OH$$

$$CO_2 + 3H_2 \leftrightarrow CH_3OH + H_2O$$

$$2CH_3OH \leftrightarrow CH_3OCH_3 + H_2O$$

The synthesis of methanol from syngas requires however some major adjustments of gas composition. The reaction requires even higher content of hydrogen to proceed than FTS, and the ratio of H_2 to CO should exceed 2:1 for the successful synthesis [24]. This raises concerns about the supply of hydrogen required to perform this reaction on large scale. Moreover, methanol synthesis is limited by the presence of other gaseous compounds such as CH_4 or N_2 [24].

3.4.6 Pyrolysis

Pyrolysis is process of endothermic conversion that produces a number of high-energy intermediates that may become valuable feedstocks for the production of biomaterial and bio-fuels. The process of pyrolysis starts with biomass drying analogous to combustion and gasification processes. Once the moisture is removed, the biomass decomposes into volatile products: gasses and tars and solid char. The reaction, however, is not allowed to proceed to the oxidation stage due to the lack of an oxidant. Instead, the gaseous streams are separated into condensable gasses that are cooled down and condensed to form bio-oil, and non-condensable high-energy gasses that are separated. These gasses undergo a series of chemical reactions among themselves largely analogous to those in gasification process. After the separation is complete, these products can be combusted to provide the energy necessary to power further endothermic reactions in the pyrolysis reactor. The final product of the transformation is char, which is analogous to the char produced in the other two conversion processes. The ratio of these products is dependent on the conditions of pyrolysis such as temperature, particle size and effects of catalysts, but the most important criterion of pyrolysis is time of reaction and heating rate. Pyrolysis technology has many modifications that yield different products. Differences between pyrolytic processes and the products they yield are influenced by the control parameters of the process and the form and chemical composition of the feedstock being processed. The summary of different variants of pyrolysis technologies and their products is presented in Table 3.2 [25].

Historically, one of the variants of slow pyrolysis (carbonisation) was a milestone technology in the development of metallurgy. Carbonisation is a process that maximises the production of char (charcoal) from biomass, usually wood. Because of its high porosity and combustion temperatures, charcoal is an excellent fuel for metallurgy. Production of charcoal still remains a major application of carbonisation technology. This conversion process can take several days at relatively low temperatures of about 400 °C and results in the conversion of biomass into solid products. An important variant of pyrolysis technology is fast pyrolysis. This variant is designed to maximise the

Table 3.2 Pyrolysis methods and their variants.

Pyrolysis technology	Residence time	Heating rate	Temperature (°C)	Major
Carbonisation	Days	Very low	400	Charcoal
Conventional	5–30 min	Low	600	Bio-oil, gas, char
Fast	0.5–5 s	Very high	650	Bio-oil
Flash-liquid	<1 s	High	<650	Bio-oil
Flash-gas	<1 s	High	<650	Chemicals, gas
Ultra	<0.5 s	Very high	1000	Chemicals, gas
Vacuum	2–30 s	Medium	400	Bio-oil
Hydro-pyrolysis	<10 s	High	<500	Bio-oil
Methano-pyrolysis	<10 s	High	>700	Chemicals

Source: Mohan *et al*. 2006 [25]. Reproduced with permission of American Chemical Society.

formation of liquid products. To achieve this goal, a higher temperature of 650 °C and very fast heating rates are applied. These conditions combined with the small size of biomass particles promote the formation of different products. The low residence time of condensable gasses limits their secondary reaction with char particles; this way these compounds can be transferred to a condenser where their condensation takes place. Usually, as a result of fast pyrolysis 60–75% of liquid fraction (bio-oil), 15–25% of char and about 10–20% of non-condensable gasses is obtained [21]. The predominant product of fast pyrolysis is a dark brown liquid called bio-oil. It contains numerous organic compounds such as phenolics, carboxylic acids, aldehydes, alcohols and others. Bio-oils have a major advantage over other biomass products, as they are organic liquids that can be stored and transported through haulage or pipelines. They are also considered as potential sources of renewable fuels and chemicals. The most straightforward application of bio-oils is their utilisation as low-heating value fuel in boilers and furnaces for heat and electricity generation [25]. Raw bio-oil is most suitable as a renewable alternative to heavy fuel oil in boilers or furnaces. Attempts were made to use bio-oil as a fuel for diesel engines, turbines and Stirling engines to increase the efficiency of energy production [26]. To date, these attempts were rather unsuccessful due to physical characteristics of bio-oil namely low heating value due to high content of oxygenated compounds and water, poor volatility, high viscosity, coking, corrosiveness, chemical instability and incompatibility with conventional fuels [25, 26]. To be able to successfully apply bio-oil as a fuel for these engines, modifications of engine designs, fuel upgrading or combination of both may be required to achieve desired performance [25]. Some of bio-oil deficiencies can be improved with relatively simple technologies, whereas others require more complex processing [26]. Typical processes of bio-oil upgrading include cracking, decarboxylation, decarbonylation, hydrocracking hydrodeoxygenation and hydrogenation, and representative reactions are presented in Figure 3.9 [27].

In principle, bio-oils are promising starting point for fractionation and subsequent upgrading of their components into fuels and platform chemicals. The biggest limitation of this process to date is very high diversity and heterogeneity of bio-oil components [27]. In a typical sample of bio-oil, more than 300 compounds are normally present [25]. Refining this mixture to useful building blocks has not been achieved to date and

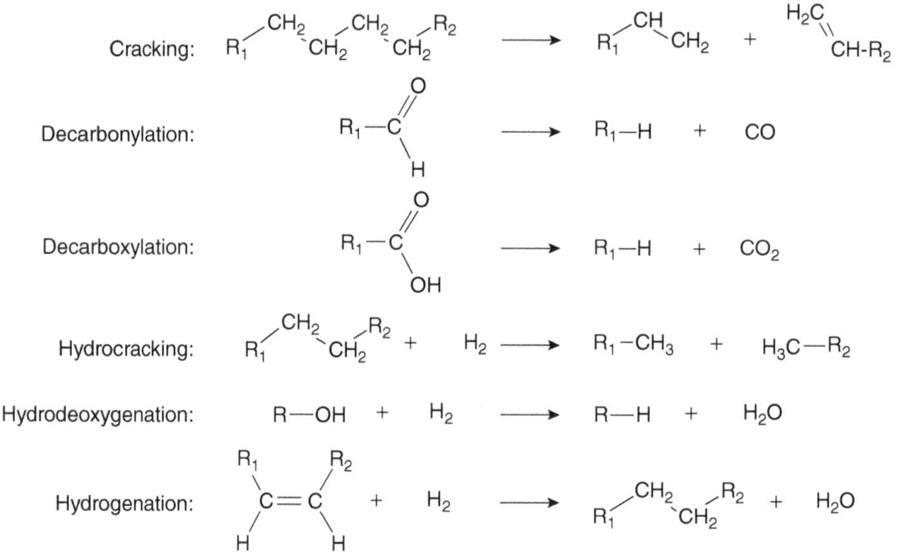

Cracking: R₁–CH₂–CH₂–CH₂–R₂ ⟶ R₁–CH=CH₂ + H₂C=CH–R₂

Decarbonylation: R₁–CHO ⟶ R₁–H + CO

Decarboxylation: R₁–COOH ⟶ R₁–H + CO₂

Hydrocracking: R₁–CH₂–CH₂–R₂ + H₂ ⟶ R₁–CH₃ + H₃C–R₂

Hydrodeoxygenation: R–OH + H₂ ⟶ R–H + H₂O

Hydrogenation: R₁(H)C=C(H)R₂ + H₂ ⟶ R₁–CH₂–CH₂–R₂ + H₂O

Figure 3.9 Examples of reactions associated with catalytic bio-oil upgrading. Reproduced with permission from: Mortensen PM, Grunwaldt JD, Jensen PA, Knudsen KG, Jensen AD. A review of catalytic upgrading of bio-oil to engine fuels. Applied Catalysis A: General. 2011;407(1–2):1–19. Copyright (2011) Elsevier.

remains a significant challenge. To address these problems, approaches to utilise bio-oil as feedstock for other processes have been studied. Utilisation of bio-oil as a feed for gasification and subsequent synthesis of hydrocarbons via FTS has been studied and proved technically feasible [25]. Other applications of bio-oil include its use in particle boards as resins, as a carrier for slow-release organic fertiliser, wood preservative, food additive and other niche applications [26]. In summary, bio-oil has an advantage of storability and transportability and has already found applications as a renewable fuel for boilers or furnaces. The most promising and valuable applications of bio-oil as a source of fuels and chemicals are still under development, and significant advances in refining and upgrading are still required to fulfil the potential of this technology.

3.4.7 Conversion of Oily Feedstocks

Oily feedstocks – triglycerides already possess long alkyl chains that are similar to those of diesel molecules. This similarity resulted in different approaches to transform lipid feedstocks than those used for whole lignocellulose biomass. An interest in using vegetable oil and its derivatives as a fuel can be dated back to the history of diesel engine itself. The first demonstration of engine during the World Fair in Paris in year 1900 used peanut oil as a fuel [28]. Currently, two approaches to derivatise oily feedstocks to fuel molecules are used: transesterification and hydroprocessing.

Transesterification – Transesterification is relatively simple process, triglycerides are reacted with a short-chain anhydrous alcohol (usually methanol or ethanol) using a catalyst. During the reaction, glycerol moiety of the fatty acid is replaced with alkyl group of an alcohol. After the reaction is complete, two products are formed: fatty acid alcohol

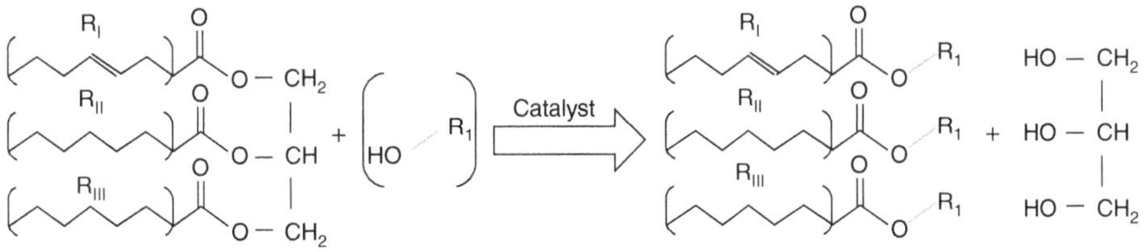

Figure 3.10 Scheme of transesterification. Biodiesel is synthesised in a chemical reaction of transesterification of an oily feedstock, triglyceride with a short-chain alcohol, usually methanol in the presence of a catalyst. Transesterification yields alkyl esters of fatty acids (biodiesel) and a by-product, glycerol. $R_{I,II,III}$ alkyl chains of fatty acid (usually C14–C22), and R_1 alkyl group of an alcohol (usually methyl or ethyl).

ester (biodiesel) and glycerol. The reaction scheme is presented in Figure 3.10. The resultant product, biodiesel, has comparable characteristics with fossil diesel and all major parameters, such as higher heating value (HHV) (39–41 MJ kg^{-1}), flash point, cetane number and kinematic viscosity, are similar to its fossil alternative [29].

Efficient transesterification requires a catalyst and/or high temperature and pressure to convert substrates to products. Type of this catalyst is fundamental as it affects both the type of feedstock that can be used and up- and down-stream biodiesel processing. There are several methods of producing biodiesel: alkali, acid, enzymatic and super-critical alcohol [30]. All of these methods have their advantages and drawbacks and currently the predominant method of industrial biodiesel production is utilisation of a two-step acid–alkali catalysed process. The first step of the conversion is used to upgrade oily feedstocks with high content of free fatty acids (esterification), and the second step results in the high efficiency of the transesterification of triglycerides to biodiesel. After the reaction is complete, both products (biodiesel and glycerol) are separated in settling tank or centrifuge due to significant density difference between both products (880 kg m^{-3} esters of fatty acids; 1050 kg m^{-3} of glycerol) and purified. Purification process is composed of repetitive heating and washing steps according to the following scheme: catalyst neutralization, deodorisation and pigment removal [30]. Production of biodiesel through a process of transesterification is a mature technology already applied by the industry worldwide.

Hydroprocessing – Hydroprocessing is a thermochemical method widely established in the petroleum industry that has been adapted to the conversion of renewable oily feedstocks to hydrocarbons [31]. One of the major advantages of hydroprocessing is a range of products that can be synthesised depending on reaction conditions that include all types of hydrocarbon-based fuels like gasoline, jet fuel and diesel. The overall goal of the reaction is to crack the alkyl chain of fatty acid chain into smaller fragments, saturate all unsaturated bond with hydrogen and remove oxygen atoms from structure. The resultant fuel is chemically indistinguishable from current fossil fuels and therefore 100% compatible with all fuel infrastructures. Major drawbacks of hydroprocessing is its huge requirement for hydrogen that is currently produced almost exclusively from non-renewable sources mainly from steam reforming of natural gas and high energy consumption during the process. One of the major challenges to make hydroprocessing a truly sustainable technology is to develop methods of renewable hydrogen production that could supplement this promising technology with renewable hydrogen.

3.4.8 Biochemical Conversion of Biomass

3.4.8.1 Aerobic and Anaerobic Metabolisms

Despite significant differences in between thermochemical and biochemical processes, there are significant parallels that can be drawn between the two conversion platforms. To understand the fundamentals of biochemical transformation of feedstocks into bio-fuels and biomaterials, basic review of cellular metabolism might be helpful. In simplest words, metabolism can be defined as a set of chemical transformations that are utilised by an organism to gather energy and to build and maintain its cellular structures. Different organisms have different ways to achieve these goals, but fundamentally all the living creatures can be assigned either to autotrophs or heterotrophs. Autotrophic organisms such as plants, algae and cyanobacteria utilise energy from light that powers cellular processes including carbon dioxide fixation that ultimately forms glucose and other more complex polymeric substances like starch cellulose or triglycerides. Heterotrophs are organisms that break down these complex substances into smaller units. Ultimately, bonds of these molecules are also broken and their chemical energy is used to power cellular functions. Heterotrophic organisms can be broadly divided into those exhibiting aerobic metabolism and those that operate under anaerobic conditions. Development of aerobic metabolisms and especially cellular respiration is considered a paramount achievement in evolutionary sense just like photosynthesis. This complex cellular system is a biological equivalent of complete oxidation (combustion) where all the energy stored in the molecules of food is transformed into energy. Carbon and hydrogen from these molecules are oxidised to carbon dioxide and water, respectively. From the energy point of view, this process is largely equivalent to complete combustion; however, because of the fragility of living organisms and their cellular structures the energy is not released in large chunks of heat but in a stepwise manner and intelligently directed towards activated energy carriers like ATP. Just like combustion processes allowed the development of the technological civilization as we know it, cellular respiration allowed the evolution of complex mammals including humans. Without the energy efficiency of cellular respiration only simple unicellular organism could develop. The availability of abundant cellular energy allowed the evolution of multicellular organisms, animals and ultimately mammals and humans. The parallels between biochemical transformation through aerobic metabolism and combustion are presented in Figure 3.11.

Activated energy carriers like ATP are excellent system of delivering energy for the organism, but they have little use outside the cells. They are not stable enough to be extracted and their chemical structure is very complex. In order to obtain compounds that could be used as fuels or platform chemicals, one needs to shift to less energy-efficient forms of microbial metabolism. These earlier evolutionary forms retain significant portion of the initial energy of the compound as high energy, stable intermediates, which can be relatively easily isolated. These metabolic transformations can be paralleled to the process of gasification in which incomplete oxidation of initial compounds results in the formation of high-energy gasses that could be used for the production of energy or platform chemicals. Various biochemical processes can yield high-energy compounds that can serve as both liquid and gaseous intermediates; most notable examples include ethanol, butanol, hydrogen and methane.

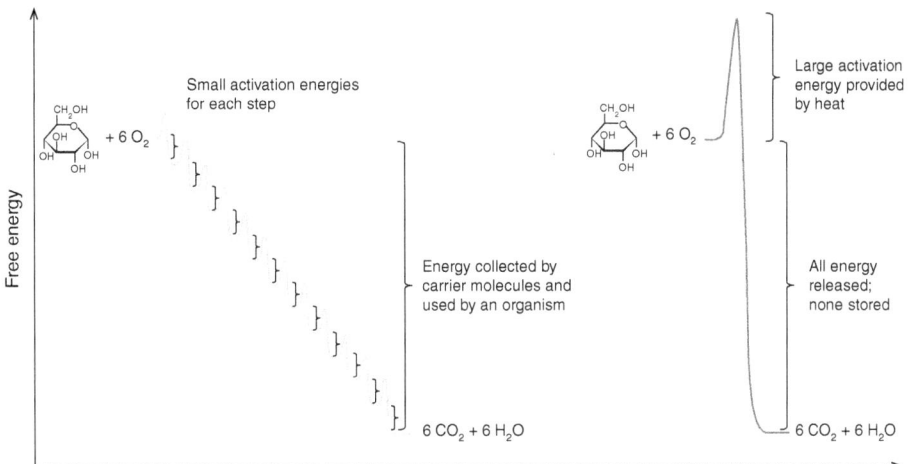

Figure 3.11 Schematic representation of stepwise oxidation of glucose in an organism compared to its combustion in oxygen. On the left biological decomposition of glucose into compounds of lower energy and ultimately into CO_2 and H_2O with most of the energy collected by carrier molecules such as ATP and NADH. On the right complete combustion of glucose in oxygen with activation energy from external heat source. Please note that substrates and products of both reactions are identical.

3.4.8.2 Central Metabolic Pathway under Anaerobic Conditions

Central metabolic pathway describes the energy generating flux of carbon in just about all organisms. The conversion of energy starts with glucose and results in the production of pyruvate, which can then enter other pathways that result in the production of energy (TCA cycle, fermentation) or building cellular components like nucleic acids or proteins. Glycolysis is the first step of energy release. During glycolysis, the six carbon sugar, glucose, is split into two molecules of pyruvate containing three carbon molecules each. During this process, chemical bond energy is released and transferred into two molecules of ATP. It is beyond the scope of this book to provide the insight into the process of glycolysis, and interested readers should refer to further reading section of this chapter for more information.

The brief summary of glycolysis is presented here; please note that glycolysis is a multi-step process and the equation does not reflect the complexity of this process.

$$C_6H_{12}O_6 + 2\,ADP + 2\,P_i + 2\,NAD^+ \leftrightarrow 2\,C_3H_4O_3 + 2\,ATP + 2\,NADH + 2\,H^+ + 2\,H_2O^* + 88\,kJ$$

* Please note that H_2O comes from elimination of water from phosphate groups (ADP to ATP synthesis) and not from glucose molecule.

The subsequent fate of pyruvate depends on the organism's enzymatic machinery and availability of oxygen. There are numerous possible routes a pyruvate molecule can take, which are summarised in Figure 3.12. We introduce two most important processes, respiration and fermentation, which could be paralleled to complete and incomplete oxidation of biomass in thermochemical routes.

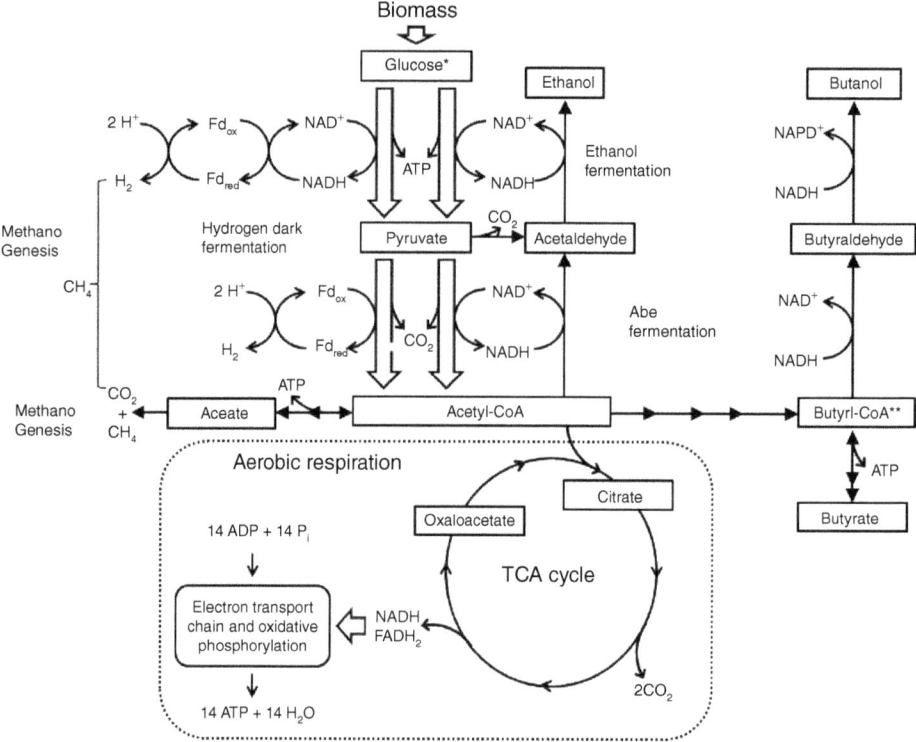

Figure 3.12 Overview of major metabolic pathways used for bioenergy production and aerobic respiration. Biomass used as a feedstock for fermentative production of bio-fuels and biochemicals is decomposed into glucose and enters glycolysis or bio-fuel-producing organism. Note: Other hexoses are converted into glucose *in vivo*, and pentoses will enter the pathway at later stages of glycolysis through pentose phosphate pathway). Oily feedstocks can enter through β-oxidation pathway as acetyl-CoA. * Glucose is split into two pyruvate molecules during glycolysis; ** Butyryl-CoA is formed through condensation of two Acetyl-CoA.

Aerobic respiration – In aerobic respiration, complete combustion of pyruvate is carried out in a multi-step process. Pyruvate is converted into acetyl CoA and enters the TCA cycle and oxidative phosphorylation – the series of enzymatic steps that perform complete combustion of these molecules to CO_2 and H_2O. This is a highly productive route as aerobic respiration produces about 14 molecules of ATP* per molecule of pyruvate [32, 33], seven times higher that glycolysis itself.

$$C_3H_4O_3 + 2.5\,O_2 + 14\,ADP + 14\,P_i \leftrightarrow 3\,CO_2 + 2\,H_2O^{**} + 14\,ATP + 14\,H_2O^{***}$$

* ATP synthesis occurs through a complex enzyme ATP synthase powered by proton gradient. The estimation of exact number of ATP molecules per molecule of pyruvate varies between organisms and does not need to be an integer.
  ** Biological combustion of pyruvate molecule.
  *** Elimination of water from phosphate groups (ADP to ATP synthesis).

Aerobic oxidation provides another important function in addition to ATP synthesis. The synthesis of energy carrier is coupled with regeneration of NAD^+ from NADH.

These NAD⁺ molecules can be then utilised in glycolsis to split another glucose molecule to maintain the flow of carbon through the process.

The process of aerobic respiration is the most energy efficient form of metabolism. This efficiency, however, makes aerobic respiration rather useless for energy extraction from biomass and its biochemical conversion to other compounds. All the energy from respiration is quickly captured, transformed into ATP and used by the organism without producing any stable intermediates that could be isolated and utilised as fuels or platform chemicals by humans.

Fermentation – Anaerobic processes also known as fermentations include several different types of metabolism that is capable of simultaneous production of ATP combined with the release of high-energy products or intermediates that can be used as fuel or platform chemicals: Fermentative production of ethanol, butanol, biohydrogen and biomethane by specialised organisms are promising routes to produce these fuels from renewable feedstocks. These four processes are described in detail in the later sections. These processes share common characteristics such as major metabolic intermediates, NADH regeneration capacity and to various extent production of additional ATP. They may differ, however, in their capacity to use different feedstocks and of course yield different products.

3.4.9 Harvesting Energy from Biochemical Processes

3.4.9.1 Ethanol Fermentation

Ethanol fermentation takes place under anaerobic conditions and is an example of anaerobic metabolism that yields high-energy intermediate that could be easily isolated. In the process of ethanol fermentation, pyruvate from the glycolysis is decarboxylated to acetaldehyde with an enzyme pyruvate decarboxylase and subsequently reduced to ethanol by alcohol dehydrogenase. The main biological function of this reaction is anaerobic regeneration of NAD⁺ from NADH, which is required for the process of glycolysis and production of ATP through this route.

$$C_3H_4O_3 \leftrightarrow C_2H_4O + CO_2$$

$$C_2H_4O + NADH + H^+ \leftrightarrow C_2H_5OH + NAD^+$$

Ethanol is one of only few chemicals that are predominantly produced by microbial fermentation. Although the process of ethanol fermentation can be carried out by numerous organisms, yeast (*Saccharomyces cerevisiae*) is the industrial organism of choice. The advantages of using yeast for ethanol production include high yield of ethanol production; capacity of using short maltodextrins, sucrose and fructose in addition to glucose; growth in slightly acidic conditions that inhibit bacterial contamination and relatively high ethanol tolerance (10–15 %). Additionally, yeast is considered as GRAS organism (Generally Regarded as Safe) that makes its biomass suitable as animal feed and has been a subject of extensive genetic studies that facilitate further strain improvement. Yeast can utilise numerous carbohydrate feedstocks to perform fermentation process. Typically, sugar and starchy feedstocks are used; utilisation of lignocellulose feedstocks has only recently entered commercial operations. Sugar feedstocks are relatively easiest to use as yeast is capable of metabolising sucrose into glucose and fructose and incorporating into glycolysis process. Starchy feedstocks require an additional step called saccharification in which starch is hydrolysed into glucose by enzymes called amylases. Lignocellulose is the most challenging source of carbohydrates. Combination of physical pre-treatment with extensive enzymatic treatment is required to release fermentable sugars from lignocellulose. The process of industrial ethanol fermentation is usually carried out in large fermenters at 30 °C at slightly acidic condition of pH 4.8–5.0 to limit the growth of undesirable bacteria that could compromise the efficiency of the fermentation process [34]. Acclimatised yeast inoculums are added to carbohydrate solution, and fermentation is initiated. Initially, rapid growth of yeast is observed until the concentration of 10^8 cells ml^{-1} is achieved. Since this stage, fermentation process takes off and ethanol concentration increases in the fermentation tank, which is coupled with the generation of heat and vigorous release of CO_2, both side products of this metabolic activity. The process of active fermentation takes about 12 h until majority of carbohydrates is utilised and is followed by a slower phase that takes 40–48 h until carbohydrates are depleted and concentration of ethanol reached 10–15% (w/w). After fermentation is completed, the broth is distilled to recover ethanol. Fractional distillation is performed to separate the ethanol at the concentration of 90–94% (azeotropic mixture with water). Further purification step is required to obtain anhydrous ethanol of fuel grade (>99%). In the industrial practice, dehydration is performed with zeolites – molecules that can entrap water molecules within their pores decreasing the content of water to almost zero.

Ethanol fermentation yields co-products that are dependent on the type of feedstock. Ethanol production from sugarcane is usually combined with production of table sugar (sucrose). Corn ethanol production yields other products including DDGS (distiller's dried grains and solubles) – dry milling process, corn oil and corn gluten – wet milling process. All ethanol fermentation processes generate high-purity CO_2 that can be collected and used in food, drink or chemical industries. Applications of ethanol extend beyond traditional uses in food and fuel industries and may become valuable feedstock for chemical industry in the synthesis of hydrocarbons, higher alcohols, acetaldehyde, acetone, diethyl ether, and ethyl acetate [35].

3.4.9.2 ABE Fermentation

Production of butanol by solventogenic bacteria, *Clostridium acetobutylicum*, and related is an example of strictly anaerobic metabolism. ABE (acetone–butanol–ethanol) fermentation is an example of mixed fermentation where numerous fermentation products are formed. In ABE fermentation, there are two stages of the process: acidogenesis and solvengenesis. In acidogenesis stage, two organic acids, acetate and butyrate, as well as two gaseous products CO_2 and H_2 are produced. Acetogenesis produces an ATP molecule in addition to those from glycolysis reaction, thanks to either acetate or butyrate kinases [36]. Production of these organic acids results in an increase of growth medium acidity that negatively impacts the producer strain, and in the second stage of the fermentation these products are reassimilated and reenter metabolic pathway. The reaction is catalysed by CoA transferase that does not require ATP utilisation; instead the reaction is driven by the formation of acetoacetate and finally acetone [36]. Once these acids are reassimilated respectively as acetyl and butyryl-CoA, the synthesis of ethanol and butanol can commence in solventogenic stage of the process.

Formation of acetyl-CoA occurs via decarboxyltion of pyruvate resulting from glycolysis and assimilation of previously secreted acetate.

$$C_3H_4O_3 + CoA\text{-}SH \leftrightarrow CoA\text{-}S\text{-}C_2H_3O + CO_2 + H_2$$

Acetyl CoA may undergo a two-step conversion into ethanol or condensation followed by saturation to form butyryl-CoA. These two substrates may form their respective alcohols in two analogous reactions:

$$CoA\text{-}S\text{-}C_2H_3O + NADH \leftrightarrow CoA\text{-}SH + C_2H_4O + NAD^+$$

$$C_2H_4O + NADH + H^+ \leftrightarrow C_2H_5OH + NAD^+$$

and

$$CoA\text{-}S\text{-}C_4H_7O + NADH \leftrightarrow CoA\text{-}SH + C_4H_8O + NAD^+$$

$$C_4H_8O + NADH + H^+ \leftrightarrow C_4H_9OH + NAD^+$$

The nature of ABE fermentation results in the formation of multiple products acetone–butanol–ethanol in the ratio of 3:6:1, respectively. ABE fermentation was one of the leading industrial processes at the beginning of the twentieth century with impacts well beyond the fields of science and technology and deep into military and politics. Although historically acetone was the most important of the three, currently most emphasis is placed on butanol, and efforts are being made to maximise its productivity. Butanol is an important industrial commodity with a primary use as a solvent and an important precursor of numerous compounds such as acrylate, and methacrylate esters, glycol ethers, butyl acetate, butylamines, and amino resins [37]. Its physicochemical properties such as higher energy content, unrestricted miscibility with traditional fuels and low vapour pressure make it a promising bio-fuel with better characteristics than ethanol [37]. ABE fermentation is a strictly anaerobic process that can use various feedstocks depending on the producer strain. Traditionally, starchy feedstocks were preferred due to amylolytic enzymes secreted by solvengenic *Clostridia*. In an effort to decrease the cost of the feedstocks, first sugar molasses and recently lignocellulose feedstocks have been utilised. The conversion of lignocellulose to butanol currently enters commercial phase. ABE fermentation is usually carried out in batch and is generally more complex than ethanol fermentation with yeast. Solvengenic *Clostridia* are strictly anaerobic organisms, sensitive to oxygen and rather slow growers. The yields of this fermentation have been traditionally rather low, and total solvent concentration rarely exceeds $20\,g\,l^{-1}$. Additionally, at about 2% concentration butanol is highly toxic to the cells and fermentation efficiency drops even further. Another bottleneck is the separation of diluted solvents from large volume of the fermentation broth, which makes the process of butanol purification expensive. It is envisaged that advances in fermentation technology and strain improvement may be required before the second wave of fermentatively produced butanol comes back to the market as fuel and platform chemical.

3.4.9.3 Biohydrogen

There are many biological routes to produce hydrogen such as direct biophotolysis, indirect biophotolysis, photofermentation and dark fermentation [38, 39]. Photolysis routes although very promising, suffer from low productivities and major engineering challenges in product separation. Fermentative routes and especially dark fermentation looks like the most technologically feasible option at the moment [38]. Among fermentative routes, hydrogen production by hyperthermophilic bacteria of the order

Thermotogales is considered particularly promising. These obligate anaerobes ferment glucose to acetate, CO_2 and hydrogen.

$$C_6H_{12}O_6 + 2H_2O \leftrightarrow 2CH_3COOH + 2CO_2 + 4H_2$$

Hydrogen production via this route is rather atypical fermentation reaction. In majority of fermentations, metabolic intermediates are reduced to regenerate NAD^+ (e.g. ethanol, butanol); in this type of fermentation, they are not. Instead, hydrogen ions are the ultimate acceptors of electrons forming hydrogen molecules [39]. The transfer of electrons from NADH to hydrogen ion is mediated by NADH: ferredoxin oxidoreductase and the final step of the hydrogen evolution by a hydrogenase. Such reaction takes place during glycolysis.

$$\text{Glyceraldehyde-3-phosphate} + NAD^+ + P_i \leftrightarrow \text{1,3-Bisphosphoglycerate}$$

$$+ NADH + H^+ NADH + H^+ + Fd_{ox} \leftrightarrow Fd_{red} + NAD^+$$

$$Fd_{red} + 2H^+ \leftrightarrow Fd_{ox} + H_2$$

Another molecule of hydrogen results from coupling pyruvate:ferredoxin oxidoreductase with hydrogenase.

$$C_3H_4O_3 + \text{CoA-SH} + Fd_{ox} \leftrightarrow \text{CoA-S-}C_2H_3O + Fd_{red} + CO_2 + 2H^+$$

$$Fd_{red} + 2H^+ \leftrightarrow Fd_{ox} + H_2$$

Hydrogen is a hugely important chemical feedstock and potentially useful bio-fuel that yields only water upon its combustion. Currently, about 95% of hydrogen is produced from reformation of natural gas and finding sustainable renewable sources of this chemical will hugely benefit other technologies of bioenergy production, especially the thermochemical. Hydrogen production with dark fermentation has numerous advantages, especially when conducted with thermophilic organisms like Thermotogales. These bacteria exhibit very large substrate flexibility and are capable of fermenting various hexoses, pentoses and even longer oligosaccharides or starch [39]. Additionally their high optimal growth temperature limits microbial contamination and simplified product recovery. Hydrogen solubility in the fermentation broth significantly drops at elevated temperatures that facilitates gas collection and removal. One of the products of fermentation process, acetate, is an economically valuable and recoverable product that could be explored commercially or as a feedstock to other processes. The advances in the development of biological hydrogen production including dark fermentation can have significant impact on the development of sustainable sources of this important chemical feedstock.

3.4.9.4 Biomethane

Biomethane production in an industrial process called anaerobic digestion is a process significantly different from other types of fermentation described earlier. Biomethane also known as (biogas) is produced by a diverse community of bacteria and archaea, which work together to produce methane and carbon dioxide through a series metabolic reactions. A unique feature of this process is that end products of one group of organisms become substrates for another until two gaseous products are obtained – methane and carbon dioxide [39, 40]. The process is composed of four steps: hydrolysis, acidogenesis, acetogenesis and methanogenesis. In the first step, large complex substances are hydrolysed into individual components like simple sugars or amino acids. In the second step, these molecules are converted into volatile fatty acids, hydrogen and carbon dioxide by another group of microorganisms. In the third stage, volatile fatty acids are converted into acetic acid, carbon dioxide and hydrogen that need to be removed by methanogens to maintain the conversion in the process called synthropy. The final stage of anaerobic digestion is methanogenesis. There are two groups of organisms that perform two separate methane synthesis reactions. The first group is capable of converting acetic acid into methane and carbon dioxide. This reaction is performed by anaerobic archea *Methanothrix* and *Methanosarcina*.

$$CH_3COOH \leftrightarrow CO_2 + CH_4$$

Another group of anaerobic archea *Methanobacteria* sp. converts gaseous products of acidogenesis into methane.

$$4H_2 + CO_2 \leftrightarrow CH_4 + 2H_2O$$

Anaerobic digestion is a very efficient process of biomass decomposition with efficiency approaching 90% [39]. Most important advantage of this process is its high feedstock flexibility. Mixed microbial communities are capable of transforming essentially any feedstock to biomethane. Most widely utilised include municipal and animal waste and wastewater, food waste, agricultural by-products and dedicated energy crops [39]. Providing that anaerobic conditions are maintained and pH controlled within 6.5–8.0, the composition of microbial communities will adjust themselves to utilise the available feedstock and decompose it into methane [40]. The productivities from different feedstocks, however, vary, and highest methane productivity is usually achieved for mixtures of manure with corn silage or fat slurries [39]. The fermentation time largely depends on the type of substrate and temperature of operation, and hydraulic retention time could range from several days to few months depending on those conditions. Methane is constantly extracted at the head of the digester and can be used in one of the two routes: initially cleaned of corrosive contaminants and used gas engine to generate heat and power; or purified from contaminants like CO_2 and H_2S and supplied to the gas grid for household and industrial use.

3.5 Metrics to Assist the Transition Towards Sustainable Production of Bioenergy and Biomaterials

3.5.1 EROI – Primary Metrics of Energy Carrier Efficiency

Inevitable depletion of fossil fuel resources forces societies to move from more efficient sources of energy and carbon that can be acquired at minimal costs into increasingly more difficult ones; those require larger investments and may come at increased environmental cost. Despite its utility to compare various methods of energy production, EROI does have its limitations. First, it does not take into consideration functionality of a particular source of carbon and energy. For example, liquid bio-fuels such as bio-ethanol, Fischer–Tropsch liquids or bio-oil have much lower EROI than wind-generated electricity. All these compounds, however, are composed of carbon and as such can have much higher applicability for the economy than electricity does. All of them can be stored, transported over long distances and after processing used as liquid fuels on demand. Moreover, all these compounds can be further used as feedstocks for a number of downstream processes to functionalise them into consumer products, wind electricity cannot. It is therefore essential to examine EROI parameters of processes that are comparable. Second, EROI does not discriminate the processes on the basis of their environmental impacts. For example, processes of extraction of shale oil may have similar EROI to the process of ethanol production. What the EROI analysis does not take into consideration is that extraction of hydrocarbons require significant inputs of freshwater, results in the production of greenhouse gasses and has generally a much bigger environmental footprint than traditional oil production processes. Production of ethanol from sugarcane on the other hand results in the production of renewable fuel (bio-ethanol), heat and power as well as valuable co product (sugar). Moreover sugarcane is a perennial crop that requires low input and is capable to sequester significant content of carbon in the soil if the plantations are established and managed properly. Therefore, although these two processes may look similar from the EROI point of view, more detailed analysis of the two shows clear differences. Therefore, sometimes additional metrics need to be taken into consideration.

3.5.2 LCA – Sustainability Determinant

Sustainable production of bioenergy and biomaterials is a complex task that needs to take into consideration a number of issues such as resource availability, local climate, technology, policy and economy. All these issues are very often region specific. Right combination of these factors can result in the production of sustainable alternatives to fossil-derived products. Sustainable production of energy and materials from biomass supports the development of rural areas, provides more balanced global economy and counteracts global climate change. On the other hand, inappropriate combination of resources, technology, management and policies can result in exacerbating food versus fuel dilemma, result in the degradation of unique ecosystems and worsening the effects of climate change. In between these two extreme cases is a multidimensional array of possibilities that result from implementations of a particular approach [41]. Therefore to implement bioenergy and biomaterial technologies on a large scale, an unprecedented in the development of any new industry scrutiny needs to be implemented [42]. This requirement resulted in the development of methods to assess environmental impacts and sustainability of a particular process called life-cycle analysis. Life-cycle assessment (LCA) is a systematic, cradle-to-grave process that evaluates the environmental impacts of products, processes and services. It is an important decision-support tool for policy makers, business analysts and researchers from public and private sectors. According to the ISO 14040.2 standard, *life cycle* is defined as 'consecutive and interlinked stages of a product or service system, from the extraction of natural resources to the final disposal'. LCA is therefore defined as 'a systematic set of procedures for compiling and examining the inputs and outputs of materials and energy and the associated environmental impacts directly attributable to the functioning of a product or service system throughout its life cycle' [43].

All products impact the environment to certain degree, and these impacts can vary significantly depending on raw materials, energy sources and waste disposal methods that were employed. Life cycle assessment identifies generation and transfer of these environmental impacts from one part of process to another and allows their quantification. LCA therefore allows to identify the most crucial steps of the production and disposal of a product or a service and to perform techno-economic analysis of alternative routes to minimise its environmental impacts without significantly affecting other parameters like performance or cost. To perform LCA, the process is broken down into its individual components and quantified. Usually, following stages of the process are distinguished: materials extraction, processing and manufacturing, use, and disposal. These stages are then broken down into fine details that become useful in identifying 'hot spots' of a process that generate most environmental impact. More sustainable alternatives or energy efficient processes can be then proposed and implemented for these stages.

3.5.3 Environmental Assessment of Bioenergy Production Processes

A number of studies evaluated the environmental impact of bioenergy production processes and identified processes that are most promising. It is beyond the scope of this book to provide detailed analysis of each of these studies, especially taking into consideration that studies of this kind should be related to local conditions, policies and availability of resources. These factors vary tremendously from region to region and as such can have huge impact on the analysis. There are however certain 'hot spots' of bioenergy production processes that have been highlighted and should be considered when developing renewable alternatives to current fossil fuel products. Four major categories of potential environmental impacts can be distinguished for bioenergy production processes:

3.5.3.1 Impacts Related to Land-Use Change

Production of dedicated feedstocks for bioenergy and biomaterials requires changes of land use. These changes can have major impact on sustainability of the process and can determine to what extent bioenergy production can make an impact on greenhouse gas emissions. Although exact quantification of the effects of land-use change is a complex issue and beyond the scope of this chapter, several examples will hopefully help to understand the issue better. Many ecosystems such as rainforests, peatlands, savannas, or grasslands contain significant amounts of sequestered carbon and act as a carbon sink. Reallocation of these resources for the production of energy crops may result in the release of this carbon to the atmosphere especially if the land-use change is performed through slash and burn practices. The land-use change from original ecosystem to bioenergy production will result in the formation of 'bio-fuel carbon debt' that would require many years of bioenergy feedstock cultivation only to restore the carbon balance before the land-use change was implemented [44]. On the other hand, land-use change that promotes accumulation of soil organic carbon can have a positive environmental effect. For example, perennial rhizomatous energy grasses like *Miscanthus* or switchgrass provide a rarely mentioned alternative that can create a 'carbon asset' shortly upon land change due to accumulation of significant amount of carbon in the soil. As long as the initial land use did not have an extraordinary carbon sequestration capacity like rainforests, the conversion of land to rhizomatous perennial grasses can bring benefits in terms of carbon sequestration.

3.5.3.2 Impacts of Feedstock Cultivation

Cultivation of bioenergy feedstocks is considered as one of the most important contributors to the overall impact of the process. First, agricultural practices such as soil preparation, biomass harvesting and transportation require significant energy inputs that usually come from fossil fuels [45]. Second, many feedstocks for bioenergy use require numerous agricultural inputs such as fertilisers, pesticides and insecticides. Especially high yielding food crops that produce abundance of easily accessible pools of starch and triglycerides, like corn or rapeseed, are particularly intensive to cultivate. Among these inputs, nitrogen fertilisers can contribute significantly to the overall sustainability of the process. Vast majority of nitrogen fertilisers used in agriculture are direct result of Haber–Bosch process, a revolutionary technology of synthetic production of ammonia from nitrogen and hydrogen. Although production of synthetic fertilisers allowed human population to grow to an unprecedented scale, it comes with significant environmental impact. First, fertilisers are produced using both energy and hydrogen derived from fossil fuels; second, the process of conversion is very energy intensive; and third, nitrogen fertilisers are usually supplied in excess that results in the release of sewage effluents and agricultural run-off carrying fertilisers into natural waters resulting in eutrophication, algal blooms and formation of anoxic zones in water bodies. Utilisation of bioenergy feedstocks that can acquire nitrogen form waste streams like microalgae or have very efficient nitrogen cycle through its remobilisation like *Miscanthus* can significantly reduce the impact of cultivation on the processes of bioenergy and biomaterial production.

3.5.3.3 Impacts of Conversion Process

Conversion of biomass to renewable fuels and materials allows introducing new functionalities to the biomass. It can enhance its utility for customer, increase energy density of the fuel and its storability and transportability. There are, however, different routes to obtain these new functionalities and broadly thermochemical routes and biochemical routes can be distinguished, as well as an emerging field of hybrid technologies. In general, thermochemical conversion processes of biomass conversion often use higher energy inputs, usually from fossil sources, more aggressive chemicals and produce waste that may be more difficult to dispose or remediate than those of biochemical processes. The advantages of thermochemical processes are their higher compatibility with current chemical industry and shorter conversion times. Biochemical processes on the other hand are generally much longer as they require cultivation of microorganisms before desired metabolites can be used. Very often they produce chemicals that are not fully compatible with current infrastructure designed for fossil fuels. Most of the side products of biochemical processes are not hazardous and often have commercial utility as food or feed components or can be relatively easily converted to methane and/or energy through anaerobic digestion technology.

3.5.3.4 Impacts of Product Use

The products resulting from conversion of biomass can have various impacts on the environment depending on their chemical properties. Their toxicity and environmental impacts could be due to the leakage or combustion. Many bio-fuels that are a direct result of microbial metabolism, that is, lower alcohols or fatty acids can be relatively easily decomposed by bacteria or fungi [31, 46]. This does not mean however that the usage of bio-fuels is free of any environmental impacts. Some biomass-derived products may in certain concentrations exhibit toxic effects to the environment – ethanol, butanol or FTS liquids. Others like biohydrogen or biogas pose risk of explosions or contain potent greenhouse gasses (methane) that need to be controlled. Various liquid bio-fuels differ in both environmental impacts and hazards. Bio-ethanol although easily biodegradable is toxic at high concentrations and miscible with water in any concentrations. Bio-butanol is more difficult to degrade by microorganisms than ethanol is [46]. Both of these bio-fuels have cleaner combustion profiles with respect to PM (particulate matter), CO and NO_x than gasoline; however, their emission profiles of carbonyl species are worse [47]. Biodiesel as a fuel has numerous advantages from the environmental point of view. It is renewable, non-toxic, biodegradable and has better handling properties than diesel fuel [48]. Combustion of biodiesel results in lower PM, hydrocarbons (HC) and CO emissions and about 10% higher NO_x emissions than those of conventional diesel fuel.

3.5.4 Sustainability Metrics in Biomass and Bioenergy Policies

In recent years, the application of metrics of environmental impact and sustainability resulted in significant shifts in bioenergy support policies worldwide. The debate about the impact of bio-fuels made from edible feedstocks can be traced back to 2008 and a surge of food prices [49]. It became apparent that bio-fuel and biomaterial production is the area where interconnected aspects of energy, security, environment, economy and social impacts need to be considered simultaneously [50]. There exist several schemes to assist the estimates of environmental impacts of processes: carbon footprint (CF) [51], ecological footprint (EF) [51, 52], environmentally extended input–output analysis [53] and many others. Early policies that were used to promote bio-fuel production were focused on volumetric mandates and subsidies for producers. Type and the origin of feedstocks for bio-fuel was not considered [54]. This resulted in the rapid conversion of land for the production of bio-fuels worldwide, very often in developing countries. These activities were profit driven and did not take into consideration any environmental impact of changing land use and greenhouse gas emissions required to transport these bio-fuels to the final users. These raised concerns about both economic efficiency and sustainability of early bio-fuel support policies especially in the EU and later the United States. These concerns resulted in introducing LCA-based sustainability criteria to bio-fuel policies in the EU and higher greenhouse gas emission reduction targets in the United States [54]. Although currently there is no obligation to include direct and indirect land-use change as one of the criteria for bio-fuel sustainability assessment, such modifications to policies are envisaged [54], and very strict sustainability criteria have been already implemented in biomass sector in the UK. In order to ensure that the transition from fossil carbon economy to renewable carbon economy in performed in a sustainable way without replacing one environmentally damaging practice with another, further advancement in creating appropriate policies are required. Utilisation of taxation of non-renewable carbon should become an important factor in the development of more sustainable fuels and materials.

3.5.5 Renewable and Non-Renewable Carbon – Taxation and Subsidies

Decision support tools such as EROI, LCA and estimations of environmental footprints should have a major impact on selection of policies to support the development of sustainable low carbon economies. Probably the easiest method to promote such change is phasing out subsidies to fossil fuels and creating a level playing ground between renewable technologies and fossil-based technologies. Despite their obvious environmental impacts fossil fuel technologies are subsidised in numerous countries directly or indirectly very often at much greater extent than renewable technologies are. These subsidies are defined by OECD as 'any government action that raises the price received by energy producers, lowers the cost of energy production, or lowers the price paid by energy consumers' [55]. The most obvious form of support for the fossil fuel economy are cash payments to producers or consumers and tax exemptions, but more often than not the subsidies have more subtle form. Common practice worldwide especially where the energy companies are state-owned is price control. This way the prices are often set below full cost of energy production, and the energy companies receive various indirect incentives to cover the difference. Sometimes it is not the energy producers or consumers but fossil fuel suppliers that are the biggest beneficiaries of the indirect subsidies. The power producers are often obliged to purchase the fossil fuel resource from (usually) domestic source at elevated prices and this way subsidise, for example, inefficient mining industry. Even more subtle ways of subsidising fossil fuel economy is utilisation of special labour laws for 'sensitive sectors', for example, coal mining where practices like an early retirement age, productivity-independent bonuses and other incentives are employed. Although widely utilised, there is little evidence in modern economies that these subsidies generate any positive outcomes [55–58]. First, subsidising fossil fuel–derived energy inevitably leads to higher consumption, waste and lower energy efficiency. Second, the environmental consequences of increased utilisation results in the deterioration of environment and human health. This leads to higher costs of health services that economies need to bear. Third, in principle subsidies are often described as a way to help poorer members of society to increase their quality of life. In reality, however the largest beneficiaries of the subsidies are the richest in the society. At the end of the day, it is rich citizens that use more energy than poor citizens do. Last but not least, these direct or indirect subsidies are a burden on public finances and drain financial resources from other vital areas. A thorough review and abolition of these subsidies should be the first step of the transition towards sustainable development.

Removal of the subsidies to fossil technologies should be followed by introducing some form of general carbon taxation to promote zero and low carbon technologies. Currently two approaches are considered: carbon trading and carbon tax. The carbon taxation component should be at least equal and preferably higher to the indirect (so-called externalised) costs that fossil technologies have on the environment, health and climate change and should increase in predictable manner to generate enough incentives to the market to invest in sustainable development. Two proposed mechanisms of carbon emission reductions carbon trading and carbon tax have different economic mechanisms. Carbon tax sets the price of carbon dioxide emissions and leaves to the market to determine the quantity of emissions reduced, carbon trading (cap and trade scheme) sets the target of emission reductions and lets the market determine the price of these reductions. As a result, their responsiveness to economic circumstances also differs. For example, cap and trade system is resistant to inflation as the price of the emissions is set by the market that adjusts prices automatically, carbon tax need to be externally indexed to take into consideration inflation. On the other hand, carbon tax sends more certain message to the markets about the price of the emissions. Once the emissions are taxed, even economic slowdown will not impact the overall cost to the emitter. Under emissions trading system, however, economic slowdown results in lower production and lower emissions, which in turn drive the cost of the emissions down sending a negative message to those who wish to invest in more environmentally friendly technologies. It is also fairly difficult to predict and administer the exact amount of carbon credits allocated to the market and even more so to adjust it when excessive allocation was made. In 2007, oversupply of carbon credits to EU Emission Trading Scheme resulted in the price to drop to zero. Subsequent economic crisis resulted in the drop of production and rendered the whole system inefficient. Any subsequent attempts to withdraw part of the carbon credits from the market were successfully blocked by the emitters. There is therefore evidence that carbon emissions trading may be more difficult to administer, more prone to political pressure and lobbying from carbon-intensive industries and require larger administration and bureaucracy than carbon tax does. Appropriately structured carbon tax could become a system that has positive environmental and social impacts. To ensure the efficiency of CO_2 emissions reduction, carbon tax should be administered as early as possible in the life cycle scheme. Ideally during first sale of a fossil resource, collected tax should be proportional to the CO_2 quantity that will be emitted during its utilisation. This will raise the costs of all processes and products in a way that is directly proportional to the environmental damage done by the resource. The tax should be rising in a predictable manner within certain also predictable timeframes to ensure that the message to the markets is clear and that the shift towards low carbon economy is a long-lasting trend and worth investing in. The carbon taxation should be introduced for the goods imported from outside the taxation region in the quantity identical to the carbon intensity of the resource used to make a product. Lastly, tax cuts should be administered to alleviate raising costs of the energy taking into consideration the less privileged groups of the society aiming for strengthening the middle class. These tax cuts could include, but are not limited to raising income tax threshold (higher tax free income), lowering personal income or VAT rates especially for the basic goods like food or public transport, lowering taxation burden on small and medium companies and so on.

Combination of these factors is likely to provide an efficient solution that can steer the development of the economies towards the sustainable future. From the environmentally efficient biomass utilisation point of view, such tax could bring significant benefits as well. First, products derived from fossil resources would become more expensive, with the increase of price corresponding to energy intensity of the process that was used to produce them. This would promote less carbon-intensive production methods and in the longer term introduction of biomass-derived replacements. Second, the energy-intensive bio-fuels like corn ethanol would make little economic sense and transition towards more efficient technologies would inevitably take place. Considering that agriculture is currently the fourth largest source of greenhouse gas emissions [59] following scheme would promote low input feedstocks such as *Miscanthus* or switchgrass due to the raise of costs of fertilisers and fuels. Therefore in the long term, introduction of a tax on fossil fuels, that is, non-renewable carbon will quickly give stimulus to develop renewable carbon technologies including biomass technologies.

3.6 Summary

- Economic growth results in the rapid depletion of easily recoverable fossil resources that power the growth of the economy. Resources of the future will inevitably be less efficient than those we are used to today. Economies and societies should prepare for this change.
- Both non-conventional fossil resources and renewable resources have lower energy return on investment than currently used energy sources. Non-conventional resources are more likely to result in serious aggravation of our current environmental problems especially in the areas of CO_2 emissions and water shortage.
- Biomass is a form of solar energy that has been captured and stored in a form of chemical molecules. Biomass shows the potential of being a sustainable, renewable source of energy and carbon for low carbon economies.
- There are many compounds that can be produced from biomass such as fuels, platform chemicals, biopolymers and others. Additionally biomass and its products can be stored and transported to various locations what gives biomass-derived fuels advantages over other renewable energy sources.
- Biomass conversion methods include thermochemical and biochemical. Thermochemical conversion methods utilise high temperature and pressure to transform biomass into heat and power as well as liquid and gaseous chemicals that could be used as fuels or chemical building blocks. Biochemical conversion methods rely on the conversion of biomass components by anaerobic microorganisms into high energy metabolites that can be used as fuels or chemicals.
- Thermochemical conversion processes include combustion, gasification and pyrolysis. Combustion releases entire chemical energy stored in biomass, whereas gasification and pyrolysis release only a part of the initial energy of the molecule and maintain most of it in high energy intermediates that could be used as fuels or chemicals.
- Biochemical conversion processes include mostly anaerobic metabolisms. Anaerobic metabolisms such as ethanol, ABE, hydrogen fermentations and anaerobic digestion result in partial oxidation of biomass components and formation of high-energy intermediates.
- Bioenergy and biomaterial production processes are strongly interconnected with other aspects such as energy, security, environment, economy and social impacts need to be considered simultaneously. This results in a need for unprecedented scrutiny of these technologies if they are to deliver truly sustainable future.
- The most straightforward method to promote sustainable development including bioeconomy could be removal of subsidies for fossil fuels worldwide and taxation of non-renewable carbon resources proportionally to the environmental damage done by utilisation of such resources.

3.7 Key References

Murphy [7].
IPCC [20].
Fargione *et al.* [44].

References

1. MacKay DJC. *Sustainable Energy – Without the Hot Air*. Cambridge: UIT Cambridge; 2008.
2. BP. BP Statistical Review of World Energy; 2014. Report No.
3. Hubbert MK. Energy Resources. In: Freeman WH, editor. *Resources and Man*. San Francisco: National Academy of Sciences; 1969. p. 157–242.
4. Bentley RW. Global oil & gas depletion: an overview. *Energy Policy*. 2002;**30**(3):189–205.
5. Hall CAS, Powers R, Schoenberg W. Peak Oil, EROI, Investments and the Economy in an Uncertain Future. In: Pimentel D, editor. *Biofuels, Solar and Wind as Renewable Energy Systems*. Springer Science + Business Media B.V.; 2008.
6. IEA. *World Energy Outlook 2012*. Paris: International Energy Agency; 2012.
7. Murphy DJ. The implications of the declining energy return on investment of oil production. *Philosophical Transactions of the Royal Society A*. 2014;**372**(2006): 20130126.
8. Mearns E. *The Global Energy Crisis and Its Role in the Pending Collapse of the Global Economy*. In: The Oil Drum, editor. Aberdeen; The Royal Society of Chemists; 2008.
9. Cleveland CJ. Net energy from the extraction of oil and gas in the United States. *Energy*. 2005;**30**(5):769–82.
10. Boden TA, Marland G, Andres RJ. *Global. Regional, and National Fossil-Fuel CO_2 Emissions. Carbon Dioxide Information Analysis Center*. Oak Ridge, TN: U.S. Department of Energy, Oak Ridge National Laboratory; 2010.
11. Le Quéré C, Peters GP, Andres RJ, Andrew RM, Boden TA, Ciais P, et al. Global carbon budget 2013. *Earth System Science Data*. 2014;**6**(1):235–63.
12. Orr JC, Fabry VJ, Aumont O, Bopp L, Doney SC, Feely RA, et al. Anthropogenic ocean acidification over the twenty-first century and its impact on calcifying organisms. *Nature*. 2005;**437**(7059):681–6. PubMed PMID: WOS:000232157900042.
13. Stern KR, Bidlack JE, Jansky SH. *Introductory Plant Biology*. 11th Edition In: Kemp MJ, editor. McGraw-Hill; 2008.
14. Carpita NC, Gibeaut DM. Structural models of primary cell walls in flowering plants: consistency of molecular structure with the physical properties of the walls during growth. *The Plant Journal*. 1993;**3**(1):1–30.
15. Delaux P-M, Nanda AK, Mathé C, Sejalon-Delmas N, Dunand C. Molecular and biochemical aspects of plant terrestrialization. *Perspectives in Plant Ecology, Evolution and Systematics* 2012;**14**(1):49–59.
16. Popper ZA, Michel G, Hervé C, Domozych DS, Willats WGT, Tuohy MG, et al. Evolution and diversity of plant cell walls: from algae to flowering plants. *Annual Review of Plant Biology*. 2011;**62**(1):567–90.
17. Komor E. The Physiology of Sucrose Storage in Sugarcane. In: Anil Kumar G, Narinder K, editors. *Developments in Crop Science*. Volume **26**. Elsevier; 2000. p. 35–53.
18. Cheng JJ. *Biomass to Renewable Energy Processes*. Boca Raton: CRC Press; 2010.
19. Eubanks LP, Middlecamp CH, Heltzel CE, Keller SW. *Chemistry in Context: Applying Chemistry to the Society*. 6th Edition. New York: McGraw-Hill; 2009.

20 IPCC. *Renewable Energy Sources and Climate Change Mitigation.* In: Edenhofer O, Madruga RP, Sokona Y, et al., editors. New York: Cambridge University Press; 2012.
21 Jameel H, Keshwani DR, Carter SF, Treasure TH. Thermochemical Conversion of Biomass to Power and Fuels. In: Cheng JJ, editor. *Biomass to Renewable Energy Processes.* Boca Raton: CRC Press; 2010. p. 517.
22 McKendry P. Energy production from biomass (part 3): gasification technologies. *Bioresource Technology.* 2002;**83**(1):55–63.
23 Wang L, Weller CL, Jones DD, Hanna MA. Contemporary issues in thermal gasification of biomass and its application to electricity and fuel production. *Biomass and Bioenergy.* 2008;**32**(7):573–81.
24 Balat M, Balat M, Kırtay E, Balat H. Main routes for the thermo-conversion of biomass into fuels and chemicals. Part 2: Gasification systems. *Energy Conversion and Management.* 2009;**50**(12):3158–68.
25 Mohan D, Pittman CU, Steele PH. Pyrolysis of wood/biomass for bio-oil: a critical review. *Energy Fuel.* 2006;**20**(3):848–89.
26 Czernik S, Bridgwater AV. Overview of applications of biomass fast pyrolysis oil. *Energy Fuel.* 2004;**18**(2):590–8. PubMed PMID: WOS:000220287400041. English.
27 Mortensen PM, Grunwaldt JD, Jensen PA, Knudsen KG, Jensen AD. A review of catalytic upgrading of bio-oil to engine fuels. *Applied Catalysis A: General.* 2011;**407**(1–2):1–19.
28 Knothe G. Historical perspectives on vegetable oil-based diesel fuels. *Inform* 2001;**12**:1103–7.
29 Demirbas A. *Biodiesel: A Realistic Fuel Alternative for Diesel Engines.* 1st Edition. Springer; 2008. p. 208.
30 Van Gerpen J, Shanks B, Pruszko R, Clements D, Knothe G. Biodiesel Production Technology, National Renewable Energy Laboratory; 2004 Contract No.: NREL/SR-510-36244.
31 Knothe G. Biodiesel and renewable diesel: a comparison. *Progress in Energy and Combustion Science.* 2010;**36**(3):364–73. PubMed PMID: WOS:000276146900003. English.
32 Hinkle PC, Kumar MA, Resetar A, Harris DL. Mechanistic stoichiometry of mitochondrial oxidative phosphorylation. *Biochemistry.* 1991;**30**(14):3576–82.
33 Berg JM, Tymoczko JL, Stryer L. *Biochemistry.* 5th Edition. W. H. Freeman; 2002. p. 1100.
34 Cheng JJ. Biological Process for Ethanol Production. In: Cheng JJ, editor. *Biomass to Renewable Energy Processes.* Boca Raton: CRC Press; 2010.
35 Sun J, Wang Y. Recent advances in catalytic conversion of ethanol to chemicals. *ACS Catalysis.* 2014;**4**(4):1078–90.
36 Jones DT, Woods DR. Acetone-butanol fermentation revisited. *Microbiological Reviews.* 1986;**50**(4):484–524. PubMed PMID: WOS:A1986F177000007.
37 Duerre P. Fermentative butanol production – bulk chemical and biofuel. *Annals of the New York Academy of Sciences* 2008:353–62. PubMed PMID: BIOSIS:PREV200800417413.

38 Kothari R, Tyagi VV, Pathak A. Waste-to-energy: a way from renewable energy sources to sustainable development. *Renewable and Sustainable Energy Reviews.* 2010;**14**(9):3164–70.
39 Drapcho CM, Nhuan NP, Walker TH. *Biofuels Engineering Process Technology.* McGraw-Hill; 2008.
40 Cheng JJ. Anaerobic Digestion for Biogas Production. In: Cheng JJ, editor. *Biomass to Renewable Energy Processes.* CRC Press; 2010.
41 Daroch M. Biodiesel Production. In: Geng S, editor. *Conversion Technologies for Biomass to Renewable Energy.* Beijing; 2015.
42 Sheehan JJ. Biofuels and the conundrum of sustainability. *Current Opinion in Biotechnology.* 2009;**20**(3):318–24. PubMed PMID: WOS:000268525800011. English.
43 ISO. *Environmental Management – Life Cycle Assesment – Principles and Framework.* Geneva: International Organization for Standardization; 2006.
44 Fargione J, Hill J, Tilman D, Polasky S, Hawthorne P. Land clearing and the biofuel carbon debt. *Science.* 2008;**319**(5867):1235–8. PubMed PMID: ISI:000253530600041. English.
45 Cherubini F, Bird DN, Cowie AL, Jungmeier G, Schlamadinger B, Woess-Gallasch S. Energy- and greenhouse gas-based LCA of biofuel and bioenergy systems: key issues, ranges and recommendations. *Resources, Conservation and Recycling* 2009;**53**:434–47.
46 Mariano AP, Tomasella RC, Di Martino C, Filho RM, Seleghim MHR, Contiero J, et al. Aerobic biodegradation of butanol and gasoline blends. *Biomass and Bioenergy.* 2009;**33**(9):1175–81.
47 Sarathy SM, Oßwald P, Hansen N, Kohse-Höinghaus K. Alcohol combustion chemistry. *Progress in Energy and Combustion Science.* 2014;**44**(0):40–102.
48 Westbrook CK. Biofuels combustion. *Annual Review of Physical Chemistry.* 2013;**64**(1):201–19. PubMed PMID: 23298249.
49 Renewable Fuels Agency. *The Gallagher Review of the Indirect Effects of Biofuels Production.* Renewable Fuels Agency; 2008.
50 Nuffield Council on Bioethics. *Biofuels: Ethical Issues.* Nuffield Council on Bioethics; 2011.
51 Matthews HS, Hendrickson CT, Weber CL. The importance of carbon footprint estimation boundaries. *Environmental Science & Technology.* 2008;**42**(16):5839–42.
52 Global Footprint Network. *National Footprint Accounts.* Oakland, CA: Global Footprint Network; 2010.
53 Minx JC, Wiedmann T, Wood R, Peters GP, Lenzen M, Owen A, et al. Input output analysis and carbon footprinting: an overview of applications. *Economic Systems Research.* 2009;**21**(3):187–216.
54 Kazamia E, Smith AG. Assessing the environmental sustainability of biofuels. *Trends in Plant Science.* 2014;**19**(10):615–8.
55 Jiang Z, Lin B. The perverse fossil fuel subsidies in China – the scale and effects. *Energy.* 2014;**70**(0):411–9.
56 IEA. *Carrots and Sticks: Taxing and Subsidising Energy.* IEA; 2006 [2015.01.02]. Available from: http://www.iea.org/publications/freepublications/publication/carrots-and-sticks-taxing-and-subsidising-energy.html.

57 Stephens JC. Time to stop investing in carbon capture and storage and reduce government subsidies of fossil-fuels. *Wiley Interdisciplinary Reviews: Climate Change.* 2014;**5**(2):169–73.

58 Global Subsidies Initiative. Untold billions: Fossil-Fuel Subsidies, their Impacts, and the Path to Reform: International Institute for Sustainable Development; 2009. Available from: http://www.iisd.org/gsi/sites/default/files/synthesis_ffs.pdf.

59 IPCC. *Climate Change 2007: Mitigation of Climate Change.* Cambridge and New York: IPCC; 2007.

4
Characterization Methods and Techniques
Noppadon Sathitsuksanoh[1] and Scott Renneckar[2]

[1] *University of Louisville, Department of Chemical Engineering, Louisville, KY, USA*
[2] *Virginia Tech, Department of Sustainable Biomaterials, Blacksburg, VA, USA*

4.1 Philosophy Statement

Humans are entering a new age when we can have significant control over the genomic make-up of plants and organisms that can operate on biomass. Understanding physiochemical characteristics of plant biomass helps us in the upstream and downstream processes. In the upstream process, information of plant biomass' characteristics aids in engineering them to have desirable traits including drought resistance or *in planta* production of metabolites such as cancer drugs. For downstream process, the characteristic information of plant biomass aids us in designing the pathways to break and transform them into valuable products. Herein, we learn what information these selected characterization methods and techniques for plant biomass provide and how to interpret and use those information.

4.2 Understanding the Characteristics of Biomass

Plants contain stored solar energy in the form of glucose that is transformed into a host of other compounds and materials such as polysaccharides, lipids, and lignins. Glucose is used to power cell metabolism and is the precursor for the structural polymers of the plant wall. The simplicity of the ingredients in photosynthesis of water and carbon dioxide results in the plant consisting of the primary elements of carbon, hydrogen, and oxygen; however, plants also require nitrogen and sulfur and other trace elements such as calcium and potassium. Nitrogen and sulfur are necessary for the formation of proteins that encode the life of the plant, while the metals form ions that assist in membrane transport and catalytic sites in cellular enzymes. During the growth of the plant, these compounds are woven into the complex tapestry of the biomass cell wall and this chapter explores methods to characterize this complex material.

Although the compounds in the cell wall start out as a simple carbohydrate of glucose, this is only a building block that is modified, polymerized, and transformed into the structural polymers that support the plant. Additional extraneous materials are formed that lend the plant defense against attacking organisms, such as terpenoids,

Introduction to Renewable Biomaterials: First Principles and Concepts, First Edition.
Edited by Ali S. Ayoub and Lucian A. Lucia.
© 2018 John Wiley & Sons Ltd. Published 2018 by John Wiley & Sons Ltd.

tannins, and other reserve materials that are inside the cell lumen, cell microfissures, and intercellular space such as resin canals. These materials are typically classified as extractives because they can be readily removed through the process of soaking the biomass in solvents with various solubility parameters that remove the compounds into the solvent. This simple extraction method is similar to the process of making a cup of coffee. It is an example of diffusion where the extractive compounds move from an area of high concentration, the cell wall, into an area of low concentration, the solvent. Depending upon the species of the plant, the location of the tissue that is being sampled within the plant, as well as the location of where it was grown there is variability of the amount of extractive content from less than a percent by weight of the mass to over 10% in some species like the heartwood of redwood [1].

Lastly, biomass has varying levels of water associated with it. Water can be tightly associated with the cell wall through hydrogen bonding to the structural framework. Because of the close association of the water, this water is usually known as bound water and has properties that are different from water that is free to associate with itself in liquid form, free water. The biomass has a limit of interaction sites, and once those spaces are occupied, any excess water is found in the cell lumen as free water. This free water can be physically squeezed out of biomass. The importance of understanding the moisture content of the biomass lies in the fact that analysis of biomass requires a known plant or biopolymer dry mass; any unaccounted moisture can cause a miscalculation of the composition or polymer size. Hence, when investigating biomass or biopolymers the amount of moisture must be first determined for the sample.

4.3 Taking Precautions Prior to Setting Up Experiments for Biomass Analysis

Before we move into the analysis of biomass using specific techniques and characterization methods, a point of caution must be made about the reliability of the analyzed data, in relation to the scientist's goals and the error that can be introduced during the evaluation of the material. Biomass is variable from species to species, plant to plant within a species, and location to location within the plant. All of the variability is dependent on a number of genetic and environmental cues, both nature and nurture! While scientists can have controlled studies where there is a comparison of two groups, by knocking out or inserting genes, or purposely changing nutrient and growing conditions of plants, it should be cautioned that samples must be representative of the study's objectives. Hence, investigating the intrinsic properties of the plant requires a wide sampling range of the biomass from the field or forest. In contrast, if the impact of a processing technique is to be measured, then most likely a narrow sampling range can help pinpoint the changes caused by processing. In both cases, a rigorous study requires that enough sample is analyzed so the investigator is confident in the data and that the results are statistically representative of the population. To do this experimental work, randomized replications are made and the variation of the results is used to understand the validity of the average.

A common issue that is found with novice experimentalists is that often replications are only made of the measurement and not the variable being investigated. Hence, when designing experimental testing protocols, the objective is to replicate the variable of interest and not necessarily the accuracy of the measurement method. The researcher should already have an idea of the degree of variability, or accuracy, of the measurement, and they should take that into account during analysis. For example, if the moisture content of a sample is being determined using gravimetric methods, the researcher should know the limit of error of the balance and should not take the same sample and weigh it three times after drying it to produce an average of the mass. This result would provide only information about the variability of the measurement and would not provide an insight into the variability of the moisture content of the biomass.

Review questions:

1. Name three classes of *nonstructural* components that are involved in the cell wall composition.
2. How could moisture content vary across an open container of wood particles?
3. Describe the impact of error of a noncalibrated balance, the error related to the resolution of the balance, and error related to copying the wrong information into their lab notebook in relation to the accuracy of the data.

4.4 Classifying Biomass Sizes for Proper Analysis

For most analytical experiments biomass samples are resized into smaller particles that range from single-digit millimeters down to single-digit micrometers depending upon the specifications of the analytical method. Various grinders and mills can be used to resize biomass ranging from a simple coffee bean grinder, available at home stores, to screened knife mills designed to control the degree of particle breakdown. Most analytical standards provide guidance to the size of the particles desired for analysis. These particle sizes are based on screen size of which the particles pass through and occasionally the screen size that particles are retained. Resizing the particles helps with randomizing the biomass as it mingles together with samples from multiple plants or multiple areas of the stem, which is obviously dependent upon what is going into the grinder. If the particles are too large during the preparation of samples for analysis, then the mass transfer is slowed by the biomass anatomical structure and potential density variations across the sample. Only partial reactions occur in the allotted time of the standard if the appropriate particle size is not selected. On the other hand, if particles are too small, then particles can easily clog filters and form colloidal suspensions that are difficult to precipitate and recover for analysis.

As indicated later, knife-milled biomass of size 20–80 mesh is used in many analytical procedures as well as chemical reactions and modifications. Knife milling requires first the resizing of large biomass pieces to scale so that these pieces can be easily fed into the processing equipment. Screen selection should occur based on the requirements of the analytical procedure. Screens are delicate and with extended use can become damaged, so screens should be inspected periodically. Additionally, extremely wet biomass does not mill into a uniform product and readily clogs screens as the biomass is smeared instead of cut. On the other hand, not all knife mills can handle biomass and caution should be taken with extremely high-density biomass. The interior shell of a coconut is composed of lignocellulosic material and has a density well above

1000 kg m^{-3}, as a result the shell will dull knife blades quickly. Other screened mills like hammer mills are used to resize high-density biomass. This machinery strikes biomass with a group of rotating bars and the particles exit the equipment through a screen of specified size. Because of their simplicity, often these mills are used in industrial settings to resize biomass for feeding into furnaces. Both hammer and knife mills cause localized temperature increases of the biomass from frictional heating, free radical formation from chain cleavage, and the researcher should be aware of this potential for temperature-sensitive samples like cellulose [2].

After particles pass through a given screen size in a knife mill, there is a range of sizes that diminish into the size of an ultrafine powder. While not always required by the standard, a more uniform sample will lead to more reproducible results [3]. Milled samples can be placed in stacked screens to obtain particles of uniform size for characterization purposes. These screens have mechanical motors that shake material back and forth or have high-frequency vibrations to help move particles across and through the screens of select sizes. A caveat is given to the student – overloading the screens with biomass samples often will limit screening effectiveness.

Occasional analytical methods require extensive milling of the material using a planetary style ball mill [4]. Cups filled with resized biomass and ceramic balls of various sizes are rotated for extended time periods collapsing the particle microstructure and turning the wood into a fine powder; this fine structure would be like transforming sugar cubes into a fine powder such as confectioner's sugar. This transformation is important in studies using ionic liquids to dissolve whole biomass or methods to isolate certain components like lignin. The former allows the use of analytical experiments like 2D (two-dimensional) nuclear magnetic resonance (NMR) spectroscopy to identify molecular structure found in plant biomass without having to extract individual components. In contrast, isolation of individual components modifies the biopolymers because intermolecular linkages between lignin and carbohydrates require cleavage. To this end, research has been performed in an attempt to limit the degree of modification during isolation. Guerra *et al.* [5] developed a method of milling wood into a fine powder, enzymatically removing much of the polysaccharide component with cellulases, and then performing a mild acidolysis reaction to obtain a relatively unmodified lignin compared to technical lignins from pulping operations.

4.5 Moisture Content of Biomass and Importance of Drying Samples Prior to Analysis

An accurate moisture content must be determined for the analysis of biomass and biopolymers and, in most cases, the moisture should be removed. While this task is relatively simple, achieved by heating the biomass and determining the moisture content gravimetrically (measured through the change in mass), an important issue must be addressed. Not all volatile compounds that can be removed through heating are water molecules; certain extractives are removed through high-temperature oven drying. While many extractives are highly polar and are solid at elevated temperatures, there are certain terpenoids, such as α- and β-pinene, that have a boiling point near 155 °C. In these cases, industrial dryers may heat samples beyond the boiling point of these compounds.

Additionally, there are two methods to calculate the moisture content of the biomass. One method uses a ratio of the water to the dry mass of the plant material and because of this ratio, moisture contents can be greater than 100%. Typical tables listing the moisture content of green wood, wood that has never been dried, will use this calculation method. The other method describes the amount of water as a fraction of the total mass of the sample. This number is used when calculating the equivalent dry mass and is directly used in most calculations as percent solids or volatiles.

Common laboratory methods to determine the dry mass, known as "oven-dry mass" of the biomass utilize lyophilization, vacuum drying, convection oven drying, or fluid-bed dryers. The choice of the different methods depends upon the starting moisture content along with the sensitivity of the structure to thermal processing. Higher temperature oven drying can cause the collapse of the ultrafine biopolymer structure resulting in a change in crystallinity and reactivity of the sample. In the pulp and paper industry, this is referred to hornification where the cellulose structure is annealed into a highly packed state [6]. Lyophilization, often commonly referred to as freeze drying, provides a sample with a more open structure without the oven drying–induced changes. One example of the sensitivity of biomass structure to drying methods is seen through the difference in crystallinity index (CrI) of freeze-dried versus air-dried nanocellulose [7]. With freeze-drying suspensions of nanocellulose, there is some reaggregation of samples into films and fibers as demonstrated by Hsieh and coworkers [8]. Some of these changes can be reversible given the right methods of activation [9]. This reversibility of structure is illustrated by a using a solvent exchange process to remove the water from the reswollen fiber. Furthermore, to dissolve cellulose in nonderivatizing solvents like dimethyl acetamide (DMAc) LiCl solutions, cellulose dissolution requires activation by first swelling the cellulose in a water solution, then stepwise removing the water by using different concentrations of methanol and then exchanging the absolute for DMAc [10]. Dissolving cellulose is important for a variety of chemical characterization methods, especially related to molecular weight determination, which is discussed.

4.6 When the Carbon is Burned

Humans have been using biomass as bioenergy for millennia in the form wood-heated stoves. One of the primary uses of biomass for bioenergy is the use of pellets for the combustion to either generate electricity directly through steam turbines or through high-temperature gasification. Biomass pellets have the advantage of using modern day carbon. The drawback of using these materials is the ash particulates that impact emissions and furnace maintenance. Hence knowledge of the ash content is critical when using wood for bioenergy production. The benefit of the excess ash from an industrial conversion process is that much of this can be recovered and used as fertilizer.

Ash content is determined gravimetrically after the combustion of the biomass at high temperatures in a muffle furnace [11]. Ceramic-based crucibles are required for the combustion process along with desiccators to store the samples as they cool. Ash content is expressed as an overall percentage of the dry matter. In some cases, ash content can be as high as 1% in some wood species and even higher in tropical wood species. There are a variety of different levels of ash content depending upon the biomass type ranging from materials such as rice straw that contains significant amount of silica upward of

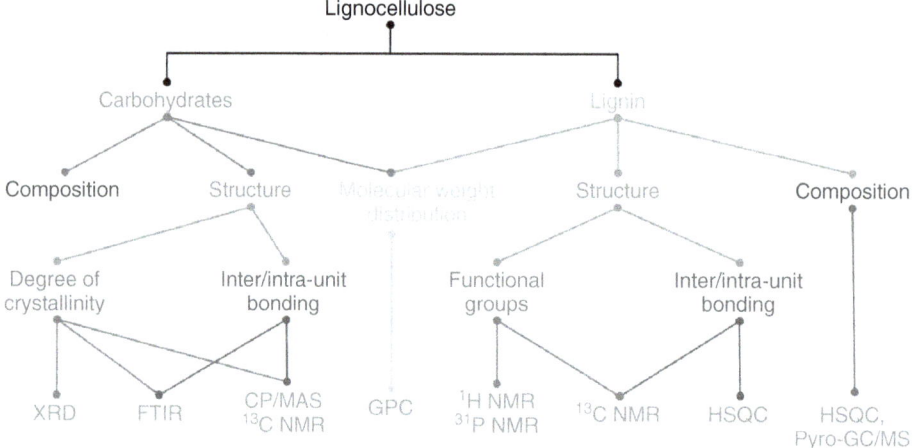

Figure 4.1 Biomass characterization diagram shows common characterization techniques for carbohydrates and lignin.

15% [12], to corn stover that is 6% [12], to some energy grasses such as switchgrass that has levels of 3–7% [12]. Mineral levels of ash do vary by stalk versus leaf material, as well as position in the stem and growing location of the plant. Another method to measure ash content is to use thermogravimetric analysis equipment, where the sample is heated on a balance pan to very high temperatures in an air environment. This method is essentially microsampling the material, as analysis is limited to 10–50 mg of material.

Figure 4.1 shows the diagram of common characterization techniques for lignocellulose. There are two major lignocellulosic components: (i) carbohydrates and (ii) lignin. Both carbohydrate and lignin can be characterized by various techniques to learn about their physiochemical characteristics. These information are useful in

- engineering plant biomass to have certain traits including (i) less recalcitrant and (ii) more drought resistant and/or
- designing biological and/or chemical pathways to degrade and/or convert plant biomass into valuable products.

4.7 Structural Cell Wall Analysis, What To Look For

The structural components of the cell wall of biomass are composed of a mixture of carbohydrates (i.e., polysaccharides) and lignin closely integrated within the cell wall. In order to fully utilize lignocellulose, robust analytical techniques for their quintessential traits (Table 4.1) are required for analysis. Changes in carbohydrate and lignin compositions, chemical structure, and functionality affect fiber properties and ultimately product quality.

Recent imaging studies of the cell wall provided insight into the arrangement of these compounds related to the biological synthesis of the cell wall [58]. Cellulose is deposited as bundles within the cell wall of various diameters, dependent upon the plant type, and their orientation controlled by microtubules. During deposition,

Table 4.1 Selected characterization methods of structural carbohydrates and lignin.

Biomass characteristics	Techniques
Composition of lignocellulose Monosaccharide and lignin contents	• Two-step acid hydrolysis [11, 13, 14] • High-pressure liquid chromatography (HPLC)
Degree of crystallinity Changes of inter-/intramolecular hydrogen bonding of polysaccharides	• Cross-polarization/magic angle spinning (CP/MAS) ^{13}C nuclear magnetic resonance (NMR) [15–17] • X-ray diffraction (XRD) [15–17] • Fourier-transform infrared (FTIR) (lateral order index (LOI)) [15–19]
p-Hydroxyphenyl (H): guaiacyl (G): syringyl (S) ratio	• FT-Raman [20–22] • Pyrolysis-gas chromatography/mass spectrometry (Pyro-GC/MS) [23–26] • 2D heteronuclear single-quantum correlation (HSQC) [27–31] • ^{13}C NMR [32, 33]
Hydroxyl functional groups of lignin	• Wet chemistries [34, 35] • ^{1}H NMR [34–38] • ^{13}C NMR [34, 36, 39] • ^{31}P NMR [34, 40–43] • FTIR [34, 35] • Ultraviolet (UV) spectroscopy [35–37]
Lignin structural information Structural types and distribution of interunit bonding patterns of lignin	• ^{13}C NMR [32, 44] • 2D HSQC [4, 45–48]
Molecular weight distribution	• Gel permeation chromatography (GPC) [49–52]
Degree of condensation of lignin	• Cross-polarization-polarization-inversion (CPPI) [53–55] • ^{13}C NMR [44, 56] • ^{31}P NMR [57]

there are heteropolysaccharides present that assemble onto the cellulose microfibril surfaces. There is a lack of any evidence for any covalent bonding between the cellulose and heteropolysaccharides, suggesting a physical association among the components. The heteropolysaccharides are thought to help space the microfibrils in the hydrogel-like network as the charged substituents like glucuronic acid give rise to longer distant repulsive forces [59]. At the end of the life of the vascular cell, monolignols are transported from the cytosol of the cell into the polysaccharide cell wall where they undergo dehydrogenative polymerization induced by a variety of protein-based enzymes. Side reactions occur during these processes linking lignin to the heteropolysaccharides, usually through benzyl ether linkages or gamma carbon esters of the propanol side chain [60].

By now it should be clear that the complex structure of the cell wall mandates careful analytical procedures to quantify the composition of the components that go into the cell wall. With an emphasis on the conversion and utilization of the polysaccharides, especially glucose, for liquid fuel production, it is important to know the amount of cellulose in order to close the total mass balance. The simplest approach would be to separate each component and weigh out exactly how much is present. Since each component cannot be fully extracted in a stepwise, perfectly quantitative manner, biomass is treated with a mineral acid that cleaves the acetal linkages among the polysaccharides reducing the biomass into a soup containing monosaccharides and heavily modified lignin, along with a few degradation by-products such as furfural compounds. The majority of the lignin remains in particulate form and can be separated via filtration through a ceramic filter and the concentration of monosaccharides determined via chromatographically.

4.8 Hydrolyzing Biomass and Determining Its Composition

Compositional analysis is one of the most common ways to understand the chemical make-up of lignocellulose. Lignocellulose is a complex biopolymer of three major lignocellulose components: cellulose, lignin, and hemicelluloses. The latter is actually a collection of heteropolysaccharides that are composed of more than one monosaccharide. In order to quantify the amount of these components, lignocellulose is required to be hydrolyzed by two-step acid hydrolysis. The first step is to transform polymers into oligomers, followed by the second step to further hydrolyze oligomers to monomers [13]. The methods have been extensively developed by the National Renewable Energy Laboratory (NREL), and an overview is given in the following.

In a typical method, extractive free biomass [14] is first hydrolyzed with concentrated acid. The biomass is ground to 20–80 mesh size and 300 mg dry lignocellulose is mixed with 3 ml of 72% (w/w) sulfuric acid. This work can be conducted in a test tube incubated water bath maintained at 30 °C. The mixture is incubated for 1 h and can be stirred intermittently with a glass rod. After 1 h, the reaction mixture is diluted with water (~84 ml water) to a final concentration of 4% (w/w) sulfuric acid solution. Caution should be used when handling the acid and mixing with water. For the second step hydrolysis, the reaction mixture is transferred into pressurized reactor vessels (e.g., pressure tubes) and autoclaved at 121 °C (15 psi) for 1 h. These two steps will cause the polysaccharides to depolymerize, cleave the acetate groups from the hemicelluloses, and dissolve a small portion of the lignin, usually less than 2% as acid-soluble lignin (ASL). The ASL can be determined by spectrophotometer at $\lambda = 240$ and 320 nm, depending on the type of biomass species [13]. The lignin undergoes acid-catalyzed cross-linking making it highly modified in the form of acid-insoluble lignin or Klason lignin. It should be noted that a small portion of the polysaccharides undergoes breakdown to furanic compounds and hydroxymethyl furan.

Subsequently, the reaction mixture is filtered using a porcelain, medium porosity ceramic filter crucible. The residue in the crucible primarily contains acid-insoluble lignin components but for certain biomass sources, it also contains some protein that must be accounted for in the residue. The acid-insoluble Klason lignin is determined by first drying the crucible at 105 °C and measuring the weight of the dried sample as well as the sample after ashing 575 °C. Residual ash (but not necessarily equivalent to total ash) is accounted for and removed from the calculation of the acid-insoluble lignin. For high protein content samples, residual protein is accounted for via nitrogen elemental analysis of the acid-insoluble residue, where the nitrogen percentage is converted into protein content by a generic multiplier of 6.25 although, a more accurate multiplier between 5.1 and 6 has been determined for a number of species [61, 62]. ASL is analyzed in the filtrate using UV–vis spectroscopy be relating the absorbance at $\lambda = 240$–320 nm to the total concentration, through the absorptivity coefficient (maintaining a given path length). Note that this analysis is based on the Beer–Lambert Law and is a highly accurate and simple method to determine the concentration. Careful selection of the absorbance range (usually below 1) and absorptivity coefficient is required based on the biomass type and the pretreatment method.

4.8.1 Analyzing Filtrate by HPLC for Monosaccharide Contents

Many of the monosaccharides are epimers of each other; hence they have the exact same molecular formula, similar ring structures but differ in conformations of the hydroxyl and hydrogen groups attached to the different carbons on the ring. This similarity makes it difficult to determine the exact amount of components without separating them in some manner. A common method of separation is liquid chromatography that partitions components between two phases, typically a mobile fluid phase and a solid stationary phase. Chromatography was first developed by M. Twsett to separate chlorophyll pigments, as their separation on a $CaCO_3$ column left bands of color (chroma is color in Greek). Nowadays high-performance liquid chromatography equipment couples a column system to an online detection system to determine the concentration. These detection systems can be simple for light absorbing compounds, such as UV–vis detectors, to detectors that measure the change in refractive index (RI) of the solution as a function of concentration, or a pulsed amperometric detector that measures the change of electrical current. The latter has gained in popularity for systems for dedicated sugar analysis as it is very sensitive for detection of sugars at very low concentrations.

4.8.2 Choosing the HPLC Column and Its Operating Conditions

Figure 4.2 shows the chromatogram of a mixture of cellobiose, glucose, xylose, and arabinose using Bio-Rad® Aminex 87**H** column and detected with an RI detector. This column can separate cellobiose, glucose, xylose, and arabinose. The column is not suitable for the analysis of lignocellulose with high mannose content, such as softwoods because mannose is co-eluted with glucose. The advantage of this column is that it is robust and will not be greatly impacted by fluctuations of the pH of the eluent. It is commonly used to analyze glucan digestibility of biomass, as glucose is released based on the specificity of the cellulase hydrolyzing cellulose. Figure 4.2 shows an example of the monosaccharide elution profile using RI detector.

To analyze samples with high mannose content, the Bio-Rad® Aminex 87**P** column is effective to separate out all the hydrolyzed monosaccharides (glucose, xylose, galactose, arabinose, and mannose); however, rigorous protocols must be used to limit pH fluctuations that will impact the life of the column.

Other components of biomass can make up a significant mass portion. In hardwoods, the hemicelluloses are highly acetylated and they can make up to 10% of the mass of the isolated xylan corresponding to 2–5% of the wood. Additionally, uronic acid branches of hemicelluloses can compose upwards of 2–6% of the wood [63]. Acetyl groups can be readily determined by measuring the acetic acid concentration in the hydrolysis liquor via HPLC (high-performance liquid chromatography) with an RI detector. The

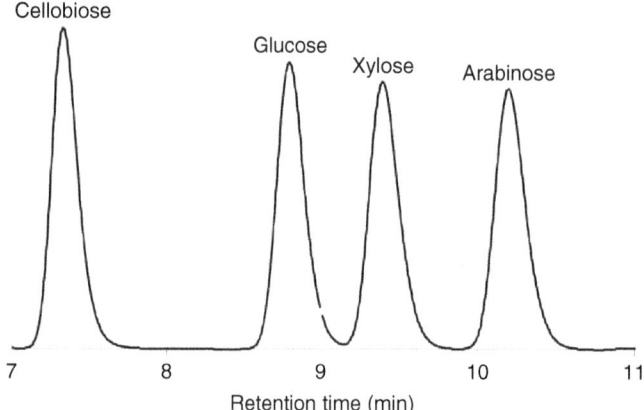

Figure 4.2 HPLC chromatogram of four monosaccharide standards separated by Bio-Rad® Aminex 87P column at 60 °C, flow rate of 06 ml min^{-1} of 4 mM H_2SO_4.

uronic acid side groups such glucuronic acid or galacturonic acid can be detected with HPLC but typically specific protocols, different from the main chain polysaccharides, must be established as their interactions with the column change their retention time significantly. While HPLC methods are common to quantify sugar components, gas chromatography (GC) methods were also developed for sugar determination. GC methods require the reduction of the monosaccharides into their nonreducing alditol form and derivatizing into acetates with acetic anhydride. The samples need to be isolated from the derivatizing solution because unreacted compounds impact peak resolution. While the method involves additional steps that would make screening 100s of samples more difficult relative to direct injection with HPLC, it is an accurate alternative for carbohydrate analysis (Table 4.2).

4.9 Determining Cell Wall Structures Through Spectroscopy and Scattering

4.9.1 Probing the Chemical Structure of Biomass

The structure of carbohydrate contributes to the fiber properties and ultimately affects the product quality. Many techniques, such as X-ray diffraction (XRD), cross-polarization/magic angle spinning (CP/MAS) ^{13}C NMR [15–17, 66, 67], Fourier-transform infrared spectroscopy (FTIR), and Raman spectroscopy [68–70] have been employed to determine chemical structure and local environment of the polysaccharides. These techniques are nondestructive and can be used with biomass in the as-received state. The degree of crystallinity is one of the commonly used parameters to probe cellulose characteristics of biomass. Herein, we discuss the resulting information from these techniques and how to correlate the obtained information with the chemical structure of biomass.

Table 4.2 Advantages and disadvantages of biomass characterization methods and techniques.

Techniques	Advantages	Disadvantages
Two-step acid hydrolysis	• Quantitative • Well established and well accepted	• Destructive • Analysis time by HPLC can be long • Labor intensive • Toxicity from acids and corrosion products
CP/MAS ^{13}C NMR	• Detailed analysis • Nondestructive • Selective	• Low throughput • Insensitive • Long analysis time • Expensive • Automation is challenging
XRD	• Fast analysis • Nondestructive • Results are qualitative and quantitative	• Low throughput • Peak convolution • Safety due to X-ray
FTIR	• High-throughput capabilities • Nondestructive • Fast analysis	• Qualitative • Peak convolution • Very sensitive to moisture • May require specific sample preparation (e.g., KBr)
Raman spectroscopy	• Not sensitive to moisture • Easy sample preparation • Multiple excitation sources including ultraviolet (UV), near-infrared (NIR), and visible (VIS), so this can tailor for various analytes • High throughput • Applicable to solid, liquid, and gas	• Weak signal • Sensitive to strayed light
Pyro-GC/MS	• Only small amounts of samples are needed • Easy sample preparation • No need to isolate lignin • Can be high throughput	• Destructive • Complex data analysis
Solution-state NMR	• Detailed structural information • Quantitation can be done without calibration standards	• Destructive • Semiquantitative • Analysis can be long • Insensitive • Expensive • Automation is challenging • Complex set-up due to inter-relationships between sample concentration, sample solubility limit, and acquisition time
GPC	• Fast analysis • A wide array of available detectors for the analysis of various polymers including multi-angled light scattering (MALS), ultraviolet (UV), and refractive index (RI)	• A number of calibration standards are needed for accuracy • Destructive • Semiquantitative • Complex analysis for polydispersed polymers

Source: Adapted from Refs [64, 65].

4.9.1.1 X-Ray Diffraction (XRD)

XRD is commonly used to analyze the degree of crystallinity of cellulose. As mentioned earlier, cellulose is always deposited as bundles or aggregates of cellulose chains. The chains have symmetry and as a result diffract X-rays according to the structure of the unit cell. Interference peaks of the diffracting electromagnetic radiation arise from the specific spacing of subatomic particles arranged in the molecular structure. In powder diffraction, there is a randomization of the scattering plans forming a concentric pattern with the angular intensity (2θ) directly related to the spacing (Å) of diffracting planes through Bragg's Law. The original indices for the diffraction of cellulose peaks have been revised for the unit cell, the most basic repeating pattern of the crystal, and it is generally accepted to report the diffraction peaks with Miller indices for cellulose I_α as (100, 010, and 110) and cellulose I_β as ($1\bar{1}0$, 110, and 200).

As highly ordered arrangements of polymers dramatically differ in mechanical properties as well as access to internal surfaces, the crystallinity of cellulose is seen as a key factor governing wood properties, fiber quality, as well as bioconversion. The degree of crystallinity can be expressed by CrI. The index, as the name implies, is not an absolute number but a comparative method to determine the relative portion of symmetry within the biomass. Sample preparation is quite easy for analysis, however, drying history has a large impact on the results, and sample preparation should be considered carefully and reported in the analysis. Because of the low bulk density of cellulose fibers, typically samples are pressed into a pellet to increase the sample per unit volume. Artifacts can sometimes occur as distortion in certain planes can occur causing orientation as samples are compressed. Samples of small amounts can be placed on quartz substrate for analysis and the sample substrate can be subtracted as background from the diffractogram. There are multiple ways to calculate the CrI from the biomass as listed. The three methods of calculating CrI are portrayed in Figure 4.3 [15]: Segal method [71]; peak deconvolution [15]; and amorphous subtraction [72]. The Segal method is the most common method for calculating CrI using the relationship between (002) and the amorphous region as follows:

$$\text{CrI} = \frac{I_{(002)} - I_{(am)}}{I_{(002)}} \times 100$$

where $I_{(002)}$ is the height of the (002) and $I_{(am)}$ is the height of the amorphous region.

The peak deconvolution method uses the idea that five crystalline planes of cellulose I_α, corresponding to (101), ($10\bar{1}$), (021), (002), and (404) are scattered on the amorphous region (Figure 4.3b). So these crystalline and amorphous planes can be deconvoluted, and the CrI value can be determined from the ratio of the crystalline area over the total area [15]. The peak deconvolution/fitting peak fitting is subjective to the user and cannot always be repeated with accuracy. Hence, the peak deconvolution has to be done with caution. The amorphous subtraction method is done by subtracting the spectrum of interest with the amorphous standard [72], which can be hemicellulose, lignin, or phosphoric acid swollen cellulose (PASC). The CrI can be calculated from the ratio of the crystalline area over the total area after all peaks are deconvoluted. Out of the three methods, the Segal method appears to provide the highest value of crystallinity.

It should be noted that most of the analysis is related to the characteristics of cellulose either in the form of the cell wall, the pulp fiber, or micro-/nanocellulose. However, hemicelluloses are amorphous in the cell wall can be crystallized through

Figure 4.3 X-ray diffraction spectra of microcrystalline cellulose (Avicel PH-101) shows three methods for calculating CrI: (a) Segal method; (b) peak deconvolution method; and (c) amorphous subtraction method.

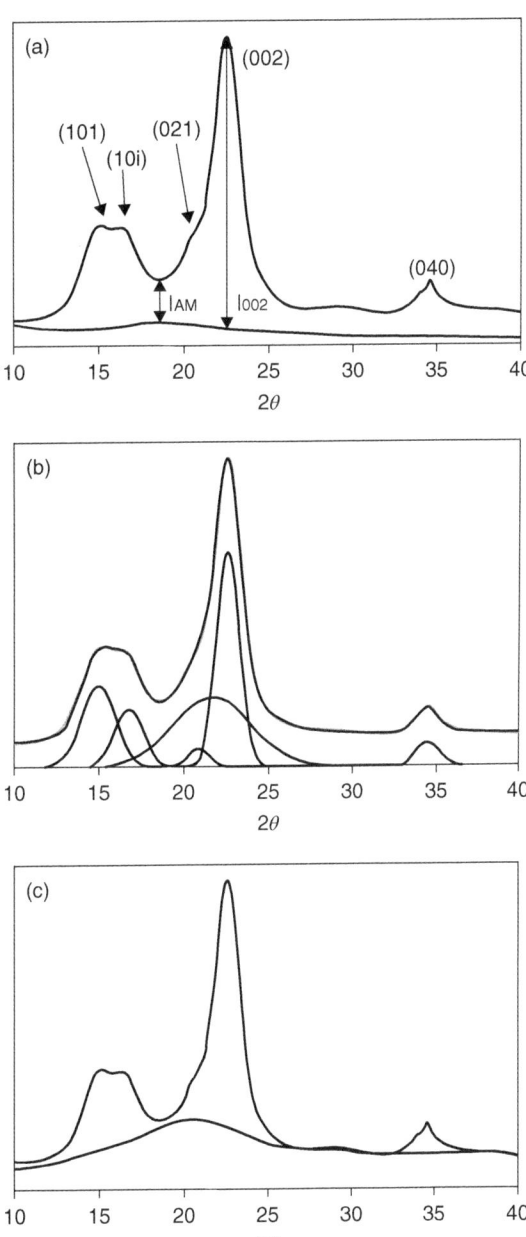

debranching and careful precipitation. Hemicelluloses can crystallize as hydrates and the diffractogram will change depending upon the amount of moisture present. Other studies have analyzed technical lignins for periodicity in their structure.

4.9.1.2 Cross-polarization/Magic Angle Spinning (CP/MAS) ^{13}C NMR

CP/MAS ^{13}C NMR or also often known as solid-state nuclear magnetic resonance (ssNMR) can be used to observe changes of inter- and intramolecular hydrogen

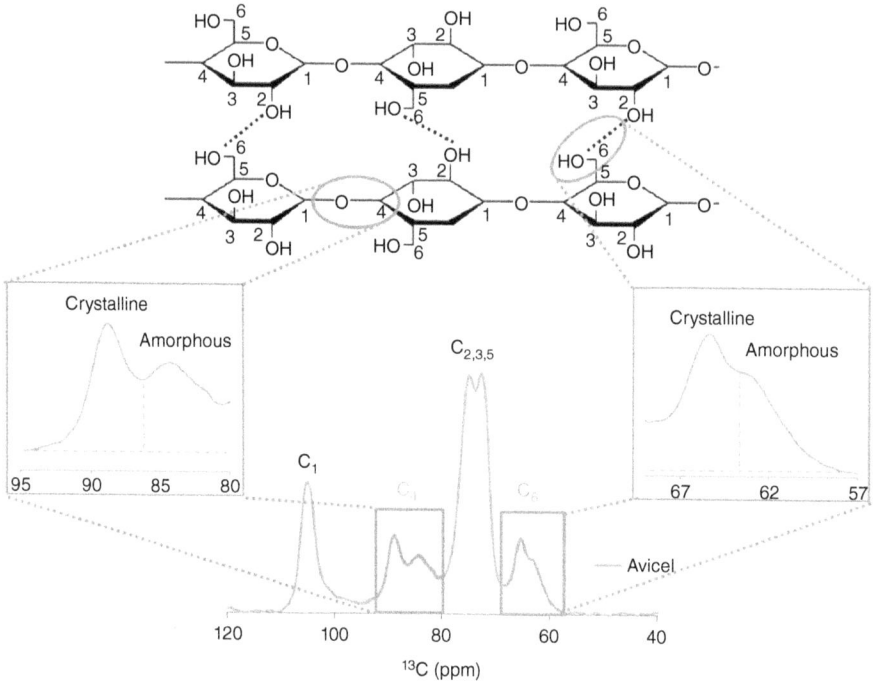

Figure 4.4 Illustration of information obtained from CP/MAS ^{13}C NMR spectrum of microcrystalline cellulose. C_4 and C_6 regions are typically used for CrI determination and evaluation of changes in inter-/intramolecular hydrogen bonding.

bonding of polysaccharide chains [15, 16, 73]. This method uses radio electromagnetic radiation to perturb spins of isotopes (with an odd number of spins) and the localized environment relaxes the perturbed spins back to its normal state. For cellulose, each carbon has a response to this perturbation. Figure 4.4 shows the spectrum of microcrystalline cellulose (Avicel). The C_1, C_4, and C_6 carbons are differentiated in the spectrum, while there is considerable overlap in the $C_{2,3,5}$ signals. C_4 region is commonly used to determine CrI [67], and C_6 region is used to observe changes in hydrogen bonding among cellulose chains. CrI can be determined by two methods: (i) C_4 peak deconvolution and (ii) amorphous subtraction.

For polymers in a highly organized state, localized interactions among cellulose chains will cause their response to differ. The core chains of cellulose microfibrils will be in a distinct local environment relative to the chains on the surface of the microfibrils. Hence, the relative amount of crystallinity in the sample can be determined using the C_4 peak deconvolution (i.e., $X/(X+Y)$ in Figure 4.4).

Similar to XRD, C_4 peak deconvolution of the NMR spectrum is subjective, and the results are user dependent to a degree. However, the method is widely used to report the relative crystallinity of cellulose as the signal intensities are directly related to the amount of species present. Caution should be taken because hemicellulose components also generate signals in the region that corresponds to the amorphous cellulose signals. To avoid falsely assigning signals to amorphous cellulose, the signal can be further deconvoluted for the individual contributions of the signal to the hemicellulose, amorphous cellulose, and crystalline cellulose [74]. Furthermore, the moisture content of the sample can cause slight increases in CrI by 5% [6]. Finally, it is interesting to note that the signal intensity of the deconvoluted peaks has been used to estimate the microfibril size of hydrolyzed samples, with all hemicelluloses removed, fitting the fractional intensity of surface chains to a rectangular cross-sectional model [74].

The amorphous subtraction method is done the same way as that by XRD. The CP/MAS ^{13}C NMR spectrum of interest is subtracted by the spectrum of the amorphous component of lignocellulose (i.e., hemicellulose, lignin, and PASC). Then, the area ratio is used to determine CrI.

4.9.1.3 Fourier-Transform Infrared Spectroscopy (FTIR)

FTIR is one of high-throughput analyses that can provide both qualitative and quantitative information about the chemical characteristics of biomass and the isolated components. FTIR works on the principle of a harmonic oscillator where the vibrational energy of bonds corresponds with the energy of certain frequencies of infrared light [75]. Hence bond strength and the mass of elements bonded together (and how the localized environment influences them) have corresponding bond energy that matches the energy of certain frequencies of infrared light. This mechanism causes certain frequencies of light to be absorbed by the corresponding bond type if the dipole moment is changed as a function of bond distance during excitement. The resulting spectrum of the Fourier-transform signal reveals aspects of the chemistry of the material. Infrared spectroscopy is typically divided into two categories based on the wavelength of the radiations used: near-infrared spectroscopy (NIR) has been developed to be used in field or industrial quality control environments, while mid-infrared (MIR) spectroscopy is typically found in most research laboratories for chemical analysis.

FTIR spectroscopic analysis of biomass requires some manipulation of the resulting spectrum. Because CO_2 and moisture can absorb electromagnetic radiation, there is always a background spectrum that needs to be subtracted from the sample spectrum. Usually, the background spectrum is acquired just prior to the analysis of a sample, and the software is designed to automatically subtract the background. Often though there is a distorted baseline because of scattering from the sample and the baseline should be adjusted by moving the spectrum to zero for regions where there is no absorbance. This shift can be automatically adjusted or user defined, but should be done with attentiveness to ensure a uniform procedure between samples. After the baseline is corrected, samples are often normalized so the intensity of absorbance is plotted in a way to see comparative changes in the spectra among samples.

Both NIR and mid-FTIR methods have been used to understand cellulose structure. For example, mid-FTIR has been correlated to the CrI [76]. The ratio of select signals that arise because of order can be correlated to the CrI determined from using other methods like XRD. Changes in cellulose crystallinity in the 850–1500 cm^{-1} range were observed by Nelson and O'Conner [69, 70]. The total crystallinity index (TCI) and lateral order index (LOI) were proposed from the ratio of 1420/893 and 1375/2900 cm^{-1}, respectively.

The molecular orientation of cellulose during the formation of wood cell wall can be analyzed by observing changes of bands at 898 (v_{as}(ring), anomeric vibration at β-glycosidic linkage) and 1160 (v_{as}(COC), COC antisymmetric stretching) [77]. Both bands are present in FTIR spectra of microcrystalline cellulose and PASC [17], suggesting that FTIR is not an absolute technique [15]. Additionally, the transformation of cellulose into cellulose II during the mercerization process, exposing cellulose fibers to aqueous alkali, causes significant peak changes in the corresponding spectra [78]. This point highlights the sensitivity of FTIR analysis to detect supramolecular changes without significant changes to the chemistry of the polysaccharide [79].

Furthermore, FTIR is widely used to confirm the derivatization of polysaccharides. During esterification, two significant spectrum changes occur that are readily notable. The prominent hydroxyl stretching is decreased while absorption in the carbonyl region is increased. These data are qualitative and can show the successful modification, but it typically lacks the quantification of substitution level. However, like CrI, a calibration curve can be created where the area of the absorbance region is plotted as a function of the degree of substitution [80]. This technique especially works well for cellulosic esters because there is not significant absorption from other functional groups in this range of 1860–1690 cm^{-1}, allowing for the quantification of the signal according to Beer–Lambert Law. Interestingly, the technique allows for the analysis of small quantities of esters compared to the other methods such as titration and ^{1}H NMR [80].

As the mechanism of IR analysis is the harmonic oscillator model where IR radiation is absorbed and converted into energy of molecular rotation [75], the molecular vibrations have harmonics at higher wavenumbers (higher frequency) in the near-field IR from 10,000 to 4000 cm^{-1}. For NIR analysis, a fiber-optic probe both delivers the IR radiation and detects it. Sampling is straightforward as reflected light from the material is detected accounting for the light that is absorbed and also scattered. Once a given set of calibration standards is made, NIR is powerful in detecting specific changes of materials. It has been used to analyze the moisture content of biomass for biofuels [81] to follow cellulose structure changes during alkali treatment [82]. With multivariate component analysis, the chemical composition of biomass can be correlated with the NIR spectra for all the major polysaccharide components, lignin, and even extractives [83].

4.9.1.4 Raman Analysis

Raman spectroscopy is another form of vibrational spectroscopy analysis. Raman signals are generated from inelastically scattered light, where bonds are excited to an elevated state of the monochromatic radiation, and light energy released from the molecule returns the bond to a lower energy level. Because energy is quantized it is analogous to a person who boarded an elevator on the second floor (an excited state) and rode the elevator to the top floor, and decided to return the first floor (a different energy state). The energy lost, inelastic scattering of light from starting and ending at two different levels, is related to the type of bond. Like FTIR, the Raman spectrum consists of signal intensity as a function of wavenumber, which is the difference in energy from the impinging light and the energy loss between starting and ending at two different energy levels. Complimentary to FTIR, where the signal is generated because of a change in dipole moment of the molecule with respect to the distance of atoms, Raman signals are generated from the polarizability of the molecule or the relative ease that an electron cloud can be disturbed. This mechanism means that carbon–hydrogen signals are much stronger than hydrogen–oxygen signals, making water relatively transparent for this method of analysis. Like FTIR, the local molecular environment greatly impacts the signals and this method can be used to examine different crystalline structures of cellulose [84].

The drawback to this analytical method is that fluorescence can saturate the signal blocking the measurement of the bonds of interest. Auto-fluorescence can be problematic with lignin-containing compounds. However, this issue can be mitigated to a point depending upon the excitation laser in the spectroscope. Longer wavelength lasers in the 700 nm region can avoid some of the issues faced with this measurement technique. The drawback of using a longer wavelength is a loss of signal intensity, so the power of laser and detectors need to be adjusted accordingly.

The method has received much attention to do two- and three-dimensional (3D) label-free mapping of chemical functional groups [85]. Hence, the spatial resolution of functional groups can be determined if the spectroscope is coupled to a confocal laser microscope. While not quantitative in nature, the method allows the determination of the relative distribution of components within the biomass cell wall of microtomed samples. Typically the spatial resolution is one micron, and this allows changes in the cell wall to be determined from the cell wall corners and middle lamellae to the innermost portion of the S3 layer [86]. However, Raman analysis has also been coupled to atomic force microscopy analysis providing detailed surface and chemical resolution [84].

4.10 Examining the Size of the Biopolymers: Molecular Weight Analysis

The molecular weight of polymers is a key component in understanding mechanical properties and processability of materials in the solution and melt state. For the strength of polymeric materials, one can imagine long thread-like molecules entangling together, and the greater the length of the molecules the more entanglements. Because polymer chain size is not uniform, an average of chain sizes is used to describe the molecular weight. For polymer characterization typically the first moment or the second moment of the distribution is used to describe the materials. The first moment is referred to as the number-average molecular weight (M_n) and is the ratio of the sums of the total number of molecules with a specific size of the polymer to the sum of the total number of polymers (providing a mean related to the summation of the mole fraction). The second moment is referred to as the weight average molecular weight (M_w) and is found by the summation of the number of molecules multiplied by the square of the molecular weight of each fraction divided by the summation of the molecular weights (summation of the number of molecules times the size of each chain). As the molecular sizes are a distribution, the ratio of the M_w over the M_n provides insight into the breadth of the distribution. This ratio is normally called the polydispersity index (PDI) [87]. M_n, M_w, and PDI can be expressed as

$$M_n = \frac{\sum N_i M_i}{\sum N_i}$$

$$M_n = \frac{\sum N_i M_i^2}{\sum N_i M_i}$$

$$\text{PDI} = \frac{M_w}{M_n}$$

The index number, i, represents the number of different molecular weights in the polymer sample and N_i is the total number of moles with the molar mass of M_i.

Gel permeation chromatography (GPC) is used to quantify the molecular weight distribution of polymers. With this technique, polymers are separated based on their hydrodynamic radius by forcing the polymer solution through a column containing particles of varied size with micropores. Large polymer chains are eluted in the column in short times because their hydrodynamic size is larger than the micropores on the bead surface in the column. As the chains elute, the concentration in the solution is determined via UV and or RI detectors. A profile of chain sizes is developed as a function time, known as the molecular weight distribution. This profile holds the key information about one of the most fundamental pieces of information about polymers that impacts polymer performance in the liquid and solid state. The problem for biopolymers like lignin is that SEC (size-exclusion chromatography) is not based on an absolute measurement, only the time it takes for the polymer chains to slip through the column. Calibrated standards can be used to get around this issue by passing materials like polystyrene or pullulan with defined molecular weights through the system because of the product of the intrinsic viscosity and molecular weight are not unique and can be directly related to each other. This method works well for many types of samples and is referred to as the "universal calibration method" [88]. For absolute measurements, other detectors can be connected to the system such as a multi-angle light scattering detector (MALS). The detection system provides an absolute molar mass and the root mean square (rms) radius of the polymer based on light scattering principles that relate the angular dependence of the scattered light to the polymer chain size and structure. MALS can be used to determine the M_w as the turbidity/scattering intensity is related to the reciprocal of the number-average molecular weight when the scattering (and concentration) is extrapolated to zero. The drawback of MALS is that fluorescence emitted from the sample can negatively impact the characterization of the sample. This issue is critical in lignin molecular weight analysis, and caution should be taken for samples that are contaminated with lignin (unbleached xylan). However, instrumentation that includes longer light wavelengths, and filters can be used to partially circumvent this issue.

The difficulty of dissolving cellulose makes molecular weight determinations problematic. In an ideal system for cellulose characterization, the GPC system utilizes a cellulose solvent, such as DMAc with 0.1 M LiCl. Several errors can occur through the dissolution process of cellulose, and molecular weights should be carefully calculated based on adjusting the mass of DMAc associated with cellulose [89]. Problems have occurred also because of the limited solubility of some high-molecular-weight celluloses. The other option to characterize the molecular weight of cellulose is based on making a cellulose derivative that is soluble in more common solvents. Tricarbanilated cellulose, where the hydroxyls are replaced by phenyl groups through a urethane linkage is one method where derivatization has limited chain degradation [90]. The derivatizing agent, phenyl isocyanate, is highly reactive and must be handled with extreme caution.

Other molecular weight determination methods involve chain end analysis or viscosity measurements that provide insight into the size but not the size distribution. The former provides information about the number-average molecular weight because it is determined from the number moles of end groups to the total number of repeat units. For polysaccharides, each chain is capped by a reducing end of the polysaccharide, which can be used to determine the number-average molecular weight [91]. The ratio of the number of reducing ends can be found through titration, which is simple; however, for cellulose in fibrous form, there must be a solubilization step that first occurs [92]. Other more soluble polysaccharides can be analyzed directly using solution-state NMR to determine the number-average degree of polymerization (DP) [93]. HPLC can also be used if the chain end has a specific saccharide termination step. For glucuronoxylans, there is a rhamnan unit associated with each end unit [94–96]. The ratio of the number of xylose units to the rhamnose units provides a simple method for molecular weight determination of xylan. The drawback to this method is that pectin-based material from the compound middle lamella may influence the results.

Using dilute viscosity measurements, the viscosity average molecular weight M_v can be determined through the relationship of the Mark–Houwink–Sakurada equation. This molecular weight falls between the first and second moment of the distribution. The viscosity of a solution relates how chains interact with each other. On a relative basis, related to the resistance to flow, the viscosity of the solution can be easily calibrated to a technique for a simpler test, named the falling ball method. Cellulose pulps are dissolved in cupriethylenediamine at 1% solution, and the measurement is based on the time it takes for an aluminum ball to drop through the solution [97]. The method is developed for quality control analysis.

Like cellulose, lignin is often derivatized in order to solubilize in common solvents like tetrahydrofuran [98]. However, polar organic solvent, such as dimethylsulfoxide (DMSO), dimethylformamide (DMF), and DMAc [99] have been used to characterize lignin without derivatization. A hybrid method has used ion-pair chromatography for lignin analysis, modifying lignin simply through adsorption, where lignin has a strong association with cationic amines [100]. Additionally, aqueous systems have been applied to hydrophilic lignin samples dissolved in an aqueous alkali. The choice of solvents, pH, and ionic strength influence the elution profile of lignin. It should be noted that polymer solutions require each polymer chain to be solvated. This idea seems straight forward, for example, with an experiment where cellulose triacetate is immersed in acetone; the sample disappears into the solution as the individual chains lose contact with each other. However, for dissolving a technical lignin-like a hardwood kraft lignin, the sample disappears into the solution, yet it is not fully dissolved. Lignin has a high degree of intermolecular associations based on its aromatic structure, and there is a significant time component to dissolution. Many treatment methods have been published on the subject, and careful attention to reporting these details should be given. This issue arises because lignin and lignin derivative solutions can be aged, and the molecular weight profiles show differences and is dependent upon the time of solution preparation and the temperature at which the material is stored. This change is also seen in the RI increment, as dn/dc values can take several days to stabilize [101]. Different additives can be introduced to solutions to limit these aggregations such as iodine [102]. To dissolve lignin in a common solvent like THF, acetylation procedures are used by reacting lignin with acetic anhydride over pyridine [103].

4.11 Intricacies of Understanding Lignin Structure

Because lignin is an amorphous biopolymer created by the oxidative coupling of monolignol(s) (i.e., *p*-coumaryl alcohol, coniferyl alcohol, and/or sinapyl alcohol), there are a number of interunit linkages that provide some level of complexity to the structure. Additionally, the structure changes dramatically during isolation and processing. Utilization and exploitation of this biopolymer hinges on understanding its structure and connecting it to its performance and some specific methods are outlined in the following sections.

4.11.1 ^{13}C NMR

As mentioned earlier, NMR relies on the detection of odd-numbered isotopes such as ^{13}C, and these signals can be quantitative from the population of atoms in the material. As such, ^{13}C NMR can be performed to examine the entire structure of soluble lignins to determine the presence of interunit linkages (e.g., β-aryl ether linkages), condensed and uncondensed aromatic and aliphatic carbons, H/G/S ratio, and the total amount of functional groups. The advantage of this technique is that one is able to study the entire soluble lignin structure intact without acetylation. The downside is that these ^{13}C nuclei are not naturally abundant, so high lignin concentration and/or long acquisition time are required. In some cases, a relaxation agent, such as chromium (III) acetylacetonate, can be added to the lignin solution to help provide a complete relaxation of nuclei. This additive reduces the time it takes for the atoms to go back to their unperturbed state (T_1), allowing better signal-to-noise ratio at shorter collection times. After collecting the data, the spectrum reveals all the different carbons in a sample (such as carbons involved in the aromatic structure, carbons in a carbonyl bond, or aliphatic carbons). The area for each of these peaks can be integrated and the ratio of peaks can be used to characterize the sample. When normalizing for the number of carbons in the signal, the ratio of peak intensities can be used to determine the functional group or interunit linkage per C_6 or C_9 [104]. The technique is sensitive to distinguishing different carbonyls attached to the lignin such as carboxylic acids and aldehydes, which are related to decomposition mechanisms. Figure 4.5 shows the ^{13}C NMR spectra of milled wood lignin (MWL) [44] isolated from loblolly pine by Bjorkman method [105]. The inset reveals the aliphatic region of the MWL and acetylated MWL (MWL-Ac). The ^{13}C NMR spectrum can be divided into regions shown in Table 4.3.

The integration of the aromatic region ($I_{160-103}$) can be set to a value of 6.12, representing all aromatic carbons plus a contribution of 0.12 per 100 aromatic units from the side-chain carbons of coniferyl alcohol and coniferaldehydes. Then integration of all moieties (e.g., methoxyl content) will be based on the aromatic ring. For example, after setting the $I_{160-103}$ to a value of 6.12 and integrating the 57–54 ppm range, the methoxyl content can be expressed as 0.95 methoxyl group per aromatic ring.

4.11.2 ^{31}P NMR

A critical aspect in lignin chemistry when synthesizing new compounds is the absolute number of functional groups per gram of isolated lignin. This information is required so the proper stoichiometric ratios are used when converting lignin into copolymers or soluble derivatives. Hydroxyl functional groups of lignins can be identified by a ^{31}P NMR technique, involving the phosphorylation of lignin with 2-chloro-4,4,5,5-tetramethyl-1,3,2-dioxaphospholane (TMDP) [106]. The reaction of TMDP with lignin hydroxyl functional groups is illustrated in Figure 4.6a. TMDP reacts with hydroxyl functional groups to give phosphite products, which are resolved by ^{31}P NMR into various regions from aliphatic hydroxyl, phenolic, and carboxylic acids

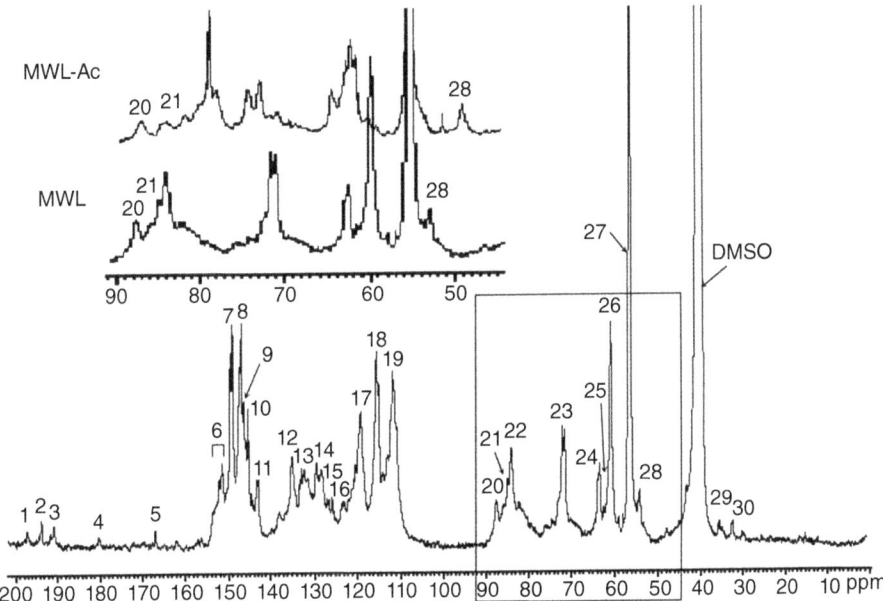

Figure 4.5 ^{13}C NMR spectrum of MWL isolated from pine shows various spectral regions.

Table 4.3 ^{13}C NMR chemical shift ranges and integration regions of all moieties.

Spectral region	Chemical shift range (ppm)
Methoxyl content	57–54
Aromatic methane carbons	125–103
Aromatic carbon–carbon structures	141–125
Oxygenated aromatic carbons	160–141
Carbon from carbonyl-type structures	195–190
Carbon from carboxyl-type structures	176–163
Degree of condensation (calculated by 3.00-$I_{125-103}$)[a]	125–103
Aliphatic hydroxyl content	171–168.5
Phenolic hydroxyl content	168.5–166

a) $I_{125-103}$ represents the integration from 125 to 103 ppm.
Source: Adapted from Ref. [44].

groups as illustrated in Figure 4.6c. The reaction occurs quickly, and the samples must be analyzed in short time periods after the reaction period because of the formation of HCl as a by-product. Peak integration of the spectrum relative to an internal standard, endo-*N*-hydroxy-5-norbornene-2,3-dicarboximide (e-HNDI) or cyclohexanol, allows the quantification of the moles of functional group per gram of sample. Different phenolic groups can be distinguished depending on their origination from different monolignols, such as sinipyl alcohol versus coniferyl alcohol, which is also helpful in

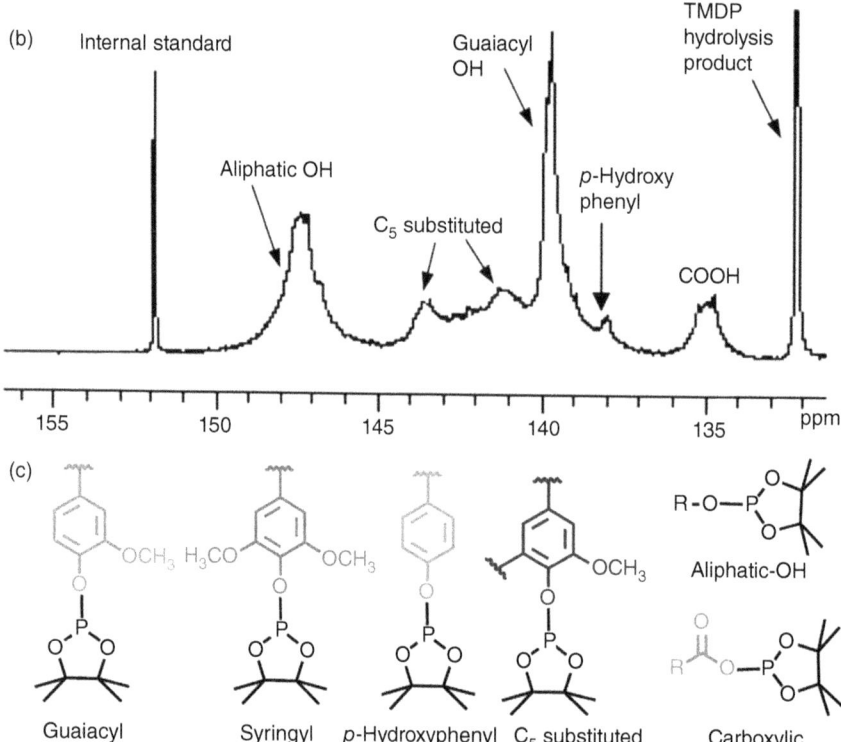

Figure 4.6 Phosphorylation of lignin with 2-chloro-4,4,5,5-tetramethyl-1,3,2-dioxaphospholane (TMDP) is fast and reaction products (c) are shown in ^{31}P NMR spectrum (b) and lignin modified (a).

examining the breakdown products. Figure 4.6b shows an example of the ^{31}P NMR spectrum of a softwood lignin derivatized with TMDP using HNDI as internal standard [107]. Typical chemical shift ranges and spectral regions for integration are shown in Table 4.4.

4.11.3 2D HSQC

2D ^{1}H–^{13}C solution-state NMR spectroscopy has been developed to reveal detailed information about lignin structure. This technique offers a semiquantitative analysis of lignin structure of ball-milled lignocellulose without extraction and acetylation of lignin. Dimethyl sulfoxide-d_6 is commonly used as a lock solvent during the procedure. In some cases, a minute amount of co-solvent, such as 1-methylimidazole (or N-methylimidazole) [48], pyridine [4, 45], and 1-ethyl-3-methyl-imidazolium acetate [27, 46], can be added to aid lignocellulose solubility, increasing solution viscosity and in turn decreasing the relaxation time. Figure 4.7a, b, and c shows the 2D HSQC

Table 4.4 ^{31}P NMR chemical shift ranges and integration regions of hydroxyl moieties.

Spectral region	Chemical shift range (ppm)
Aliphatic	145.4–150.0
Phenols	137.6–144.0
C_5 substituted	140.0–144.5
β-5	~143.5
Syringyl	~142.7
4-O-5	~142.3
5-5	~141.2
Guaiacyl	139.0–140.2
Catechol	~138.9
p-Hydroxyphenyl	~137.8
Carboxylic acid	133.6–136.0

Source: Adapted from Ref. [107].

(heteronuclear single-quantum correlation) spectra of enzymatic mild acidolysis lignin (EMAL) isolated from hardwood [108]. These spectra show the C–H coupling in aliphatic, aromatic, and anomeric regions. Aliphatic region reveals lignin interunit linkages. Important correlations include methoxyl groups, β-aryl ether (β-O-4), phenylcoumaran (β-5), resinol (β-β), and dibenzodioxocin (5-5/4-O-β). Aromatic region reveals the difference in the H:G:S distributions in the lignins. Anomeric region shows polysaccharide anomerics; however, overlapping of peaks makes it hard for interpretation. For EMAL, most carbohydrates were hydrolyzed by enzymes and acid. Hence, only a trace amount of carbohydrates were observed. Figure 4.7d shows the anomeric region of switchgrass, consisting of cellulose and hemicelluloses.

HSQC analysis is semiquantitative, and it can be used for both structural identification and estimation of the relative abundance of interunit linkages, H:G:S ratios of lignin, profiling cellulose and hemicellulose [4, 109, 110]. The relative abundance of each respective interunit linkage is then calculated as the percentage of integrals of total linkages. For determining H:G:S ratios of lignin from integrating cross-peaks from HSQC spectra, it should be noted that the G_5 cross-peak is not suitable to use for quantification of the G content because G_5 cross-peak overlaps with the $pCA_{3,5}$ and $H_{3,5}$ correlations. G_6 is *para* to the methoxy group, increasing the chance to participate in condensation reactions. Hence, G_2 cross-peak is typically used for quantification of the H:G:S ratio [111, 112].

The C_2 and C_6 positions of syringyl units in lignin rarely get substituted [113]. Hence, the overall C_9 units in lignin can be estimated by

$$C_9 \text{ units} = \frac{1}{2}S_{2,6} + G_2$$

The C_9 unit value can be used to normalize the interunit linkages of lignin, which can be expressed as percent of C_9 units. For example, β-O-4 content is 49.1% of C_9 units.

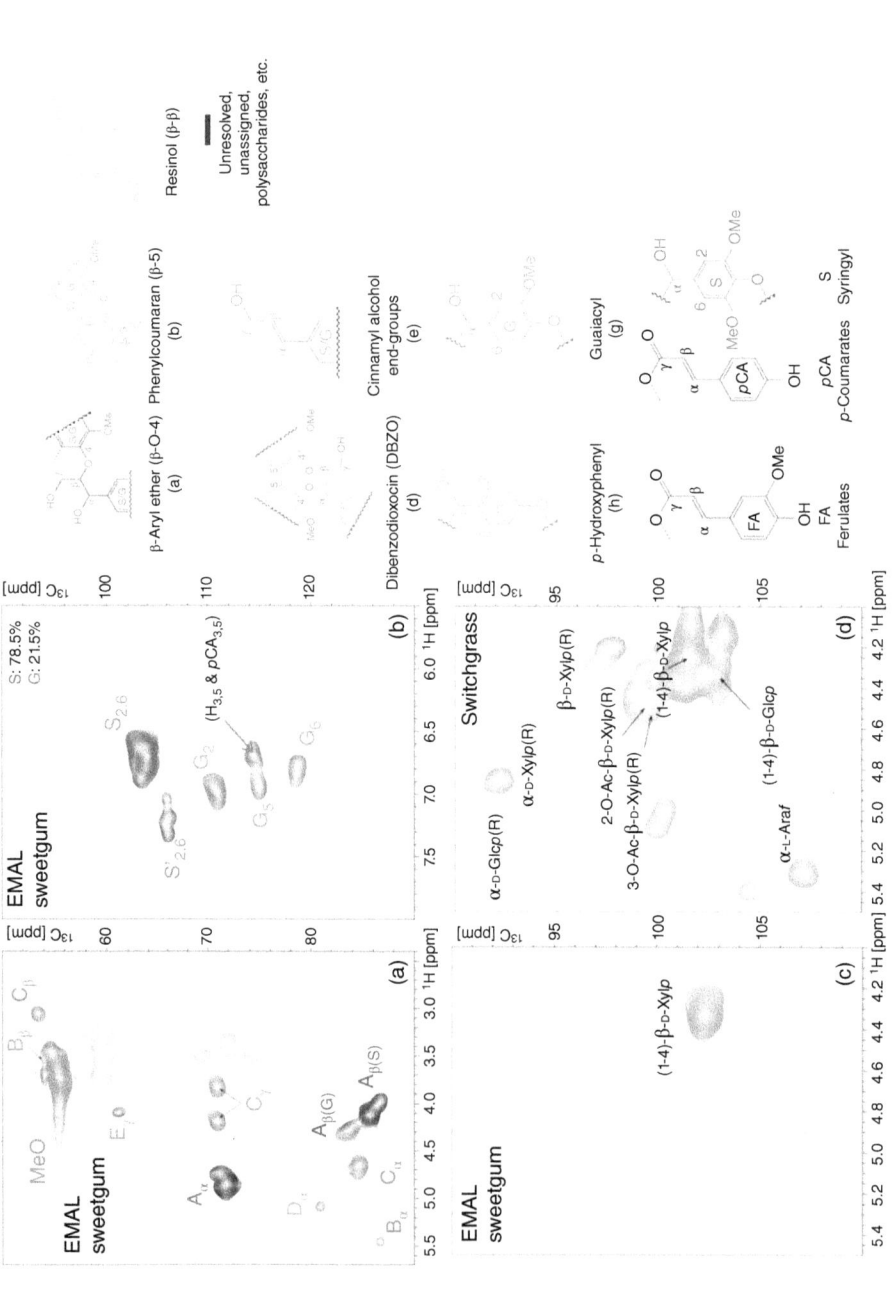

Figure 4.7 The 2D HSQC spectra of enzymatic mild acidolysis lignin (EMAL) isolated from hardwood [108].

It should be noted that all of the integration of cross-peaks should be done at the same contour level.

Table 4.5 summarizes important chemical shifts and assignments of cross-peaks of interunit linkages and/or subunits in HSQC spectra of lignin.

The 2D HSQC has proven to be a powerful tool to evaluate changes in transgenic plants [28, 29, 115, 116] and changes due to chemical and biological modifications of plant biomass [46, 47, 52, 117–123].

Table 4.5 Assignments of carbohydrates/lignin $^{13}C-^{1}H$ correlation peaks in the 2D HSQC.

Region	Label	δ_C/δ_H (ppm)	Assignment
Aliphatic	A_α	71.8/4.83	$C_\alpha–H_\alpha$ in β-O-4' substructures (**A**)
	$A_{\beta(G)}$	83.4/4.27	$C_\beta–H_\beta$ in β-O-4' substructures (**A**) linked to a G unit
	$A_{\beta(S)}$	85.9/4.10	$C_\beta–H_\beta$ in β-O-4' substructures linked (**A**) to a S unit
	B_α	86.8/5.43	$C_\alpha–H_\alpha$ in α-5 phenylcoumaran substructures (**B**)
	C_α	84.8/4.65	$C_\alpha–H_\alpha$ in α–α' resinol substructures (**C**)
	C_β	53.5/3.05	$C_\beta–H_\beta$ in β–β' resinol substructures (**C**)
	C_γ	71.1/3.81 and 4.16	$C_\gamma–H_\gamma$ in β–β' resinol substructures (**C**)
	D_α	83.5/4.98	$C_\alpha–H_\alpha$ in 5–5' dibenzodioxocin substructures (**D**)
	E_γ	61.3/4.08	$C_\gamma–H_\gamma$ in cinnamyl alcohol end groups (**E**) overlaps with carbohydrate signals
	MeO (-OCH$_3$)	55.6/3.73	C–H in methoxyls
	X_2'	99.4/4.50	2-O-Ac-β-D-Xyl$p$(R)
	X_3'	101.5/4.39	3-O-Ac-β-D-Xyl$p$(R)
Aromatic	$H_{2,6}$	127.1/7.17	$C_2–H_2$ and $C_6–H_6$ p-hydroxyphenyl units (H)
	G_2	110.9/6.99	$C_2–H_2$ in guaiacyl units (**G**)
	G_5	114.9/6.72 and 6.94	$C_5–H_5$ in guaiacyl units (**G**)
	G_6	118.7/6.77	$C_5–H_5$ in guaiacyl units (**G**)
	$S_{2,6}$	103.8/6.69	$C_2–H_2$ and $C_6–H_6$ in etherified syringyl units (**S**)
	$pCA_{2,6}$	130.1/7.45	$C_2–H_2$ and $C_6–H_6$ in p-coumarate (pCA)
	$pCA_\beta + FA_\beta$	111.0/7.32	$C_2–H_2$ in ferulate (FA)
Anomeric	(1–4)-β-D-Glc$p$	102.8/4.40	Cellulose
	D-α-Glc$p$(R)	92.2/5.10	$C_\alpha–H_\alpha$ reducing end of glucose
	D-β-Glc$p$(R)	97.1/4.44	$C_\beta–H_\beta$ reducing end of glucose
	(1–4)-β-D-Xyl$p$	102.4/4.50	Xylan
	D-α-Xyl$p$(R)	92.4/5.10	$C_\alpha–H_\alpha$ reducing end of xylose
	D-β-Xyl$p$(R)	97.5/4.38	$C_\beta–H_\beta$ reducing end of xylose
	2-O-Ac-β-D-Man$p$	98.9/4.86	O-2 Acetylated mannan
	3-O-Ac-β-D-Man$p$	99.9/4.78	O-3 Acetylated mannan

Source: Adapted from Refs [4, 114].

4.11.4 Methoxyl Content Determination

Three selected methods to assess methoxyl contents are described in the following sections.

4.11.4.1 ^{1}H NMR

^{1}H NMR can be used to quantify the methoxyl content. The lignin samples are required to be acetylated before dissolution in CDCl$_3$. The signal for methoxyl protons at ~4.5–3.6 ppm can be quantified [124]. Peak integration is rapid and there is an advantage of a relatively rapid analysis.

4.11.4.2 Hydriodic Acid

In contrast, mixing lignin with the concentrated hydriodic acid (HI) and refluxing [125] can be used to quantify methoxyl content as a wet chemistry method. This method leads to the cleavage of the alkoxyl groups with formation of methyl or ethyl iodide. The volatile iodide, passing through a solution of silver nitrate, yields a precipitate of silver iodide, which is used to determine quantitatively the methoxyl content of the lignin.

4.11.4.3 Direct Methanol

The direct methanol determination has been employed to quantify the methoxyl content [126–128]. In short, lignin is mixed with concentrated sulfuric acid (96% (w/w)). The reaction is exothermic and methanol is formed as a result. Water is added to stop the reaction and methanol is distilled and quantified by GC.

Out of these three methods, ^{1}H NMR method is commonly used because it is simple and does not involve toxic chemicals. The analysis of the resulting NMR spectrum is straightforward. The spectral range and integration region are known.

4.12 Questions for Further Consideration

1. If the actual volume of crystallinity is the same for the cellulose in wood fiber as well as a delignified pulp fiber, will the CrI be equal for the two samples? Will the measurement technique influence this result?
2. After composition analysis of biomass using HPLC your total acid-insoluble lignin mass and monosaccharides adds up to be 91% of the total mass. Explain what other components that may not be accounted for in the total mass.
3. For ^{1}H and ^{31}P NMR analysis of lignin, the total mass of lignin required for analysis is much less than that used in ^{13}C NMR experiments. Why would the same instrument require different concentrations of sample for these different nuclei?
4. Explain why the 2D NMR analysis of lignin is an important breakthrough in the analysis of lignin's structure within wood?

References

1 Resch H, Arganbright DG. Variation of specific gravity, extractive content, and tracheid length in redwood trees. *For Sci.* 1968;**14**(2):148–55.
2 Stefanovic B, Pirker KF, Rosenau T, Potthast A. Effects of tribochemical treatments on the integrity of cellulose. *Carbohydr Polym.* 2014;**111**(0):688–99.

3 Silva GGD, Guilbert S, Rouau X. Successive centrifugal grinding and sieving of wheat straw. *Powder Technol.* 2011;**208**(2):266–70.
4 Kim H, Ralph J. Solution-state 2D NMR of ball-milled plant cell wall gels in DMSO-d6/pyridine-d5. *Org Biomol Chem.* 2010;**8**(3):576–91.
5 Guerra A, Filpponen I, Lucia LA, Argyropoulos DS. Comparative evaluation of three lignin isolation protocols for various wood species. *J Agric Food Chem.* 2006;**54**(26):9696–705.
6 Park S, Johnson DK, Ishizawa CI, Parilla PA, Davis MF. Measuring the crystallinity index of cellulose by solid state ^{13}C nuclear magnetic resonance. *Cellulose (Dordrecht, Neth).* 2009;**16**(4):641–7.
7 Peng Y, Gardner DJ, Han Y, Kiziltas A, Cai Z, Tshabalala MA. Influence of drying method on the material properties of nanocellulose I: thermostability and crystallinity. *Cellulose (Dordrecht, Neth).* 2013;**20**(5):2379–92.
8 Jiang F, Hsieh Y-L. Chemically and mechanically isolated nanocellulose and their self-assembled structures. *Carbohydr Polym.* 2013;**95**(1):32–40.
9 Jiang F, Hsieh Y-L. Assembling and redispersibility of rice straw nanocellulose: effect of tert-butanol. *ACS Appl Mater Interfaces.* 2014;**6**(22):20075–84.
10 McCormick CL, Callais PA, Hutchinson Jr, BH. Solution studies of cellulose in lithium chloride and N,N-dimethylacetamide, *Macromolecules* 1985;**18**(12):2394–401.
11 Determination of ash in biomass. Laboratory Analytical Procedure (LAP). [Internet]. 2008.
12 Liu X, Liu Z, Fei B, Cai Z, Jiang Z. Comparative properties of bamboo and rice straw pellets. *BioResources.* 2012;**8**(1):638–47.
13 Determination of structural carbohydrates and lignin in biomass. Laboratory Analytical Procedure (LAP). Available: http://www.nrel.gov/biomass/pdfs/42618.pdf. Accessed 2013 July 7. [Internet]. 2011.
14 Determination of extractives in biomass. http://www.nrel.gov/biomass/pdfs/42619.pdf [Internet]. 2005.
15 Park S, Baker JO, Himmel ME, Parilla PA, Johnson DK. Cellulose crystallinity index: measurement techniques and their impact on interpreting cellulase performance. *Biotechnol Biofuels.* 2010;**3**:10.
16 Park S, Johnson D, Ishizawa C, Parilla P, Davis M. Measuring the crystallinity index of cellulose by solid state 13 C nuclear magnetic resonance. *Cellulose.* 2009;**16**(4):641–7.
17 Sathitsuksanoh N, Zhu Z, Wi S, Zhang Y-HP. Cellulose solvent based biomass pretreatment breaks highly ordered hydrogen bonds in cellulose fibers of switchgrass. *Biotechnol Bioeng.* 2011;**108**(3):521–9.
18 Hurtubise FG, Krässig H. Classification of fine structural characteristics in cellulose by infrared spectroscopy. Use of potassium bromide pellet technique. *Anal Chem.* 1960;**32**(2):177–81.
19 Schultz TP, McGinnis GD, Bertran MS. Estimation of cellulose crystallinity using Fourier transform-infrared spectroscopy and dynamic thermogravimetry. *J Wood Chem Technol.* 1985;**5**(4):543–57.
20 Sun L, Varanasi P, Yang F, Loqué D, Simmons BA, Singh S. Rapid determination of syringyl: guaiacyl ratios using FT-Raman spectroscopy. *Biotechnol Bioeng.* 2012;**109**(3):647–56.

21 Papa G, Varanasi P, Sun L, Cheng G, Stavila V, Holmes B, et al. Exploring the effect of different plant lignin content and composition on ionic liquid pretreatment efficiency and enzymatic saccharification of *Eucalyptus globulus* L. mutants. *Biores Technol.* 2012;**117**:352–9.

22 Lupoi JS, Singh S, Davis M, Lee DJ, Shepherd M, Simmons BA, et al. High-throughput prediction of eucalypt lignin syringyl/guaiacyl content using multivariate analysis: a comparison between mid-infrared, near-infrared, and Raman spectroscopies for model development. *Biotechnol Biofuels.* 2014;**7**(1):93.

23 Rencoret J, Gutiérrez A, Nieto L, Jiménez-Barbero J, Faulds CB, Kim H, et al. Lignin composition and structure in young versus adult *Eucalyptus globulus* plants. *Plant Physiol.* 2011;**115**(2):667–82.

24 Del Rio JC, Rencoret J, Marques G, Li J, Gellerstedt G, Jiménez-Barbero J, et al. Structural characterization of the lignin from jute (*Corchorus capsularis*) fibers. *J Agric Food Chem.* 2009;**57**(21):10271–81.

25 Del Río JC, Marques G, Rencoret J, Martínez ÁT, Gutiérrez A. Occurrence of naturally acetylated lignin units. *J Agric Food Chem.* 2007;**55**(14):5461–8.

26 Del Rio JC, Gutierrez A, Martinez AT. Identifying acetylated lignin units in non-wood fibers using pyrolysis-gas chromatography/mass spectrometry. *Rapid Commun Mass Spectrom.* 2004;**18**(11):1181–5.

27 Cheng K, Sorek H, Zimmermann H, Wemmer DE, Pauly M. Solution-state 2D NMR spectroscopy of plant cell walls enabled by a dimethylsulfoxide-d 6/1-ethyl-3-methylimidazolium acetate solvent. *Anal Chem.* 2013;**85**(6):3213–21.

28 Eudes A, Sathitsuksanoh N, Baidoo EK, George A, Liang Y, Yang F, et al. Expression of a bacterial 3-dehydroshikimate dehydratase reduces lignin content and improves biomass saccharification efficiency. *Plant Biotechnol J.* 2015;**13**(9):1241–50.

29 Eudes A, Zhao N, Sathitsuksanoh N, Baidoo EEK, Lao J, Wang G, et al. Expression of S-adenosylmethionine hydrolase in tissues synthesizing secondary cell walls alters specific methylated cell wall fractions and improves biomass digestibility. *Front Bioeng Biotechnol.* 2016;**4**:doi: 10.3389/fbioe.2016.00058.

30 Wen J-L, Sun S-L, Xue B-L, Sun R-C. Quantitative structural characterization of the lignins from the stem and pith of bamboo (*Phyllostachys pubescens*). *Holzforschung.* 2013;**67**(6):613–27.

31 Martínez ÁT, Rencoret J, Marques G, Gutiérrez A, Ibarra D, Jiménez-Barbero J, et al. Monolignol acylation and lignin structure in some nonwoody plants: a 2D NMR study. *Phytochemistry.* 2008;**69**(16):2831–43.

32 Landucci LL. Quantitative ^{13}C NMR characterization of lignin 1. A methodology for high precision. *Holzforschung.* 1985;**39**(6):355–60.

33 Evtuguin DV, Neto CP, Silva AMS, Domingues PM, Amado FML, Robert D, et al. Comprehensive study on the chemical structure of dioxane lignin from plantation *Eucalyptus globulus* wood. *J Agric Food Chem.* 2001;**49**(9):4252–61.

34 Faix O, Argyropoulos DS, Robert D, Neirinck V. Determination of hydroxyl groups in lignins evaluation of ^{1}H-, ^{13}C-, ^{31}P-NMR, FTIR and wet chemical methods. *Holzforschung.* 1994;**48**(5):387–94.

35 Faix O, Bottcher JH. Determination of phenolic hydroxyl group contents in milled wood lignins by FTIR spectroscopy applying partial least-squares (PLS) and principal components regression (PCR). *Holzforschung.* 1993;**47**(1):45–9.

36 El Mansouri N-E, Salvadó J. Analytical methods for determining functional groups in various technical lignins. *Ind Crops Prod*. 2007;**26**(2):116–24.
37 Mansouri N-EE, Salvadó J. Structural characterization of technical lignins for the production of adhesives: application to lignosulfonate, kraft, soda-anthraquinone, organosolv and ethanol process lignins. *Ind Crops Prod*. 2006;**24**(1):8–16.
38 Kubo S, Kadla JF. Hydrogen bonding in lignin: a Fourier transform infrared model compound study. *Biomacromolecules*. 2005;**6**(5):2815–21.
39 Robert DR, Brunow G. Quantitative estimation of hydroxyl groups in milled wood lignin from spruce and in a dehydrogenation polymer from coniferyl alcohol using ^{13}C NMR spectroscopy. *Holzforschung*. 1984;**38**(2):85–90.
40 Argyropoulos DS, Bolker HI, Heitner C, Archipov Y. ^{31}P NMR spectroscopy in wood chemistry part V. Qualitative analysis of lignin functional groups. *J Wood Chem Technol*. 1993;**13**(2):187–212.
41 El Hage R, Brosse N, Chrusciel L, Sanchez C, Sannigrahi P, Ragauskas A. Characterization of milled wood lignin and ethanol organosolv lignin from miscanthus. *Polym Degrad Stab*. 2009;**94**(10):1632–8.
42 Crestini C, Argyropoulos DS. Structural analysis of wheat straw lignin by quantitative ^{31}P and 2D NMR spectroscopy. The occurrence of ester bonds and α-O-4 substructures. *J Agric Food Chem*. 1997;**45**(4):1212–9.
43 Argyropoulos DS. Quantitative phosphorus-31 NMR analysis of lignins, a new tool for the lignin chemist. *J Wood Chem Technol*. 1994;**14**(1):45–63.
44 Holtman KM, Chang HM, Jameel H, Kadla JF. Quantitative ^{13}C NMR characterization of milled wood lignins isolated by different milling techniques. *J Wood Chem Technol*. 2006;**26**(1):21–34.
45 Kim H, Ralph J, Akiyama T. Solution-state 2D NMR of ball-milled plant cell wall gels in DMSO-d6. *BioEnerg Res*. 2008;**1**(1):56–66.
46 Sathitsuksanoh N, Holtman KM, Yelle DJ, Morgan T, Stavila V, Pelton J, et al. Lignin fate and characterization during ionic liquid biomass pretreatment for renewable chemicals and fuels production. *Green Chem*. 2014;**16**(3):1236–47.
47 Yelle DJ, Kaparaju P, Hunt CG, Hirth K, Kim H, Ralph J, et al. Two-dimensional NMR evidence for cleavage of lignin and Xylan substituents in wheat straw through hydrothermal pretreatment and enzymatic hydrolysis. *Bioenergy Res*. 2013;**6**(1):211–21.
48 Yelle DJ, Ralph J, Frihart CR. Characterization of nonderivatized plant cell walls using high-resolution solution-state NMR spectroscopy. *Magn Reson Chem*. 2008;**46**(6):508–17.
49 Glasser WG, Dave V, Frazier CE. Molecular weight distribution of (semi-) commercial lignin derivatives. *J Wood Chem Technol*. 1993;**13**(4):545–59.
50 Colombini MP, Orlandi M, Modugno F, Tolppa E-L, Sardelli M, Zoia L, et al. Archaeological wood characterisation by PY/GC/MS, GC/MS, NMR and GPC techniques. *Microchem J* 2007;**85**(1):164–73.
51 Tejado A, Pena C, Labidi J, Echeverria JM, Mondragon I. Physico-chemical characterization of lignins from different sources for use in phenol–formaldehyde resin synthesis. *Biores Technol*. 2007;**98**(8):1655–63.
52 Sathitsuksanoh N, Sawant M, Truong Q, Tan J, Canlas CG, Sun N, et al. How alkyl chain length of alcohols affects lignin fractionation and ionic liquid recycle during lignocellulose pretreatment. *BioEnerg Res*. 2015;**8**(3):973–81.

53 Holtman KM, Chang HM, Kadla JF. An NMR comparison of the whole lignin from milled wood, MWL, and REL dissolved by the DMSO/NMI procedure. *J Wood Chem Technol.* 2007;**27**(3–4):179–200.

54 Hatcher PG. Dipolar-dephasing ^{13}C NMR studies of decomposed wood and coalified xylem tissue: evidence for chemical structural changes associated with defunctionalization of lignin structural units during coalification. *Energy Fuels.* 1988;**2**(1):48–58.

55 Park J, Meng J, Lim KH, Rojas OJ, Park S. Transformation of lignocellulosic biomass during torrefaction. *J Anal Appl Pyrolysis* 2013;**100**:199–206.

56 Holtman KM, Chang H-M, Kadla JF. Solution-state nuclear magnetic resonance study of the similarities between milled wood lignin and cellulolytic enzyme lignin. *J Agric Food Chem.* 2004;**52**(4):720–6.

57 Granata A, Argyropoulos DS. 2-Chloro-4, 4, 5, 5-tetramethyl-1, 3, 2-dioxaphospholane, a reagent for the accurate determination of the uncondensed and condensed phenolic moieties in lignins. *J Agric Food Chem.* 1995;**43**(6):1538–44.

58 Ding S-Y, Himmel ME. The maize primary cell wall microfibril: a new model derived from direct visualization. *J Agric Food Chem.* 2006;**54**(3):597–606.

59 Reis D, Vian B. Helicoidal pattern in secondary cell walls and possible role of xylans in their construction. *Comptes Rendus Biologies.* 2004;**327**(9–10):785–90.

60 Balakshin M, Capanema E, Gracz H, Chang H-M, Jameel H. Quantification of lignin–carbohydrate linkages with high-resolution NMR spectroscopy. *Planta.* 2011;**233**(6):1097–110.

61 Determination of protein content in biomass. Laboratory Analytical Procedure (LAP) [Internet]. 2008.

62 Mosse J. Nitrogen-to-protein conversion factor for ten cereals and six legumes or oilseeds. A reappraisal of its definition and determination. Variation according to species and to seed protein content. *J Agric Food Chem.* 1990;**38**(1):18–24.

63 Fengel D, Wegener G. *Wood: chemistry, ultrastructure, reactions.* Walter de Gruyter, Berlin; 1983.

64 Lupoi JS, Singh S, Simmons BA, Henry RJ. Assessment of lignocellulosic biomass using analytical spectroscopy: an evolution to high-throughput techniques. *Bioenerg Res.* 2014;**7**(1):1–23.

65 Lupoi JS, Singh S, Parthasarathi R, Simmons BA, Henry RJ. Recent innovations in analytical methods for the qualitative and quantitative assessment of lignin. *Renew Sustainable Energy Rev.* 2015;**49**:871–906.

66 Teeäär R, Serimaa R, Paakkarl T. Crystallinity of cellulose, as determined by CP/MAS NMR and XRD methods. *Polym Bull.* 1987;**17**(3):231–7.

67 Newman R. Estimation of the lateral dimensions of cellulose crystallites using ^{13}C NMR signal strengths. *Solid State Nucl Magn Reson.* 1999;**15**(1):21–9.

68 Oh S, Yoo D, Shin Y, Seo G. FTIR analysis of cellulose treated with sodium hydroxide and carbon dioxide. *Carbohydr Res.* 2005;**340**(3):417–28.

69 Nelson M, O'Connor R. Relation of certain infrared bands to cellulose crystallinity and crystal latticed type. Part I. Spectra of lattice types I, II, III and of amorphous cellulose. *J Appl Polym Sci.* 1964;**8**(3):1311–24.

70 Nelson M, O'Connor R. Relation of certain infrared bands to cellulose crystallinity and crystal lattice type. Part II. A new infrared ratio for estimation of crystallinity in celluloses I and II. *J Appl Polym Sci.* 1964;**8**(3):1325–41.

71 Segal L, Creely J, Martin Jr, A, Conrad C. An empirical method for estimating the degree of crystallinity of native cellulose using the X-ray diffractometer. *Text Res J.* 1959;**29**(10):786.

72 Ruland W. X-ray determination of crystallinity and diffuse disorder scattering. *Acta Crystallogr.* 1961;**14**(11):1180–5.

73 You C, Chen H, Myung S, Sathitsuksanoh N, Ma H, Zhang XZ, et al. Enzymatic transformation of nonfood biomass to starch. *PNAS.* 2013;**110**(18):7182–7.

74 Wickholm K, Larsson PT, Iversen T. Assignment of non-crystalline forms in cellulose I by CP/MAS ^{13}C NMR spectroscopy. *Carbohydr Res.* 1998;**312**(3):123–9.

75 Silverstein RM, Webster FX, Kiemle D, Bryce DL. *Spectrometric identification of organic compounds*: John Wiley & Sons; 2014.

76 Åkerholm M, Hinterstoisser B, Salmén L. Characterization of the crystalline structure of cellulose using static and dynamic FT-IR spectroscopy. *Carbohydr Res.* 2004;**339**(3):569–78.

77 Kataoka Y, Kondo T. FT-IR microscopic analysis of changing cellulose crystalline structure during wood cell wall formation. *Macromolecules.* 1998;**31**(3):760–4.

78 Nelson ML, O'Connor RT. Relation of certain infrared bands to cellulose crystallinity and crystal latticed type. Part I. Spectra of lattice types I, II, III and of amorphous cellulose. *J Appl Polym Sci.* 1964;**8**(3):1311–24.

79 Li Q, Renneckar S. Supramolecular structure characterization of molecularly thin cellulose I nanoparticles. *Biomacromolecules.* 2011;**12**(3):650–9.

80 Casarano R, Fidale LC, Lucheti CM, Heinze T, Seoud OAE. Expedient, accurate methods for the determination of the degree of substitution of cellulose carboxylic esters: application of UV–vis spectroscopy (dye solvatochromism) and FTIR. *Carbohydr Polym.* 2011;**83**(3):1285–92.

81 Lestander TA, Rhén C. Multivariate NIR spectroscopy models for moisture, ash and calorific content in biofuels using bi-orthogonal partial least squares regression. *Analyst.* 2005;**130**(8):1182–9.

82 Lindgren T, Edlund U, Iversen T. A multivariate characterization of crystal transformations of cellulose. *Cellulose.* 1995;**2**(4):273–88.

83 Kelley S, Rials T, Snell R, Groom L, Sluiter A. Use of near infrared spectroscopy to measure the chemical and mechanical properties of solid wood. *Wood Sci Technol.* 2004;**38**(4):257–76.

84 Eronen P, Österberg M, Jääskeläinen A-S. Effect of alkaline treatment on cellulose supramolecular structure studied with combined confocal Raman spectroscopy and atomic force microscopy. *Cellulose.* 2009;**16**(2):167–78.

85 Gierlinger N, Keplinger T, Harrington M. Imaging of plant cell walls by confocal Raman microscopy. *Nat Protoc.* 2012;**7**(9):1694–708.

86 Agarwal UP. Raman imaging to investigate ultrastructure and composition of plant cell walls: distribution of lignin and cellulose in black spruce wood (*Picea mariana*). *Planta.* 2006;**224**(5):1141–53.

87 Rogošić M, Mencer HJ, Gomzi Z. Polydispersity index and molecular weight distributions of polymers. *Eur Polym J.* 1996;**32**(11):1337–44.

88 Grubisic Z, Rempp P, Benoit H. A universal calibration for gel permeation chromatography. *J Polym Sci B: Polym Lett.* 1967;**5**(9):753–9.

89 Ishii D, Isogai A. The residual amide content of cellulose sequentially solvent-exchanged and then vacuum-dried. *Cellulose (Dordrecht, Neth).* 2008;**15**(4):547–53.

90 Evans R, Wearne RH, Wallis AF. Molecular weight distribution of cellulose as its tricarbanilate by high performance size exclusion chromatography. *J Appl Polym Sci.* 1989;**37**(12):3291–303.

91 Hiller LA, Jr., Pacsu E. Cellulose studies. V. Reducing end-group estimation. A new method using potassium permanganate. *Text Res J.* 1946;**16**:318–23.

92 Zhang YHP, Lynd LR. Determination of the number-average degree of polymerization of cellodextrins and cellulose with application to enzymatic hydrolysis. *Biomacromolecules.* 2005;**6**(3):1510–5.

93 Kim Y-T, Kim E-H, Cheong C, Williams DL, Kim C-W, Lim S-T. Structural characterization of β-D-(1 → 3, 1 → 6)-linked glucans using NMR spectroscopy. *Carbohydr Res.* 2000;**328**(3):331–41.

94 Johansson M, Samuelson O. Reducing end groups in brich xylan and their alkaline degradation. *Wood Sci Technol.* 1977;**11**(4):251–63.

95 Brown DM, Goubet F, Wong VW, Goodacre R, Stephens E, Dupree P, et al. Comparison of five xylan synthesis mutants reveals new insight into the mechanisms of xylan synthesis. *Plant J.* 2007;**52**(6):1154–68.

96 Peña MJ, Zhong R, Zhou G-K, Richardson EA, O'Neill MA, Darvill AG, et al. Arabidopsis irregular xylem8 and irregular xylem9: implications for the complexity of glucuronoxylan biosynthesis. *Plant Cell.* 2007;**19**(2):549–63.

97 Hatch RS. Cupriethylene diamine as solvent for precise determination of cellulose viscosity. *Ind Eng Chem Anal Ed* 1944;**16**(2):104–7.

98 Gellerstedt G. Gel permeation chromatography. In: Lin SY, Dence CW, editors. *Methods in lignin chemistry*: Springer; 1992. p. 487–97.

99 Chum HL, Johnson DK, Tucker MP, Himmel ME. Some aspects of lignin characterization by high performance size exclusion chromatography using styrene divinylbenzene copolymer gels. *Holzforschung.* 1987;**41**(2):97–108.

100 Majcherczyk A, Hüttermann A. Size-exclusion chromatography of lignin as ion-pair complex. *J Chromatogr A.* 1997;**764**(2):183–91.

101 Contreras S, Gaspar AR, Guerra A, Lucia LA, Argyropoulos DS. Propensity of lignin to associate: light scattering photometry study with native lignins. *Biomacromolecules.* 2008;**9**(12):3362–9.

102 Guerra A, Gaspar AR, Contreras S, Lucia LA, Crestini C, Argyropoulos DS. On the propensity of lignin to associate: a size exclusion chromatography study with lignin derivatives isolated from different plant species. *Phytochemistry.* 2007;**68**(20):2570–83.

103 Glasser WG, Dave V, Frazier CE. Molecular weight distribution of (semi-) commercial lignin derivatives. *J Wood Chem Technol.* 1993;**13**(4):545–59.

104 Robert D. Carbon-13 nuclear magnetic resonance spectrometry. In: Lin SY, Dence CW, editors. *Methods in lignin chemistry*: Springer; 1992. p. 250–73.

105 Björkman A. Studies on finely divided wood. Part 1. Extraction of lignin with neutral solvents. *Svensk Papperstidning – Nordisk Cellulosa.* 1956;**59**(13):477–85.

106 Argyropoulos DS. Quantitative phosphorus-31 NMR analysis of lignins, a new tool for the lignin chemist. *J Wood Chem Technol*. 1994;**14**(1):45–63.
107 Pu Y, Cao S, Ragauskas AJ. Application of quantitative ^{31}P NMR in biomass lignin and biofuel precursors characterization. *Energy Environ Sci*. 2011;**4**(9):3154–66.
108 Zhang W, Sathitsuksanoh N, Simmons BA, Frazier CE, Barone JR, Renneckar S. Revealing the thermal sensitivity of lignin during glycerol thermal processing through structural analysis. *RSC Adv*. 2016;**6**(36):30234–46.
109 Kim H, Ralph J. A gel-state 2D-NMR method for plant cell wall profiling and analysis: a model study with the amorphous cellulose and xylan from ball-milled cotton linters. *RSC Adv*. 2014;**4**(15):7549–60.
110 Rencoret J, Marques G, Gutiérrez A, Nieto L, Santos JI, Jiménez-Barbero J, et al. HSQC-NMR analysis of lignin in woody (*Eucalyptus globulus* and *Picea abies*) and non-woody (*Agave sisalana*) ball-milled plant materials at the gel state. *Holzforschung*. 2009;**63**(6):691–8.
111 Sette M, Lange H, Crestini C. Quantitative HSQC analyses of lignin: a practical comparison. *Comput Struct Biotechnol J*. 2013;**6**(7):1–7.
112 Brandt A, Chen L, van Dongen BE, Welton T, Hallett JP. Structural changes in lignins isolated using an acidic ionic liquid water mixture. *Green Chem*. 2015;**17**(11):5019–34.
113 Sette M, Wechselberger R, Crestini C. Elucidation of lignin structure by quantitative 2D NMR. *Chem Eur J*. 2011;**17**(34):9529–35.
114 Dolan JA, Sathitsuksanoh N, Rodriguez K, Simmons BA, Frazier CE, Renneckar S. Biocomposite adhesion without added resin: understanding the chemistry of the direct conversion of wood into adhesives. *RSC Adv*. 2015;**5**(82):67267–76.
115 Stewart JJ, Akiyama T, Chapple C, Ralph J, Mansfield SD. The effects on lignin structure of overexpression of ferulate 5-hydroxylase in hybrid poplar1. *Plant Physiol*. 2009;**150**(2):621–35.
116 Gille S, de Souza A, Xiong G, Benz M, Cheng K, Schultink A, et al. O-acetylation of Arabidopsis hemicellulose xyloglucan requires AXY4 or AXY4L, proteins with a TBL and DUF231 domain. *Plant Cell*. 2011;**23**(11):4041–53.
117 Yelle DJ, Wei D, Ralph J, Hammel KE. Multidimensional NMR analysis reveals truncated lignin structures in wood decayed by the brown rot basidiomycete *Postia placenta*. *Environ Microbiol*. 2011;**13**(4):1091–100.
118 Bauer S, Sorek H, Mitchell VD, Ibáñez AB, Wemmer DE. Characterization of *Miscanthus giganteus* lignin isolated by ethanol organosolv process under reflux condition. *J Agric Food Chem*. 2012;**60**(33):8203–12.
119 Eichorst SA, Joshua C, Sathitsuksanoh N, Singh S, Simmons BA, Singer SW. Substrate-specific development of thermophilic bacterial consortia by using chemically pretreated switchgrass. *Appl Environ Microbiol*. 2014;**80**(23):7423–32.
120 Yelle DJ, Ralph J, Lu F, Hammel KE. Evidence for cleavage of lignin by a brown rot basidiomycete. *Environ Microbiol*. 2008;**10**(7):1844–9.
121 Shuai L, Yang Q, Zhu J, Lu F, Weimer P, Ralph J, et al. Comparative study of SPORL and dilute-acid pretreatments of spruce for cellulosic ethanol production. *Biores Tech* 2010;**101**:3106–14.
122 Van den Bosch S, Schutyser W, Vanholme R, Driessen T, Koelewijn S-F, Renders T, et al. Reductive lignocellulose fractionation into soluble lignin-derived phenolic

monomers and dimers and processable carbohydrate pulps. *Energy Environ Sci.* 2015;**8**(6):1748–63.

123 Liu Z, Padmanabhan S, Cheng K, Xie H, Gokhale A, Afzal W, et al. Two-step delignification of miscanthus to enhance enzymatic hydrolysis: aqueous ammonia followed by sodium hydroxide and oxidants. *Energy Fuels.* 2014;**28**(1):542–8.

124 Li S, Lundquist K. A new method for the analysis of phenolic groups in lignins by ^{1}H NMR spectroscopy. *Nord Pulp Pap Res J.* 1994;**9**(3):191–5.

125 Chen C-L. Determination of methoxyl groups. In: Lin SY, Dence CW, editors. Methods in lignin chemistry: Springer; 1992. p. 465–72.

126 Balogh D, Curvelo A, De Groote R. Solvent effects on organosolv lignin from *Pinus caribaea* hondurensis. *Holzforschung.* 1992;**46**(4):343–8.

127 Vazquez G, Antorrena G, Gonzalez J, Freire S. FTIR, ^{1}H and ^{13}C NMR characterization of acetosolv-solubilized pine and eucalyptus lignins. *Holzforschung.* 1997;**51**(2):158–66.

128 Ligero P, Villaverde JJ, de Vega A, Bao M. Delignification of *Eucalyptus globulus* saplings in two organosolv systems (formic and acetic acid): preliminary analysis of dissolved lignins. *Ind Crop Prod.* 2008;**27**(1):110–7.

5

Introduction to Life-Cycle Assessment and Decision Making Applied to Forest Biomaterials

Jesse Daystar and Richard Venditti

North Carolina State University, Department of Forest Biomaterials, Raleigh, NC, USA

> *All models are wrong; some are useful*
>
> Albert Einstein

5.1 Introduction

5.1.1 What is LCA?

Demands on the earth's resources are escalating with increasing population and standards of living. These new demands are causing the increased extraction of raw materials harvested from the environment and emissions and wastes that are ultimately introduced into the environment. It is mankind's responsibility to actively search for the best solutions to meet the society's needs and to be the best steward of the planet.

For every demand that society has, there are an infinite set of possible solutions including different products, behaviors, and services to meet the demand. Life-cycle assessment (LCA) is a comprehensive life-cycle approach that quantifies ecological and human health impacts of a product or system over its complete life cycle. LCA uses credible scientific methods to model steady-state, global environmental and human health impacts. It can provide quantitative measures of multiple environmental impacts that can help us to choose more sustainable pathways. LCA helps decision makers understand the scale and trade-offs of many environmental and human health impacts for competing products, services, policies, or actions.

There are many definitions of sustainability but the major idea is safeguarding our natural resources today so that future generations can continue to use the same resources indefinitely. The University of California, Los Angeles, sustainability group states that

> the physical development and institutional operating practices that meet the needs of present users without compromising the ability of future generations to meet their own needs, particularly with regard to use and waste of natural resources. Sustainable practices support ecological, human, and economic health and vitality. Sustainability presumes that resources are finite, and should be used

conservatively and wisely with a view to long-term priorities and consequences of the ways in which resources are used.

UCLA, 2016

5.1.1.1 History

The process of evaluating a product or service with a holistic view of all the life cycle stages evolved from activities in the mid-1900s with respect to purchasing (Novick, 1959). In that report, not only the cost to purchase weapon systems was considered in total cost, but other life stages including use, development, and end of life provide the true cost over the entire lifetime. This kind of life cycle thinking for cost is now very prevalent. Most people have heard about the "cost of ownership" of automobiles, which includes purchase price in addition to other costs like fuel, maintenance, and trade-in value. The principle of life-cycle analysis in the 1970s was a starting point for life-cycle analysis applied to other issues such as environmental impacts of products. Dissatisfaction with some of the early LCA studies included ambiguous methods and results that had questionable motivations. LCAs performed with poor methods and hidden motivations unfortunately discredited LCAs for some time. The issues surrounding methods and motivations needed to be addressed to enable creditable studies to inform product designers and decisions makers. This motivated the standardization of LCA methods using a predefined framework and detailed methodological guidance. The International Organization of Standards created two guidance documents for performing LCAs that are defined in standards ISO 14040 (2006) and ISO 14044 (2006). More recently, an additional ISO method was developed to provide guidance on accounting for water use and consumption in LCAs – ISO 14046 (2014).

Triple Bottom Line Triple bottom line refers to the evaluation of environmental, economic, and social considerations for a product or service (Figure 5.1). Product or service need to consider and address all three pillars of sustainability to be considered sustainable from the perspective of triple bottom line. Environmental LCA contributes to the triple bottom line reporting by quantifying the ecological and human health performance of competing products and services. Of the three measures, the economic evaluation is the most certain and objective. Environmental and health LCA analyses are often less certain and less objective, with the social analysis the most difficult to produce objective, clear, and certain results.

5.1.2 LCA for Decision Making

As stated earlier, the purpose of environmental LCAs is to quantify the inputs and outputs to the environment resulting from a product or service and how these flows impact the environment. This information is used to support decisions made by consumers, industry, and policy makers that will hopefully result in reduced environmental impacts. Everyone makes these decisions, ranging from the leaders of government, to businesses, to individual consumers. In particular, the impact of an LCA can be immense when used by entities that have great influence on products or service provided. These significant influencers can include business product managers and planners, company procurement and purchasing agents, industrial sector consortia,

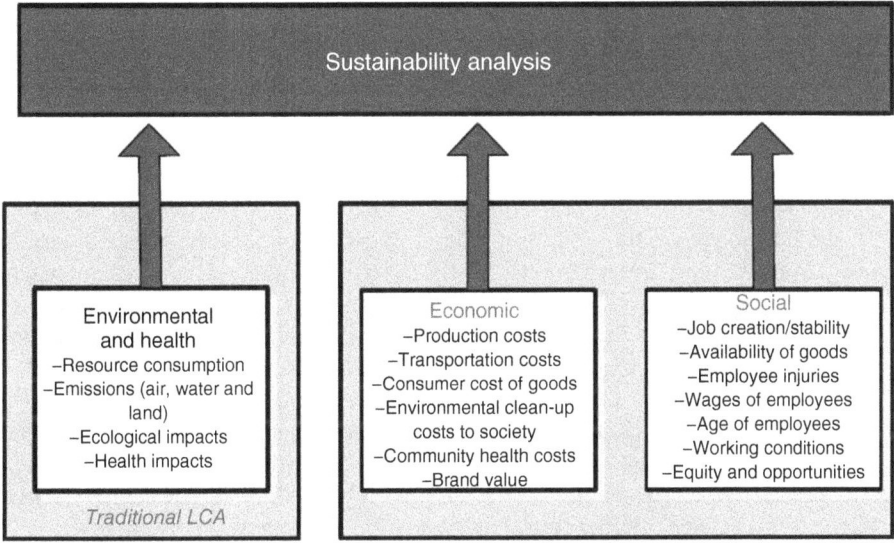

Figure 5.1 The three pillars of the triple-bottom-line perspective of sustainability.

and regional or national policy makers. However, consumer and customers, because of their large numbers, can use LCAs as a total group of independent decision makers to influence product and services consumed.

There is a wide range of types of decisions that can be influenced by LCAs. Strategic planning and capital investments in things such as green buildings, waste management infrastructure, and transportation infrastructure can have large effects over long periods of time. Product development and the eco-design of new products can be shaped by LCAs. How businesses, governments, and other institutions operate and manage activities can be influenced by LCA studies. Purchasing of products and services can be influenced by eco-labeling and clear communication of LCA-based environmental performance. Collectively, all of these decisions together can have a great and positive influence. However, the quality and benefits of these decision hinges on the quality of the LCAs that are considered.

To be a useful tool for decision making, environmental LCAs need to be effective in several dimensions. Primarily, the environmental LCA should be able to communicate the environmental performance of products and services. It should be able to identify critical operations, flows, or life-cycle stages within a product or service that are critical in the overall environmental impact and have high potential for improvement of the overall system. When alternates exist, the LCA should also be able to allow for the understanding of differences between products or services. An effective LCA may also provide other benefits, including minimizing production and regulatory costs. Overall, the LCA should assist in making decisions that minimize environmental impacts, resource depletion, and human health damages.

5.1.2.1 Eco-labels

In general, consumers collectively have great influence on products and services that are prevalent. It is logical to think that different people from different countries, cultures, religions, and economic status would have different opinions and preferences. These groups would have different opinions on what is important with respect to the environment. In the end, we all share the planet and collectively should care for it. To the average consumer, products and service environmental performance is performed mainly through environmental labels. An eco-label is a statement that discloses the environmental performance of products based on an environmental LCA. These are important in that similar to a nutrition label, an eco-label is how a consumer may evaluate alternative products. These eco-labels can be powerful tools in gaining larger shares of a market. With data-based, fair, and objective eco-labels, consumers can make better choices.

Eco-labels are classified into three types, each with different levels of rigorous LCA backing. Type 2 eco-labels are environmental self-declaration claims, usually focusing on a single claim. Examples might be the labeling of a product as natural, biodegradable, or recyclable. These often are not independently verified and so there is a risk of "green-washing," the deliberate misrepresentation of an environmental performance. These labels should be looked at with a critical eye. Type 1 eco-labels are third-party-certified multicriteria environmental labeling that ensure that a set of predetermined requirements are met. Within a product category, this can allow a consumer to understand which products are preferable. Examples of this are the Energy Star program that promotes energy efficiency and the Forestry Stewardship Council that promotes sustainable management of forests.

The most rigorous eco-label is type 3, in which quantified environmental information on the life cycle of a product is performed using standard LCA procedures so that fair and objective comparison of products can be made. For a type 3 eco label or environmental product declaration, a program operator must define LCA protocols for a product category that is reviewed and commented on by interested stakeholders. With this consistent methodology and reporting, competing entities with products can perform LCAs on their products using the defined methods enabling comparable results that can ultimately provide consumers with creditable information to make purchasing decisions.

5.2 LCA Components Overview

LCA is a standardized procedure used to determine the environmental impacts of products, services, or goods. The standardized procedure can be described by four-part framework as outlined by the ISO 14044, which includes

1. goal and scope definition
2. life-cycle inventory
3. life-cycle impact assessment
4. interpretation.

This integrated framework was inspired by earlier forms of life cycle thinking originating in life cycle financial analysis. Examining a product from origination of materials, to use, and disposal provides more holistic analysis of systems that can identify where environmental impacts originate and guide efforts in reducing these impacts.

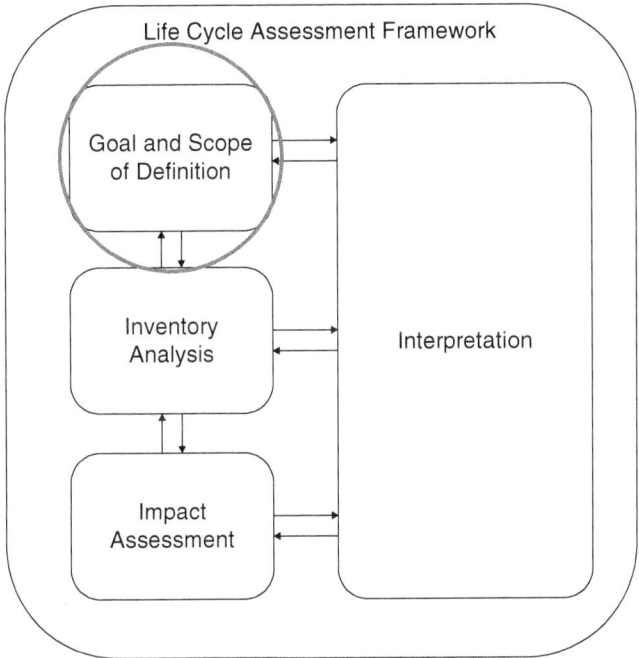

The ISO standards provide guidance on the structure framework and reuse requirements of data, study assumptions, and methods. With more consistent LCA methodologies, studies can be more comparable and of more scientific rigor. A standardized method helps LCA practitioners manage complex data sets consistently, enable comparisons between different products, and allow benchmarking. Without a standardized method, the results of LCA studies would be even more variable depending on study assumptions and methods. The ISO standards help reduce the influence of practitioner influence on study results.

A brief description of the four steps is provided here before presenting an in-depth description of each process in the following section.

5.2.1 Goal and Scope Definition

The assumptions surrounding an LCA study can heavily influence the analysis of results and conclusion. There are many different types of studies requiring different levels of data collection and analysis. The goal and scope of a LCA defines the purpose, audience, and intended use of the study. The intended use guides the further decisions surrounding the scope, functional unit of comparison, and data collection methods. For instance, if an LCA study is to be used internally within a company, a full review panel of LCA experts is not required; however, when making publically facing environmental claims about a competing product, this review is required.

5.2.2 Inventory Analysis

The life-cycle inventory (LCI) represents the most laborious step of an LCA where data are collected and organized for further analysis. This step often involves contacting companies, literature review, and building models in LCA software. Material flows in and out of processes, types of materials, product life time, and product energy requirements are examples of data typically collected in the LCI phase.

5.2.3 Life-Cycle Impact Assessment

The life-cycle impact assessment (LCIA) step of the analysis process takes LCI data and computes values that represent some form of environmental impacts. This process simplifies the data set from hundreds of flows into 10 or less impact categories that can then be used for decision making. There are many different methods for LCIA based on location, goal, and scope of the study.

5.2.4 Interpretation

The interpretation step of LCA reflects on what was found in the other steps to create new knowledge. It should be noted that the interpretation step is not the last step; rather it is continually done throughout each process. When this is done in each stage, study assumptions, goals and scopes, and methods are often refined to create to better suit the needs of the study commissioner.

5.3 Life-Cycle Assessment Steps

5.3.1 Goal, Scope, System Boundaries

5.3.1.1 Goal Definition

The first step of an LCA is defining the goal of the study. In this part of the LCA, the aim of the study and breadth and depth of the study are communicated. There are two types of LCA purposes: descriptive and change-oriented. A descriptive LCA generally looks at broader aspects of an issues, for example, how much of the world's global warming impact can be attributed to transportation. These larger environmental questions are answered by descriptive LCAs. The second purpose of an LCA is a change-oriented LCA where two decision options for fulfilling a function are compared. Some typical examples of change-oriented LCAs are paper versus plastic bags, and flying versus driving your car. These types of studies can guide the audience in ways to reduce environmental impacts through changing behaviors based on the findings of the LCA.

The intended audience is another aspect of the goal and scope, which is important to communicate. The audience can include interest groups such as policy makers, company marketing groups, or product development teams. Additionally, the involved interest groups and parties should be identified. These include companies, funding sources, target audiences, and expert reviewers. It is noted that the intended use of the LCA can be different from the end use as the information may be relevant to other decisions and analysis beyond the original intent.

One specific type of LCA used to compare two different products is a "comparative assertion disclosed to the public." In this type of study, "environmental claim regarding the superiority or equivalence of one product vs a competing product which performs the same function" are communicated to the public. These types of studies must follow the ISO 14044 standards including the nine steps required for a "comparative assertion."

5.3.1.2 Scope Definition

The scope definition serves the purpose of communicating to the audience what is included and what is excluded from the study. Depending on the goal of the study, there are several types of scopes including cradle-to-gate, cradle-to-grave, and gate-to-gate. There are other words that are commonly used to describe these scopes but the ideas are similar.

- Cradle-to-grave includes all flows and impacts from raw material extraction to disposal and reuse.
- Cradle-to-gate includes all flows and impacts from raw material to production and excludes product use and end of life.
- Gate-to-gate only includes flows from production or material processing steps of a product life cycle.

The scope should be carefully selected considering the potential implications of not including product stages or phases in the scope of the work. For example, a product may have lower production emissions but has a shorter lifetime than an alternative product that would not be considered if a cradle-to-gate boundary was selected. The scope of the study is often best communicated in a process stage diagram as seen in Figures 5.2 and 5.3. These types of diagrams list the major unit steps that are considered within a study and clearly show what is not included in the study.

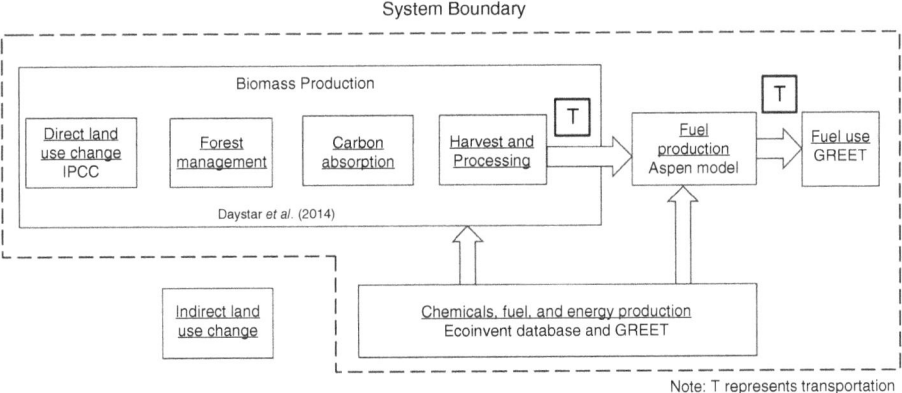

Figure 5.2 System boundary diagram of a cradle-to-grave bio-fuels process but excludes indirect land-use change.

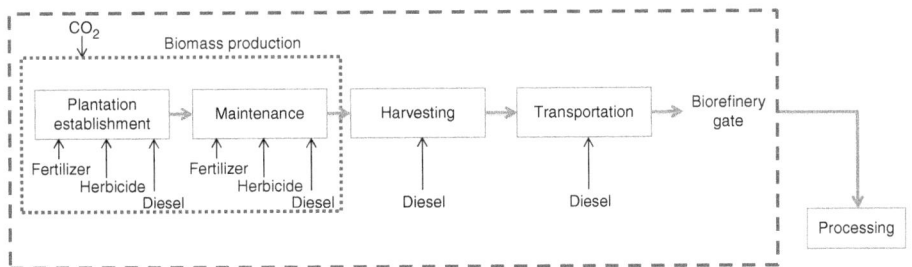

Figure 5.3 System boundary diagram of a cradle-to-gate biomass production system excluding processing.

Temporal boundaries are also set in the scope definition. Temporal assumptions, or assumptions relating to time, can have large influences on the results of a study. It is important to pick a study timeframe that will best capture the impacts of the product or processes being studied. Impacts occurring in 100 years is a common temporal boundary for global warming potential (GWP). In a 100-year temporal boundary, impacts occurring after 100 years would not be calculated and included in the results. There is an emerging field of dynamic LCA, which can more accurately model emissions through time for product systems lasting over many years (Daystar *et al.*, 2016; Levasseur *et al.*, 2010). This new method improves LCA temporal consistency that enables a better determination of the global impacts over a given time period.

Other aspects to be included in the scope are technology types and geographical regions. Many studies are spatially dependent, and the overall results and conclusions may not be broadly applied to other regions. Product manufacturing technology can also be important to the study results. Products or services from older technologies often have different impacts than the most current technology. For this reason, it is important to clearly communicate the type and stage of the technology under analysis. In addition to the aspect listed, allocation procedures impact assessment methods used should be reported in the scope of an LCA.

5.3.1.3 Functional Unit

A functional unit is the primary measure of the product, service, or good you are studying. ISO states "the functional unit defines the quantification of the identified functions (performance characteristics) of the product. The primary purpose of a functional unit is to provide a reference to which the inputs and outputs are related. This reference is necessary to ensure comparability of LCA results" (ISO 14044, 2006). This can be a service, mass of material, or an amount of energy. Selecting appropriate functional units is critical to creating an unbiased analysis. For example, comparing paper milk cartons to glass milk bottles may not be the best option due to the different possible sizes of each container. The real purpose of the container would be to deliver a quantity of fresh milk to a consumer. For this example, a better functional unit may be impacts of a container delivering 8 ounces of milk to a consumer. The results will then be normalized to a quantity of milk, which is what the consumer really wants not the container it comes from.

5.3.1.4 Cutoff Criteria

Data collection for an LCA is the most time-intensive and laborious step. To expedite this process, cutoff criteria are often used. A cutoff criteria defines a level of product content or other parameter which the study will not consider. For example, material contents less than 1% of the total product mass are often not considered in the LCA. This allows the LCA practitioner to focus on data from the main flows while systematically eliminating flows that may not influence the results. An LCA practitioner should perform cutoff decisions carefully as some materials can produce emissions and environmental impacts disproportionate to their component weight.

5.3.1.5 Problems Set – Goal and Scope Definition

Problem 1 Design a goal and scope for a life-cycle analysis that compares two heating methods to produce steam in a factory: a wood pellet burning furnace versus a conventional heating oil furnace.

Problem 2 The following lists the inputs and outputs in a wood beam processing operation – data from the US LCI database. Use *mass* cutoff criteria of 0.1% to determine which inputs would be included in an LCA. Use *mass* cut off criteria of 0.001% to determine which outputs would be included in an LCA.

Estimate that all liquids and gels have a density of $1.0\,\mathrm{g\,cm^{-3}}$ (a gross assumption), all solids have a density of $1.5\,\mathrm{g\,cm^{-3}}$, and all powders have a density of $0.5\,\mathrm{g\,cm^{-3}}$. Show all your work including unit conversions. Make a table with columns listing the amount of each input or output in the units presented, the converted quantity in mass in grams, the % mass of the entire product, and whether or not it is included.

Details for Glue laminated beam processing, at plant, US PNW				
Flow	Category	Type	Unit	Amount
Outputs				
2-Propanol	Air/unspecified	ELEMENTARY_FLOW	kg	4.34E-04
Co-products of glue laminated beam production, at plant, unspecified, US PNW	Wood product manufacturing/sawmills	PRODUCT_FLOW	kg	2.24E-01
Ethanol	Air/unspecified	ELEMENTARY_FLOW	kg	1.04E-05
Formaldehyde	Air/unspecified	ELEMENTARY_FLOW	kg	7.12E-07
Glue laminated beam processing, at plant, US PNW	Wood product manufacturing/sawmills	PRODUCT_FLOW	kg	1.00E + 00
Methanol	Air/unspecified	ELEMENTARY_FLOW	kg	2.15E-05
Particulates, unspecified	Air/unspecified	ELEMENTARY_FLOW	kg	1.28E-03
Phenol	Air/unspecified	ELEMENTARY_FLOW	kg	2.17E-05
Resorcinol	Air/unspecified	ELEMENTARY_FLOW	kg	2.07E-08
VOC, volatile organic compounds	Air/unspecified	ELEMENTARY_FLOW	kg	6.03E-04
Inputs				
CUTOFF Hogfuel-Biomass (50% MC). combusted in industrial boiler	Null/CUTOFF Flows	PRODUCT_FLOW	kg	1.75E-01
CUTOFF Kerosene, combusted in industrial boiler	Null/CUTOFF Flows	PRODUCT_FLOW	l	3.40E-07
Diesel, combusted in industrial boiler	Utilities/steam and air-conditioning supply	PRODUCT_FLOW	l	7.49E-04
Dry rough lumber, at kiln, US PNW	Wood product manufacturing/sawmills	PRODUCT_FLOW	kg	9.17E-01
Electricity, at grid, Western US, 2000	Utilities/electric power distribution	PRODUCT_FLOW	kWh	1.74E-01
Gasoline, combusted in equipment	Utilities/steam and air-conditioning supply	PRODUCT_FLOW	l	1.90E-04
Liquefied petroleum gas, combusted in industrial boiler	Utilities/steam and air-conditioning supply	PRODUCT_FLOW	l	3.34E-03
Melamine urea formaldehyde hardener, at plant	Chemical manufacturing	PRODUCT_FLOW	kg	1.82E-04

| Details for Glue laminated beam processing, at plant, US PNW |||||
Flow	Category	Type	Unit	Amount
Melamine urea formaldehyde resin, at plant	Chemical manufacturing	PRODUCT_FLOW	kg	1.64E-03
Natural gas, combusted in industrial boiler	Utilities/steam and air-conditioning supply	PRODUCT_FLOW	m3	8.23E-03
Rough green lumber, softwood, at sawmill, US PNW	Wood product manufacturing/sawmills	PRODUCT_FLOW	kg	3.08E-01
Transport, combination truck, average fuel mix	Truck transportation/general freight trucking	PRODUCT_FLOW	t*km	2.51E-01
Water, unspecified natural origin	Resource/in water	ELEMENTARY_FLOW	L	7.38E-02

Problem 3 Identify a reasonable functional unit for different wall constructions. Clearly define the specifications that the wall must meet to provide the necessary function. Then describe several wall constructions including one that is a log cabin wall, a normal wall with drywall and 2 by 4 wooden studs, and a concrete block wall. For the desired function, define for each of these cases the reference flow that would satisfy the functional unit.

5.3.2 Life-Cycle Inventory

Learning objectives

- Be able to describe the different types of flows
- Understand the concept of and draw unit processes
- Understand the concept of mass and elemental component balance
- Distinguish between primary and secondary data
- Be able to describe product allocation and system expansion
- Describe the difference between biogenic and anthropogenic carbon

LCI data describes the material flows into and out of a system or to and from the environment as a result of a product or service. These data alone provide useful information such as water use; however, it is not directly correlated to how the emissions or resource uses impact the environment. Some studies only determine the LCI and do not proceed further to the LCIA step, such as water footprint or energy analysis. The goal and scope of the study will define what parameters are tracked and calculated for the studied system. For example, greenhouse gas emissions and energy use are common metrics that are examined without other impact categories. In these studies, data surrounding energy and GHG are collected while other data describing other resource uses and emissions are not collected. Alternatively, when the goal and scope are set for a cradle-to-gate analysis, the product use and end of life steps are not included in the inventory analysis. The goal and scope define what data will be included and excluded.

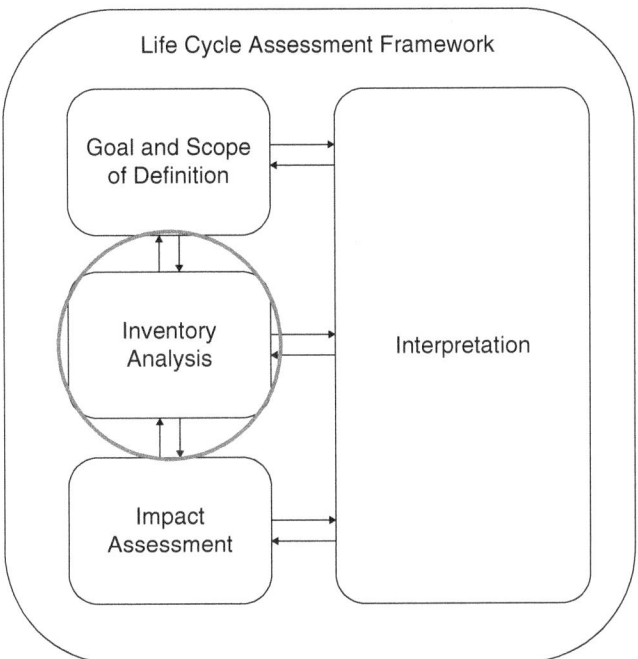

The data collected in the LCI should account for materials used in production, resources consumed in production, products and coproducts produced, waste streams sent to waste treatment, and emissions released to the environment. Collecting data and modeling a robust LCI is the most laborious aspect of performing an LCA, however, the quality of the overall LCA is heavily dependent on the quality of the LCI data. In this section, we describe the basic steps of creating an LCI, the data used to generate LCI and the software that is used to leverage preexisting data that can save time in modeling.

As an overview, the basic steps of performing an LCI as described by ISO 14044:2006 are listed here.

- Preparation for data collection based on goal and scope.
- Data collection.
- Data validation.
- Data allocation.
- Relating data to the unit process.
- Relating data to the functional unit.
- Data aggregation.

These steps are briefly discussed in the following sections; however, for an extensive description of these steps and performing an LCI, refer to www.lcatextbook.com.

5.3.2.1 Preparation of Data Collection Based on Goal and Scope

The goal and defined system boundary for an LCI will define which life-cycle stages and unit processes data are collected for. For life-cycle stages within the defined system boundary, unit processes should be identified. A unit process is a transformation of material or service performed. Many unit processes can make up a larger process or

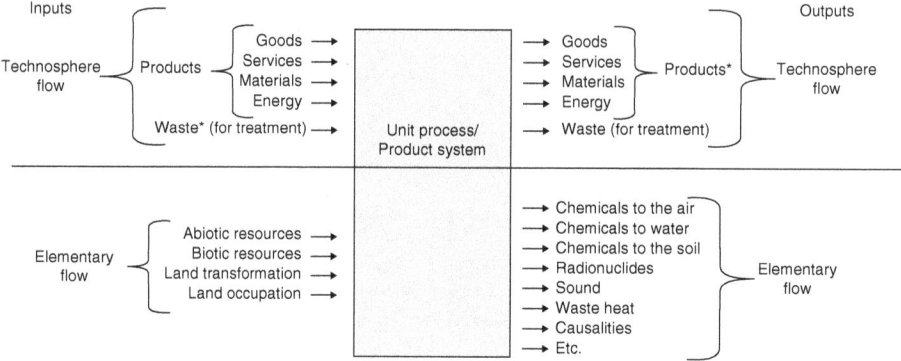

Figure 5.4 Life-cycle inventory data from product systems.

life-cycle stage. Identifying the material flows into and out of the defined system is the first step of an LCI performed by the LCA practitioner.

Elementary flows originate in the environment and are mined or retrieved to be used in a process or flows that are released from processes that are released to the environment and are not used by other processes. One can think of these elementary flows as the actual material used and materials released to the environment as a result of the studied product system. In Figure 5.4, the two types of LCI data can be seen. On the top half of the figure, technosphere flows such as products, services, and other goods are listed. The lower half of figure lists the elementary flows such as chemicals released to soil or air.

After the materials flows have been determined through interviews, literature searches, and measurement, LCA software can be used to track the material process flows back elementary flows to and from the environment.

5.3.2.2 Data Collection

Primary Data   When collecting data needed to perform an LCA, there are multiple types of data that are collected in different ways. Primary data and secondary data describe the two data types that are collected. Primary data are data specific to the studied product or service that are collected by the practitioner or someone working with the practitioner. For instance, if an LCA of plywood was being conducted, the material weights of wood, glue, and other materials would be measured as well as the energy required in the kiln. Primary data could be collected from natural gas flow meters from the kiln furnace, which would be classified as primary data. Primary data are unique to the process under study, which often results in more accurate data describing the system under study. Often there is generic data in literature describing the studies system; however, primary data should be used when it is available, specifically when the measurements are describing the major components of the studies system.

When collecting data from flow or electricity meters, it is important to collect data with statistical descriptors to ensure data that represents the processes. This can be achieved by collecting data from multiple pieces of similar equipment and from different times in different production runs to achieve multiple measures describing the processes. The median, mean, and standard deviation of the measured primary data should be reported as part of the primary data collection.

Consideration should also be given to the precision and accuracy of the equipment that process direct measurement. Precision can be thought of as the ability of a measurement device to measure a specified quantity, for example, mass of wood entering a kiln, in a repeatable way. For instance, if a mass was measured five times and the data from each measurement was 32.0 kg, the measurement device would be precise. Measurement precisions are different than accuracy. Accuracy of a measurement device describes the closeness of the measurement to the known value that it is measuring. For instance, if the measurement device reported a value of 8.7 kg for a known mass of 10 kg, the scale would not be considered accurate. However, if this same device reported 8.7 kg for ten different measurements of the 10 kg mass, this would be a precise but inaccurate measurement. Both precise and accurate measurements are important for collecting primary LCI data. To ensure both accurate and precise data collection, it is important to use measurement equipment with proper capabilities as well as using equipment with recent and proper calibration.

Many different types of data can be termed "primary data" and what data are collected depends on the product being studied. Some common primary data measurements are listed here; however, this is only a few of the types of data that could be collected in an LCI.

Example of data collected in an LCI:

- Raw material use
- Energy use
- Transportation distances
- Chemical use
- Waste treatment information
- Process yields
- Life times
- Water use
- Product and coproduct flows
- Other flows in or out of the system that are within the defined cutoff criteria.

Biogenic and Anthropogenic Carbon   In performing LCIs for forest biomaterials and other materials that include some form of plant materials, it is important to understand the differences in accounting methods of carbon derived from biological resources and petroleum resources. Carbon contained in a piece of wood is typically termed biogenic carbon as it was taken in from the environment during tree growth through photosynthesis. The natural uptake of carbon during tree growth and eventual decay of wood carbon during natural decomposition in forests or in a landfill is often termed the carbon cycle. This uptake and release of carbon within a relatively short time period, often less than 100 years, is in contrast to anthropogenic carbon emissions that are produced from oil and other carbon sources that have been affixed in some type of material for thousands of years. When oil or coal is burned, additional CO_2 is released and added to the environment, which over times increases the net concentration of CO_2 as the coal or other source did not capture the carbon in recent time periods.

To determine the CO_2 absorbed during a biomaterials growth, the following equation can be used.

$$CO_2 \text{ uptake} = \text{Mass of Material} \times \% \text{ Carbon in Material} \times \frac{44}{12}$$

In this equation, the percent carbon in the material, for instance, piece of loblolly pine with a carbon content of ~50%, can be multiplied by both the mass of the material and the molecular weight of CO_2 then divided by the molecular weight of carbon. This calculation can determine the CO_2 taken in from the atmosphere during growth and the potential for the material to release CO_2 to the environment during decomposition.

There are different methods that can be used to track both biogenic and anthropogenic carbon. Some studies do not track biogenic carbon as they assume that biogenic CO_2 is "carbon neutral" meaning has no effect on global climate change. This standpoint is controversial and is not as transparent as tracking the uptake and the later potential releases. Tracking biogenic carbon can add additional work, however, is more a more robust method and provides additional transparency.

Secondary Data Data that are not directly collected from measurement within the studies systems are often referred to as secondary data. Secondary data are often collected from LCI databases, literature, or other previous studies. A useful LCI database is provided by the National Renewable Energy Laboratory, the US LCI, available on the web (http://www.nrel.gov/lci/).

The use of secondary data are often required in studies; however, there are often discrepancies between the studied system and the system the secondary data describe. One of the most common discrepancies is the location of production. For instance, an LCI secondary data point from a database may describe loblolly pine production in Georgia; however, the studied product is in reality made in North Carolina. The data do not describe the system under study, however, could be representative of tree growth in North Carolina. When secondary data are used, it is important to document the differences in regionality, technology used in production, and other characteristics that would result in secondary data not accurately representing the process of interest in context of the studied LCI. In reality, secondary data can be as good as primary data, as secondary data are at some point primary data of another study. Reusing other study data within your own study is unavoidable in many cases, as time and funding constraints often limit the time spent on data collection. Additionally, there are times when collecting data on processes is just not feasible for other various reasons.

LCA Software There are many tools to assist an analyst in performing LCAs. There are software and data packages specifically designed for performing LCAs, and tools made in other software focused on certain aspects of LCA. No matter the form of the software, the use of some sort of LCA software and data management system is nearly needed in all LCAs. The LCI step of an LCA often requires a large data set listing hundreds of emissions to the environment. Keeping track of these flows manually would be too arduous, and LCA software is designed to manage these flows and perform specific functions such as impact assessments based on the inventory as well as uncertainty analysis. In this section, we discuss common LCA software as well as several LCA tools which hopes help the reader find the best software and tools for their specific needs.

There is a large list of LCA software emerging onto the market all with various selling features. This review is not exhaustive to all LCA software, and no preference is given any software provider. What is learned about the reviewed software packages can also be helpful in understanding how other tools and software work as well. A basic overview of how data and LCA software will first be provided then a list of software packages.

Basic LCA Software Structure LCA software can be split into several components: (i) the software package, (ii) data sets, and (iii) LCIA methods (which is explained in detail in the next section). The software package such as SimaPro, openLCA, and Gabi can be thought of as a framework or a calculator that keeps track of data and performs intensive numerical calculations. With the many flows and detailed data, much effort has been invested in creating efficient calculation methods to speed up analysis time. This framework, however, is not useful without inventory data. There are many premade secondary data sets provided from sources such as Ecoinvent, Gabi, and United States Department of Agriculture (USDA) that contain previous LCI results for various chemicals, materials, energy, services, and waste treatment processes. LCA software can access this previously developed data and allow an LCA practitioner to include a chemical or other process from a data set in their LCA without the need to perform an entire LCA on that particular material or process. This fundamental aspect of LCA, the leveraging of previous study results for new studies, is a key benefit of LCA software and can save countless hours on the LCI step. LCIA methods are procedures and conversions that are used in performing an LCIA such as GWP characterizations and weighting methods. There are many accepted LCIA methods that calculate LCA results using different impact categories, types of impacts, and weighting methods. Further discussion surrounding LCIA is provided later.

Figure 5.5 visually depicts how the different components of LCA software and data interact. The LCI step requires data from data sets (e.g., Ecoinvent) and primary data gathered by the LCA practitioner surrounding the process or product under analysis. An LCI is calculated with the combination of these two types of data and the use of LCA software calculations. The LCI data can then be used to perform an impact assessment using the LCIA methods (e.g., TRACI) (Table 5.1).

5.3.2.3 Data Quality

Mass and Component Balance Data Check Much of the LCI data comes from the product manufacturing process. These data can be collected directly from a company, from literature, or from process conversion models. When examining a process, a mass balance should be performed to ensure that all process streams are accounted captured.

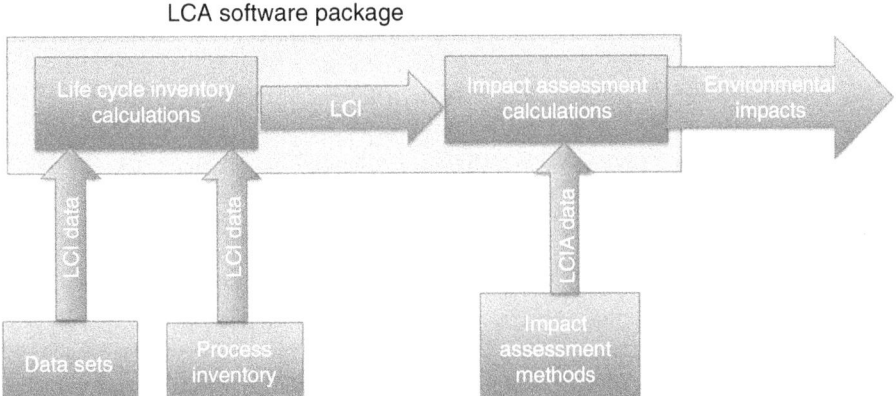

Figure 5.5 Life-cycle assessment software structure.

Table 5.1 Three common LCA software package options.

Software	Licensing	Data sets	Software features	Website
openLCA	Open source and free	Ecoinvent, Gabi, USLCI, CML, and others	Fast calculation engine, easily shares models, no yearly subscription, process based on transparent data, used for USDA digital commons LCA data development	www.openlca.org
SimaPro	Paid licensing	Ecoinvent, USLCI, CML, and others	Process based on transparent data, good customer support, robust uncertainty analysis	www.pre-sustainability.com/simapro
Gabi	Paid licensing	Gabi Dataset, Ecoinvent, USLCI	Robust data set, visual process flow-based modeling, ease of use, data frequently updated	www.gabi-software.com

A mass balance should equal zero when adding all of the material inputs to a system subtracted by all the materials exiting in the system. In reality and especially in complicated production processes, the difference of the in–out flows may not be zero. The percent mass closure can be calculated by

$$\% \text{ Mass Closure} = \frac{(\text{mass in} - \text{mass out})}{\text{mass in}} \times 100$$

Providing mass balances and listing percent closure is a good practice that leads to data transparency. Ideally, the percent closure system should be 100%; however, in reality this is often not the case due to measurement errors, fugitive emissions, and other modeling errors. In practice, mass balances above 95% can often provide meaningful data suitable for use.

Similar to a total mass balance a component balance can be performed to ensure proper tracking of an element within a system. For instance, carbon balances track the mass of carbon flowing into and out of a process. The in and out should be equal or the percent closure should be near 100%. If this is not the case, the data should be reexamined to find the error or missing data.

When using secondary data from LCI databases or literature, it is also important to perform data quality checks. Mass balance is also a valid approach to checking secondary data surrounding unit processes. Further analysis should examine the spatial data surrounding the secondary data. For instance, a material produced in China using an average electrical grid that relies heavily on coal power will have different environmental impacts than a product produced in the Northwest United States where hydroelectric power is more dominant in the average electrical grid. One way to overcome these regional differences is to change the electricity type used for the secondary data and recalculate the overall impacts. Doing this can provide a better representation of the impacts of the product under study. Technology is another important factor. Often, secondary data are out of date and are based on older technology than what is currently employed in the industry. The record date for the secondary data can give some indication of whether the technology is current or not, but further analysis of the documentation should be performed to determine if the technology used is representative of the technology used to produce the product under study.

5.3.2.4 Coproduct Treatment – Allocation

Some production processes produce more than one product, and the emissions of the process cannot be easily attributed to a single product from the process. Coproduct treatment methods as defined by ISO 14044 are used to properly account for emissions from the production of multiple products. The ISO standard states that wherever possible, allocation should be avoided by

1. dividing the unit process to be allocated into two or more subprocesses and collecting the input and output data related to these subprocesses;
2. expanding the product system to include the additional functions related to the coproducts.

If allocation cannot be avoided by these two methods, the allocation method should

1. partition inputs and outputs of the system between its different products or functions in a way that reflects the underlying physical relationships between them;
2. partition input and output data between coproducts in proportion to the economic value of the products.

The first route to avoiding allocation by process subdivision can remove the need to account for multiple products from the overall system by splitting it into additional subprocesses. With a more detailed process flow diagram and data, some products may be produced through separate subprocess and thus can be accounted for individually.

Process subdivision is not always possible and expanding the product system may be necessary to account for the coproducts. Using this method, all impacts associated with production are assigned to the primary product of interest and a credit or negative emission is used to account for the displacement of the coproduct production in other manufacturing processes. The system expansion method works only when the manufacturing process is not the primary route to the coproduct. The life-cycle handbook (Curran, 2012) gives the example of a hydrocracking unit that produces ethylene, propylene, other hydrocarbons, fuel gas, and heat. In this example, energy and gas can be accounted for using system expansion by giving a displacement credit. However, system expansion does not work on the other products, as hydrocracking is the primary commercial route to these products. The other products of the hydrocracking process are accounted for using allocation.

When allocation is to be performed, physical parameters and relationships should be used to attribute the total impacts to individual products. Such parameters may include mass of final product, raw material ratio required to produce the final products, energy content of products, or other physical relationships. If a physical relationship cannot be determined, economic allocation can be performed by attributing the impacts in accordance with the revenue associated with each product. For instance, in a process producing products A, B, and C, if product A generates 98% of the revenue from a process, 98% of the total impacts would be assigned to product A. Economic allocation should be avoided as product prices can change and thus changes the overall LCA results even when the production process and overall emissions remain constant.

158 | Introduction to Renewable Biomaterials

5.3.2.5 Relating Data to the Unit Process

Much of the data collected for an LCI will not be in correct or the most meaningful units. Often, there may be inventories based on a month's production and could have units such as energy use per number of units produced in a month. On most occasions, data surrounding one unit are needed, and thus some data manipulation is required.

As an example, Table 5.2 lists the inputs and outputs of dry rough lumber, at kiln, US PNW as developed by the US LCI database. The outputs from this process are listed in the top portion of the table and the outputs on the bottom. Note that there are product flows and elementary flows as discussed earlier in this chapter. Under the amount column, the total quantity of product is 23 units. This would indicate that all the emissions listed in the amount column are per 23 units of product, which in this case is 1 kg of dry rough lumber at kiln in the US Pacific Northwest. To make these

Table 5.2 US LCI inventory for wood product manufacturing/sawmills.

Flow	Category	Type	Unit	Amount	Adjusted to one unit
Outputs					
Dry rough lumber, at kiln, US PNW	Wood product manufacturing/sawmills	PRODUCT_FLOW	kg	2.30E+01	1.00E+00
Particulates, unspecified	Air/unspecified	ELEMENTARY_FLOW	kg	2.46E−04	1.07E−05
VOC, volatile organic compounds	Air/unspecified	ELEMENTARY_FLOW	kg	2.51E−03	1.09E−04
Inputs					
CUTOFF disposal, inert solid waste, to inert material landfill	Null/CUTOFF flows	PRODUCT_FLOW	kg	2.88E−04	1.25E−05
CUTOFF hogfuel-biomass (50% MC), combusted in industrial boiler	Null/CUTOFF flows	PRODUCT_FLOW	kg	7.80E+00	3.39E−01
Diesel, combusted in industrial boiler	Utilities/steam and air-conditioning supply	PRODUCT_FLOW	l	2.67E−03	1.16E−04
Electricity, at grid, Western US, 2000	Utilities/electric power distribution	PRODUCT_FLOW	kWh	1.44E+00	6.25E−02
Natural gas, combusted in industrial boiler	Utilities/steam and air-conditioning supply	PRODUCT_FLOW	m^3	1.01E+00	4.39E−02
Rough green lumber, softwood, at sawmill, US PNW	Wood product manufacturing/sawmills	PRODUCT_FLOW	kg	2.30E+01	1.00E+00
Transport, combination truck, average fuel mix	Truck transportation/general freight trucking	PRODUCT_FLOW	t*km	1.04E+00	4.50E−02

secondary data more useful, it is helpful to relate all the flows to one unit, 1 kg, of output so that in later processes this can be scaled to meet the needs of one product. In this example, all the numbers in the "Amount" column are divided by 23 to get the inputs and outputs per one unit of product.

5.3.2.6 Relating Data to the Functional Unit

The next step of LCI is similar to relating to unit process step, but instead this time the data are related to the functional unit defined in the goal and scope. For instance, if the functional unit of the LCA was a rustic chair, that chair might require several kilograms of wood as well as other materials. In relating the data to the functional units, all the inputs and outputs are scaled to the quantity of material/product that is required to fulfill the requirements of the functional unit; this flow is called the reference flow. This step can be performed in Microsoft Excel worksheet or in an LCA software package. The results after this step may include numbers such as energy use per functional unit or CO_2 emission per functional unit. Elementary, waste, and product flows may be listed at this point; however, they would all be listed in relation to the required amount per functional unit.

5.3.2.7 Data Aggregation

When performing an LCI, many calculations are required for the different life-cycle stages (remember: product production, product use, end of life) that may be useful to analyze separately before combining. Often, the final results of both LCIs and LCAs are reported by life-cycle stages as well as the total impacts. Since the final total number is required, the LCI data are summed across all the life-cycle stages. For instance, if electricity was used by five different processes, the total electric usage may be summed for all these processes and reported.

5.3.2.8 LCI Data Interpretation

Inventory data interpretation is an important step within the larger interpretation of the whole study. Throughout the LCI development process, some level of interpretation must be performed. For example, when collecting data, the practitioner must interpret the available data and make a judgment call on the quality and relevance to the goal and scope. Often when performing an LCI, it will become clear that the goal and scope are at times not appropriate given the availability of data, time, and resources available to complete the study. Developing a high-quality LCI is the most time-consuming part of an LCA, which often experiences hang-ups and delays that are in many cases beyond the control of the practitioner. In these cases, where data are just not available, the goal and scope can be adjusted so that the available data can support the goal and scope and eventually the overall study conclusions.

Another aspect of interpretation is uncertainty in data and modeling assumptions. Though the use of a sensitivity analysis, a variety of study assumptions, and data can be tested to determine the influence on the overall LCI results. For instance, an assumption on a process yield where incoming material is converted to a product material can be varied depending on incoming material composition that varies with time. To determine how this yield that can often change influences the energy or other LCI parameters, the yield could be adjusted up or down a set percentage, for example,

25%, or adjusted according to some statistical measure associated with the value such as standard deviation. Providing a range of values and understanding how different values influence the results and eventually conclusions are far more valuable than providing a static one-number answer without deeper insights into what is driving the overall results.

In some studies, the goal may be to compare one product to another. This type of study is referred to as a comparative LCA, and when a company wants to publically communicate such results they must be first certified by through a peer-reviewed process as defined by the ISO 14044 standards. In these types of studies, it is important in the LCI phase to determine if the available data and models can reasonably calculate the differences in environmental flows and impacts. To reasonably claim a difference between flows for competing products, a general rule of thumb is that the values are not significantly different unless they are at least 25% different. This 25% different rule is often used as there is inherently uncertainty in the data, modeling, and other factors that are not possible to completely model. This 25% rule at times could be too high and a robust uncertainty analysis could be performed to further determine the certainty and the probability that one product will produce lower flows and environmental impacts than another.

Another important part of LCI interpretation is determining and communicating the "hotspots" or process areas and process flows that influence the overall LCI the most. For instance, in dried rough lumber production, the electricity used during drying would be an energy use hotspot as well as a large contributor to GHG emissions. Insights surrounding the hotspots are often some of the most important and actionable information that is attainted by performing an LCA, and time should be dedicated to understanding the driving factors behind the values of the most important environmental flows.

For more information surrounding LCI methods, please refer to www.lcatextbook.com and navigate to Chapter 5. This textbook provides additional details and resources that could not be included in this brief introduction to LCA and is a free book available to all.

5.3.2.9 Problems Set – Life-Cycle Inventory

Problem 1 A wood products production facility has emissions that contribute to global warming (GWP) of 50,000 kg CO_2eq day^{-1}. The production facility makes 1000 2×4 in.^{2} boards a day, 500 1×4 in.^{2} boards a day and 125 2×6 in.^{2} boards per day.

	Prices ($)	Weights (lb)
2×4 in.^{2}	2.36	17
1×4 in.^{2}	6.79	9
2×6 in.^{2}	5.89	27

Using mass allocation methods, what is the GWP per kg of each product? Using economic allocation methods, what is the GWP per board for each board type? Final units are to be reported in kg CO_2 eq.

Problem 2 Use the USLCI data below for inputs and outputs, for Cotton, whole plant, at field. For 8 metric tonnes of cotton product produced, determine the following quantities:

The total amount of cotton straw coproduct.
The total amount of outputs in kilogram of air pollutants.
The total amount of outputs in kilogram of water pollutants.
The amount of total land needed.
The amount of total fertilizers needed.
The amount of total pesticides needed.
The amount of total electricity needed.
The amount of total water needed.

Inputs	Unit	Amount	Comments
Carbon dioxide	kg	1.69	Carbon fixation through plant growth
Diesel, combusted in industrial equipment	l	0.231	Farm tractor, energy consumption
Dummy, agrochemicals, at plant	kg	0.009	Pesticides – consumption
Dummy, phosphorous fertilizer (TSP as P_2O_5), at plant	kg	0.0503	Fertilizers NPK Ca and method of application
Dummy, potash fertilizer (K_2O), at plant	kg	0.0556	Fertilizers NPK Ca and method of application
Electricity, at grid, United States, 2000	kWh	0.247	Energy consumption
Liquefied petroleum gas, combusted in industrial boiler	l	0.0257	Energy consumption
Natural gas, combusted in industrial boiler	m^3	0.0000661	Energy consumption
Nitrogen fertilizer, production mix, at plant	kg	0.124	Fertilizers NPK Ca and method of application
Occupation, arable, conservation tillage	m^2*a	2.75	Planted surface, tillage practices
Occupation, arable, conventional tillage	m^2*a	10.4	Planted surface, tillage practices
Occupation, arable, reduced tillage	m^2*a	1.55	Planted surface, tillage practices
Quicklime, at plant	kg	0.158	Fertilizers NPK Ca and method of application
Transport, single unit truck, diesel powered	t*km	0.113	Transportation of fertilizers (200 km)
Transport, train, diesel powered	t*km	0.302	Transportation of fertilizers (400 km)
Water	l	287	Water consumption – consumptive use for
Water, well	l	1020	Water consumption – consumptive use for

Outputs	Category	Unit	Amount	Comments
2,4-D	Air/low population density	kg	4.72E−07	Pesticides – consumption, Pesticides – application method, Pe
Acephate	Air/low population density	kg	4.36E−05	Pesticides – consumption, Pesticides – application method, Pe
Aldicarb	Air/low population density	kg	1.10E−04	Pesticides – consumption, Pesticides – application method, Pe
Ammonia	Air/low population density	kg	7.48E−03	Air emissions of N_2O and NH_3 and NO_x
Azinphos-methyl	Air/low population density	kg	9.73E−06	Pesticides – consumption, Pesticides – application method, Pe
Bromoxynil	Air/low population density	kg	5.21E−06	Pesticides – consumption, Pesticides – application method, Pe
Carbofuran	Air/low population density	kg	7.85E−06	Pesticides – consumption, Pesticides – application method, Pe
Chlorpyrifos	Air/low population density	kg	4.63E−05	Pesticides – consumption, Pesticides – application method, Pe
Clomazone	Air/low population density	kg	2.45E−05	Pesticides – consumption, Pesticides – application method, Pe
Cyanazine	Air/low population density	kg	1.27E−04	Pesticides – consumption, Pesticides – application method, Pe
Dinitrogen monoxide	Air/low population density	kg	5.69E−03	Air emissions of N_2O and NH_3 and NO_x
Disulfoton	Air/low population density	kg	2.97E−05	Pesticides – consumption, Pesticides – application method, Pe
Diuron	Air/low population density	kg	5.63E−05	Pesticides – consumption, Pesticides – application method, Pe
Endosulfan	Air/low population density	kg	1.84E−05	Pesticides – consumption, Pesticides – application method, Pe
Ethephon	Air/low population density	kg	2.16E−04	Pesticides – consumption, Pesticides – application method, Pe
Fluometuron	Air/low population density	kg	2.20E−04	Pesticides – consumption, Pesticides – application method, Pe
Glyphosate	Air/low population density	kg	7.72E−05	Pesticides – consumption, Pesticides – application method, Pe

Outputs	Category	Unit	Amount	Comments
Hydrocarbons, unspecified	Air/low population density	kg	5.40E−03	Pesticides – consumption, Pesticides – application method, Pe
MSMA	Air/low population density	kg	1.94E−04	Pesticides – consumption, Pesticides – application method, Pe
Malathion	Air/low population density	kg	1.74E−04	Pesticides – consumption, Pesticides – application method, Pe
Mancozeb	Air/low population density	kg	5.19E−07	Pesticides – consumption, Pesticides – application method, Pe
Metam sodium	Air/low population density	kg	5.52E−05	Pesticides – consumption, Pesticides – application method, Pe
Metolachlor	Air/low population density	kg	2.06E−05	Pesticides – consumption, Pesticides – application method, Pe
Nitrogen oxides	Air/low population density	kg	2.02E−02	Air emissions of N_2O and NH_3 and NO_x
Norflurazon	Air/low population density	kg	6.42E−05	Pesticides – consumption, Pesticides – application method, Pe
Oxamyl	Air/low population density	kg	4.41E−05	Pesticides – consumption, Pesticides – application method, Pe
Paraquat	Air/low population density	kg	4.34E−05	Pesticides – consumption, Pesticides – application method, Pe
Parathion methyl	Air/low population density	kg	1.31E−04	Pesticides – consumption, Pesticides – application method, Pe
Particulates, unspecified	Air/low population density	kg	4.08E−05	Particulate emissions during harvesting
Pendimethalin	Air/low population density	kg	1.50E−04	Pesticides – consumption, Pesticides – application method, Pe
Phorate	Air/low population density	kg	4.00E−05	Pesticides – consumption, Pesticides – application method, Pe
Prometryn	Air/low population density	kg	6.78E−05	Pesticides – consumption, Pesticides – application method, Pe
Propene, 1,3-dichloro-	Air/low population density	kg	2.50E−04	Pesticides – consumption, Pesticides – application method, Pe
S,S,S-tributyl phosphorotrithioate	Air/low population density	kg	2.04E−04	Pesticides – consumption, Pesticides – application method, Pe

Outputs	Category	Unit	Amount	Comments
Sodium chlorate	Air/low population density	kg	2.71E−04	Pesticides − consumption, Pesticides − application method, Pe
Trifluralin	Air/low population density	kg	3.93E−04	Pesticides − consumption, Pesticides − application method, Pe
Cotton straw, at field	None	kg	4.50E+00	Residue to Crop Ratio, C content of residues − same value as soybe
Cotton, at field	None	kg	1.00E+00	Planted surface, Production, Seeds
2,4-D	Water/unspecified	kg	2.02E−08	Pesticides − consumption, Pesticides − application method, Ru
Acephate	Water/unspecified	kg	1.87E−06	Pesticides − consumption, Pesticides − application method, Ru
Aldicarb	Water/unspecified	kg	4.73E−06	Pesticides − consumption, Pesticides − application method, Ru
Azinphos-methyl	Water/unspecified	kg	4.17E−07	Pesticides − consumption, Pesticides − application method, Ru
Bromoxynil	Water/unspecified	kg	1.35E−07	Pesticides − consumption, Pesticides − application method, Ru
Carbofuran	Water/unspecified	kg	3.36E−07	Pesticides − consumption, Pesticides − application method, Ru
Chlorpyrifos	Water/unspecified	kg	1.99E−06	Pesticides − consumption, Pesticides − application method, Ru
Clomazone	Water/unspecified	kg	1.05E−06	Pesticides − consumption, Pesticides − application method, Ru
Cyanazine	Water/unspecified	kg	5.43E−06	Pesticides − consumption, Pesticides − application method, Ru
Disulfoton	Water/unspecified	kg	7.69E−07	Pesticides − consumption, Pesticides − application method, Ru
Diuron	Water/unspecified	kg	2.41E−06	Pesticides − consumption, Pesticides − application method, Ru
Endosulfan	Water/unspecified	kg	7.88E−07	Pesticides − consumption, Pesticides − application method, Ru
Ethephon	Water/unspecified	kg	9.25E−06	Pesticides − consumption, Pesticides − application method, Ru

Outputs	Category	Unit	Amount	Comments
Fluometuron	Water/unspecified	kg	9.44E−06	Pesticides – consumption, Pesticides – application method, Ru
Glyphostate	Water/unspecified	kg	3.31E−06	Pesticides – consumption, Pesticides – application method, Ru
MSMA	Water/unspecified	kg	8.31E−06	Pesticides – consumption, Pesticides – application method, Ru
Malathion	Water/unspecified	kg	7.47E−06	Pesticides – consumption, Pesticides – application method, Ru
Mancozeb	Water/unspecified	kg	1.34E−08	Pesticides – consumption, Pesticides – application method, Ru
Metam sodium	Water/unspecified	kg	2.37E−06	Pesticides – consumption, Pesticides – application method, Ru
Metolachlor	Water/unspecified	kg	8.85E−07	Pesticides – consumption, Pesticides – application method, Ru
Nitrogen, total	Water/unspecified	kg	1.52E−02	Runoff/leaching of fertilizers
Norflurazon	Water/unspecified	kg	2.75E−06	Pesticides – consumption, Pesticides – application method, Ru
Oxamyl	Water/unspecified	kg	1.14E−06	Pesticides – consumption, Pesticides – application method, Ru
Paraquat	Water/unspecified	kg	1.86E−06	Pesticides – consumption, Pesticides – application method, Ru
Parathion methyl	Water/unspecified	kg	5.63E−06	Pesticides – consumption, Pesticides – application method, Ru
Pendimethalin	Water/unspecified	kg	6.45E−06	Pesticides – consumption, Pesticides – application method, Ru
Phorate	Water/unspecified	kg	1.03E−06	Pesticides – consumption, Pesticides – application method, Ru
Phosphorus compounds, unspecified	Water/unspecified	Kg	2.96E−04	Runoff/leaching of fertilizers
Prometryn	Water/unspecified	kg	2.91E−06	Pesticides – consumption, Pesticides – application method, Ru
Propene, 1,3-dichloro-	Water/unspecified	kg	6.46E−06	Pesticides – consumption, Pesticides – application method, Ru

Outputs	Category	Unit	Amount	Comments
S,S,S-tributyl phosphorotrithioate	Water/unspecified	kg	8.74E−06	Pesticides – consumption, Pesticides – application method, Ru
Sodium chlorate	Water/unspecified	kg	1.16E−05	Pesticides – consumption, Pesticides – application method, Ru
Suspended solids, unspecified	Water/unspecified	kg	1.02E+01	Erosion
Trifluralin	Water/unspecified	kg	1.02E−05	Pesticides – consumption, Pesticides – application method, Ru

"*" denotes the meaning 'per'.
"a" denotes the meaning 'year'.

Life-Cycle Impact Assessment
LCIA is the third sequential step of an LCA. The purpose of an LCIA "is to provide additional information to assess LCI result and help users better understand the environmental significance of natural resource use and environmental releases". The LCIA helps provide significance and simplify results for easier decision making; however, it is important to understand that it does not directly measure the impacts of chemical releases to the environment as an environmental risk assessment does. The third step of LCIA follows sequentially after the LCI using the many flows to and from the environment developed in the LCI. These LCI flows without an impact assessment step are not easily interpreted, and understanding the significance of emissions can be impossible (Figure 5.6).

The LCIA as previously mentioned is different from a risk assessment measuring absolute values of environmental impacts; rather, the LCIA helps determine the significance of emissions and impacts in relation to the study scope. The absolute value of the impacts cannot be determined by the LCIA due to (Margni and Curran, 2012)

- the relative expression of potential environmental impacts to a reference unit;
- the integration of environmental data over space and time;
- the inherent uncertainty in modeling environmental impact;
- the fact that some possible environmental impact occur in the future.

Even though the LCIA has limitations, it is useful in determining what impacts matter the most, what unit processes are contributing most through hot spot analysis, and help identify best-scenario options when environmental trade-offs occur.

According to ISO, there are three mandatory processes of an LCIA including selection of impact categories, classification, and characterization (Figure 5.7).

5.3.2.10 Mandatory Elements

Selection of Impact Methods The selection of impact methods should reflect the intent and methods outlined in the goal and scope of the study. The impact indicators of the LCIA method must reflect the purpose of the study and examine the resources or

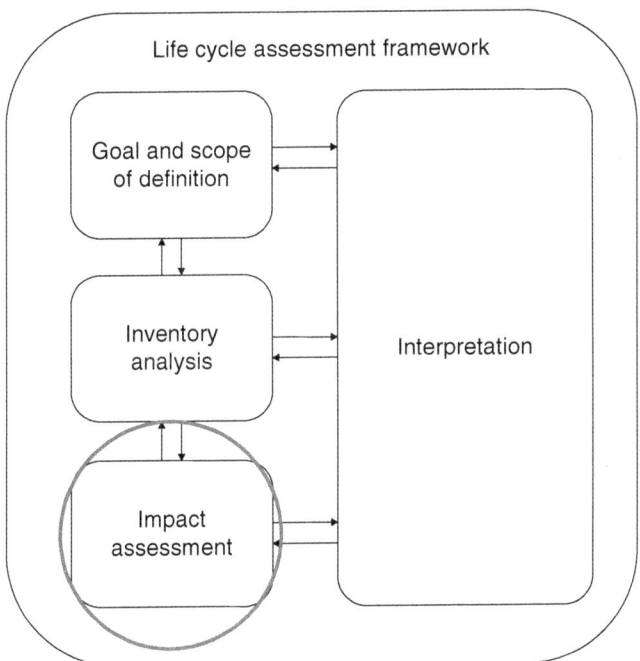

Figure 5.6 Life-cycle assessment stages.

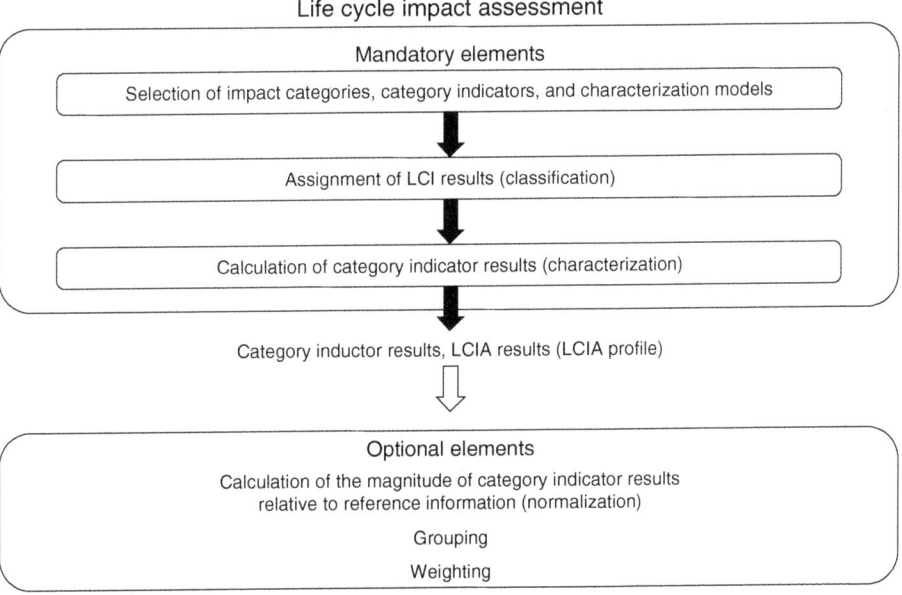

Figure 5.7 Impact assessment ISO mandatory and optional steps (ISO 14044, 2006).

168 *Introduction to Renewable Biomaterials*

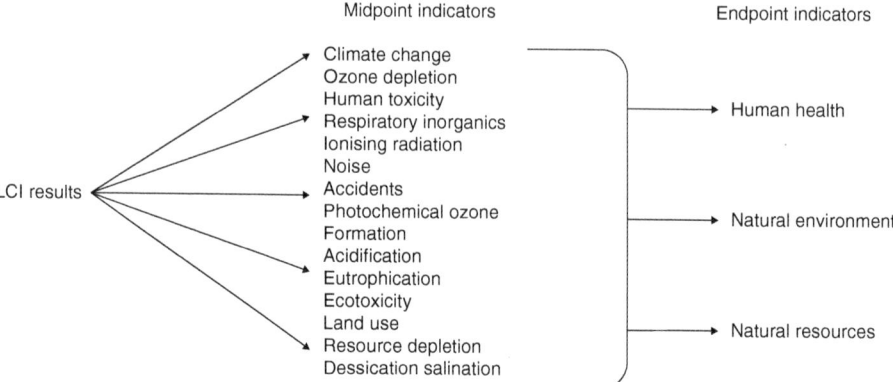

Figure 5.8 The relationship between end point and midpoint impacts as proposed by the ILCD Handbook (Wolf et al., 2012).

impacts to the environment that the study is addressing. For instance, if quantifying fossil fuel usage of a product is a stated goal, the impact assessment method must calculate this to enable the interpretation of results.

There are primarily two main types of impact assessment methods: midpoint and end point impact assessment methods (Goedkoop and Spriensma, 2001; Bare et al., 2006). The midpoint indicator methods are closely tied to science and are based on more exact models. As there are fewer assumptions associated with midpoint indicators, they generally have less uncertainty than end point indicators. End point indicators are useful in that they are easier to understand and are more appealing to a general audience (Figure 5.8).

5.3.2.11 Classification

Classification is the second of the ISO mandatory LCIA steps where emissions are sorted into groups that have an impact on a midpoint indicator. Figure 5.9 lists LCI data of different elemental flows and then shows arrows grouping the emissions to

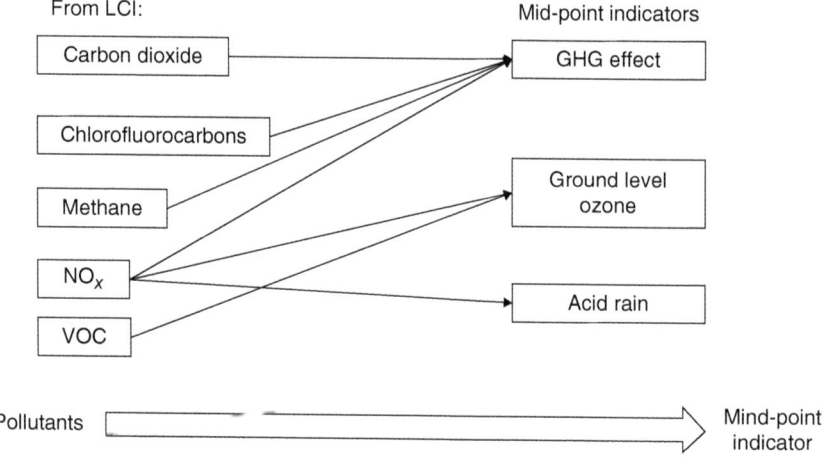

Figure 5.9 Classification of LCI data into midpoint indicators.

Table 5.3 Midpoint indicators with associated emissions and scale (Bare et al., 2006).

Impact category	Scale	Examples of LCI data (i.e., classification)
Global warming	Global	Carbon dioxide (CO_2), nitrous oxide (N_2O), methane (CH_4), chlorofluorocarbons (CFCs), hydrochlorofluorocarbons (HCFCs), methyl bromide (CH_3Br)
Stratospheric ozone depletion	Global	Chlorofluorocarbons (CFCs), hydrochlorofluorocarbons (HCFCs), halons, methyl bromide (CH_3Br)
Acidification	Regional, local	Sulfur oxides (SO_x), nitrogen oxides (NO_x), hydrochloric acid (HCl), hydrofluoric acid (HF), ammonia (NH_4)
Eutrophication	Local	Phosphate (PO_4), nitrogen oxide (NO), nitrogen dioxide (NO_2), nitrates, ammonia (NH_4)
Photochemical smog	Local	Non-methane hydrocarbon (NMHC)
Terrestrial toxicity	Local	Toxic chemicals with a reported lethal concentration to rodents
Aquatic toxicity	Local	Toxic chemicals with a reported lethal concentration to fish
Human health	Global, regional, local	Total releases to air, water, and soil
Resource depletion	Global, regional, local	Quantity of minerals used, quantity of fossil fuels used
Land use	Global, regional, local	Quantity disposed off in a landfill or other land modifications
Water use	Regional, local	Water used or consumed

the corresponding impact categories. Often, an emission can impact more than one category (Table 5.3).

5.3.2.12 Characterization

The characterization step relates the emission flow to the potential impact in the impact category in a common unit relating multiple flows to a reference flow. Environmental models are used to determine the potential impact each flow has to the corresponding impact category. One common example of this is the relationship between global warming gases of carbon dioxide and methane. The impact of 1 kg of methane, according to the Intergovernmental Panel on Climate change 2013, is 34 times greater than 1 kg of carbon dioxide over a 100-year analytical time horizon. This value is calculated using models to predict the warming effect of greenhouse gasses based on insulating capacity of each gas as well as gas degradation patterns. Table 5.4 provides a list of contributors to GWP and the characterization factors for both a 20-year time horizon and a 100-year time horizon. Note the characterization factor is listed in kg CO_2 equivalents per kg of substance. The kg of the emission times the characterization factor provides the equivalent emission in CO_2 to the emission of the other compound. These and many other emissions have a different potential environmental impact and characterization factor, depending on the time in which they are considered, or analytical time horizon. The characterization factors are often standardized for a set impact assessment method

Table 5.4 Greenhouse gas lifetime before decomposition and corresponding global warming potential (GWP) for a 20-year time horizon and a 100-year time horizon (Myhre et al., 2013).

	Lifetime years	GPW20	GPW100
CH_4	12.4	86	34
HFC-134a	13.4	3790	1550
CF-11	45	6900	5350
N_2O	121	268	298
CF_4	50,000	4950	7350

such as Ecoinvent or the TRACI Method. Both of these methods are further discussed in the following sections.

$$\text{Characterized flow} = \text{flow (inventory unit)} \times \text{Char. factor} \frac{\text{characterized units}}{\text{inventory units}}$$

5.3.2.13 Optional Elements

Normalization Normalization is an optimal step of LCIA according to the ISO standards; however, it can provide some additional insights into the relevancy of certain emissions. Since midpoint indicators represent different measure and use different units, relating the quantity of a midpoint indicator to another midpoint indicator can be difficult. Furthermore, it is not always possible to know how significant a 50 kg of CO_2 eq. emitted is compared to 50 kg H^+ equivalents. You might ask is 50 kg of CO_2 and H^+ huge emission in comparison to what is currently being emitted in the world?

Relating midpoint indicators to emissions of an exterior reference system is referred to as external normalization. For instance, dividing the emission of 50 kg CO_2 by the CO_2 emissions of a human in the United States over the course of a year allows a comparison of the current system to the total impact that a person may have in a year. This external normalization, defined as impact of study scenario divided by an external reference value in the same units, provides a unitless value that can be further examined in the next optional steps of impact assessment. External normalization is commonly used; however, it has a major weakness. As an example, when the CO_2 emissions of a product are normalized by the CO_2 emissions of a US citizen over a year, the impacts of the product are often trivial in comparison to the yearly total. This yearly total emission that is used for normalization does not have any relation to what the earth or natural environment can sustainably accommodate; rather it is the current and often unstainable level of emissions. By using often large and unsustainable emissions number to normalize midpoint indicators, the influence of certain impact categories is often diminished in the final single score. In this example, 50 kg CO_2 divided by 24,000 kg (emissions of a US citizen for CO_2 per year,) yields 0.021, somewhat trivial and small number (Table 5.5).

Another method of normalization for comparative LCA studies is internal normalization. In such a comparative study, the midpoint indicators between an option A and

Table 5.5 Normalization factors based on a US citizen's impact over the course of a year in 2008 (Ryberg et al., 2014).

Impact category	US normalization factors reference year 2008	
	Annual (impact per year)	Per capita (impact per person year)
Ecotoxicity – nonmetals (CTUe)	2.2E+10	7.6E+1
Ecotoxicity – metals (CTUe)	3.3E+12	1.1E+04
Carcinogens – nonmetals (CTUcanc.)	1.7E+03	5.5E−06
Carcinogens – metals (CTUcanc.)	1.4E+04	4.5E−05
Noncarcinogens – nonmetals (CTUnoncanc.)	1.1E+04	3.7E−05
Noncarcinogens – metals (CTUcanc.)	3.1E+05	1.0E−03
Global warming (kg CO_2 eq)	7.4E+12	2.4E+04
Ozone depletion (kg CFC-11 eq)	4.9E+07	1.6E−01
Acidification (kg SO_2 eq)	2.8E+10	9.1E+01
Eutrophication (kg N eq)	6.6E+09	2.2E+01
Photochemical ozone formation (kg O_3 eq)	4.2E+11	1.4E+03
Respiratory effects (kg PM2.5 eq)	7.4E+09	2.4E+01
Fossil fuel depletion (MJ surplus)	5.3E+12	1.7E+04

option B are divided by scenario with the highest value for an individual impact. For instance, if product A produced 20 kg eq. of CO_2 and option B produced 50 kg CO_2 eq., scenario A values would be divided by option B, 20 kg/50 kg. In a comparative study, this method has some advantages as this goal is often to provide information that leads to making a decision that will produce the lowest impacts. Since this goal is based on deciding between one of multiple options, the scenarios can be normalized between each other.

Weighting Weighting is a subjective methodology where the relative importance of impact categories is determined. This can be useful in simplifying the results of an LCA especially when there are trade-offs between scenarios. For instance, in the production of a paper product two bleaching processes are used and compared in an LCA; however, option A has lower impacts in three categories while option B has lower impacts in four categories. It is often unclear how to compare GWP to acidification potential midpoint indicators. Through weighting normalized midpoint indicators can provide further insights that can be useful in decision making and can be used in a single environmental score.

Weighting values are often determined by stakeholder groups involved in the project, such as a company commissioning the LCA study. Alternatively, standardized weighting values established in other studies can be used. One example of standardized weighting values is the Eco-indicator 99 impact assessment method. For the Eco-indicator 99 method (Ministry of Housing, 2000), a panel of 365 people was asked to rank the importance of ecosystems health, resource use, and human health. The results from this Swiss LCA interest group panel indicated that the human health and ecosystems

Table 5.6 Ecoindicator weighting values and survey responses (Goedkoop and Spriensma, 2001).

Impact category	Mean (%)	Rounded (%)	St. deviation (%)	Median (%)
Human health	36	40	19	33
Ecosystem quality	43	40	20	33
Resources	21	20	14	23

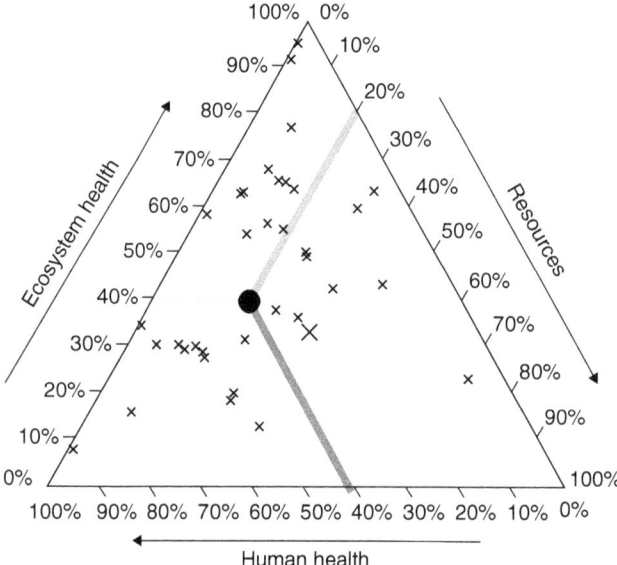

Figure 5.10 LCA interest group response to weighting survey used for Eco-indicator 99 weighting method (Goedkoop and Spriensma, 2001).

quality were twice as important as recourse use (Table 5.6). Figure 5.10 shows the wide range of responses of this panel, which further illustrates the subjective nature of weighting. United States-based weighting factors were also determined through survey panel by Gloria *et al.* (2007).

Since the act of weighting is subjective, various scientists have sought to create methods that reduce the subject nature of a single score. One method employed by Daystar *et al.* (2016)) tested the scenario outcomes using 16 different weighting methods established by LCA experts, product producers, and product users. In some results, the weighting factor can play a major role in influencing the results; however, in Daystar *et al.* (2016), the results were generally the same when different weighting methods were applied. A more robust approach to weighting and single score is the stochastic multiattribute analysis (SMAA) (Prado-Lopez *et al.*, 2014) where all possible weighting factors are used in combination with internal normalization and uncertainty data. This helps provide insight into whether the differences are significant between options as well as the probability of one option resulting in a more favorable outcome than another based on the range of weighting factors tested.

Single Score A single environmental score result is calculated using both the normalized midpoint results and weighting methods as described by the following equation. Communicating the single score results has the advantage of one data point that is easy to compare across multiple scenario option; however, this one value is inherently subjective and has more uncertainty than midpoint indicators. When communicating results in a single score, it is necessary to also provide midpoint indicator score and properly document the normalization and weighting factors used to determine the single score. Study transparency is a critical to the integrity of any LCA study and requires much documentation of these and other methods and data sources.

$$\text{Single Score} = \sum \frac{\text{Midpoint Indicator}}{\text{Normilization Factor}} \times \text{Weighting Value}$$

5.3.2.14 Life Cycle Impact Assessment Interpretation

The impact assessment methods and impact categories selected in the goal and scope can shape the study conclusions. Omitting or including different impact categories is one way in which the study can be drastically altered. Additionally, there are times in which the study set out to examine a wide list of impacts, but available data are not sufficient to calculate one or several of the impact categories. It is important to again interpret the results from each of the LCA steps and to realign new insights with the goal and scope.

With regard to impact assessment, it is important to evaluate if the LCIA results provide enough data to make decisions or to take action. At times, there can be two options that present environmental trade-offs or where one options will have higher values for some impacts and lower values for others. In these scenarios, LCIA left at a midpoint indicator can do little to help a decision maker. Additional analysis such as single scores or an SMAA can be performed to provide further guidance into the most advisable action or decision (Prado-Lopez *et al.*, 2014).

5.3.2.15 Problems Set –Life-Cycle Impact Assessment

Problem 1 Use the NREL US LCI database (discussed later) and determine the emissions that contribute to GWP for 1 person*km and also for a transport scenario of 100 people transported 600 km. Then calculate the GWP using the characterization factors shown in this chapter for the 100-year horizon and for the 100 people transported 600 km. Use the following transportation operations.

a. Transport, intercity bus, diesel powered, south
b. Transport, passenger car, gasoline powered

Transport, intercity bus, diesel powered, south

Flow	Category	Type	Unit	Amount
Outputs				
Ammonia	Air/unspecified	ELEMENTARY_FLOW	kg	5.49E−07
Carbon dioxide, fossil	Air/unspecified	ELEMENTARY_FLOW	kg	4.04E−02
Carbon monoxide, fossil	Air/unspecified	ELEMENTARY_FLOW	kg	9.37E−05

Flow	Category	Type	Unit	Amount
Dinitrogen monoxide	Air/unspecified	ELEMENTARY_FLOW	kg	5.40E−08
Hydrocarbons (other than methane)	Air/unspecified	ELEMENTARY_FLOW	kg	1.52E−05
Methane	Air/unspecified	ELEMENTARY_FLOW	kg	3.59E−07
Nitrogen dioxide	Air/unspecified	ELEMENTARY_FLOW	kg	2.61E−05
Nitrogen oxide	Air/unspecified	ELEMENTARY_FLOW	kg	3.13E−04
Nitrogen oxides	Air/unspecified	ELEMENTARY_FLOW	kg	3.39E−04
Particulates, <10 um	Air/unspecified	ELEMENTARY_FLOW	kg	1.90E−06
Particulates, <10 um	Air/unspecified	ELEMENTARY_FLOW	kg	4.64E−07
Particulates, <10 um	Air/unspecified	ELEMENTARY_FLOW	kg	1.95E−05
Particulates, <2.5 um	Air/unspecified	ELEMENTARY_FLOW	kg	1.11E−07
Particulates, <2.5 um	Air/unspecified	ELEMENTARY_FLOW	kg	1.89E−05
Particulates, <2.5 um	Air/unspecified	ELEMENTARY_FLOW	kg	4.97E−07
Sulfur dioxide	Air/unspecified	ELEMENTARY_FLOW	kg	6.76E−07
Transport, intercity bus, diesel powered, South	Transit and ground pa	PRODUCT_FLOW	p*km	1.00E+00
VOC, volatile organic compounds	Air/unspecified	ELEMENTARY_FLOW	kg	1.57E−05
Inputs				
Diesel, at refinery	Petroleum and coal pr	PRODUCT_FLOW	l	1.54E−02
Transport, barge, average fuel mix	Water transportation/	PRODUCT_FLOW	t*km	4.79E−03
Transport, combination truck, average fuel mix	Truck transportation/	PRODUCT_FLOW	t*km	4.96E−03
Transport, ocean freighter, average fuel mix	Water transportation/	PRODUCT_FLOW	t*km	3.39E−02
Transport, pipeline, unspecified petroleum products	Utilities/fossil fuel ele	PRODUCT_FLOW	t*km	5.00E−02
Transport, train, diesel powered	Rail transportation/r	PRODUCT_FLOW	t*km	9.25E−03

Transport, passenger car, gasoline powered

Flow	Category	Type	Unit	Amount
Outputs				
Ammonia	Air/unspecified	ELEMENTARY_FLOW	kg	1.55E-05
Carbon dioxide, fossil	Air/unspecified	ELEMENTARY_FLOW	kg	1.47E-01
Carbon monoxide, fossil	Air/unspecified	ELEMENTARY_FLOW	kg	2.52E-03
Dinitrogen monoxide	Air/unspecified	ELEMENTARY_FLOW	kg	5.44E-06

Flow	Category	Type	Unit	Amount
Hydrocarbons (other than methane)	Air/unspecified	ELEMENTARY_FLOW	kg	2.16E-04
Methane	Air/unspecified	ELEMENTARY_FLOW	kg	8.38E-06
Nitrogen dioxide	Air/unspecified	ELEMENTARY_FLOW	kg	3.19E-05
Nitrogen oxide	Air/unspecified	ELEMENTARY_FLOW	kg	2.75E-04
Nitrogen oxides	Air/unspecified	ELEMENTARY_FLOW	kg	3.07E-04
Particulates, <10 um	Air/unspecified	ELEMENTARY_FLOW	kg	4.81E-06
Particulates, <10 um	Air/unspecified	ELEMENTARY_FLOW	kg	2.26E-06
Particulates, <10 um	Air/unspecified	ELEMENTARY_FLOW	kg	5.78E-06
Particulates, <2.5 um	Air/unspecified	ELEMENTARY_FLOW	kg	5.41E-07
Particulates, <2.5 um	Air/unspecified	ELEMENTARY_FLOW	kg	1.26E-06
Particulates, <2.5 um	Air/unspecified	ELEMENTARY_FLOW	kg	5.33E-06
Sulfur dioxide	Air/unspecified	ELEMENTARY_FLOW	kg	2.74E-06
Transport, passenger car, gasoline powered	Transit and ground pa	PRODUCT_FLOW	p*km	1.00E + 00
VOC, volatile organic compounds	Air/unspecified	ELEMENTARY_FLOW	kg	2.22E-04
Inputs				
Gasoline, at refinery	Petroleum and coal pr	PRODUCT_FLOW	l	6.30E-02
Transport, barge, average fuel mix	Water transportation/	PRODUCT_FLOW	t*km	1.96E-02
Transport, combination truck, average fuel mix	Truck transportation/	PRODUCT_FLOW	t*km	2.03E-02
Transport, ocean freighter, average fuel mix	Water transportation/	PRODUCT_FLOW	t*km	1.39E-01
Transport, pipeline, unspecified petroleum products	Utilities/fossil fuel ele	PRODUCT_FLOW	t*km	2.05E-01
Transport, train, diesel powered	Rail transportation/r	PRODUCT_FLOW	t*km	3.79E-02

Problem 2 A cellulose super absorbent production plant is producing absorbent at a rate of 50,000 kg per day. The factory burns nonrenewable natural gas at a rate of 0.06 million BTU per kg of product. During the combustion, the following are released per kg of product: 0.05 g CH_4, 0.005 g N_2O, and 1.6 kg of CO_2. All other sources of GWP for the factory are negligible compared to this combustion. What is the gate-to-gate carbon footprint in kg CO_2e per kg of super absorbent? Characterization factor CO_2: 1 kgCO_2e/kg CO_2, Characterization factor CH_4: 34 kgCO_2e/kg CH_4, Characterization factor N_2O: 298 kgCO_2e/kg N_2O.

Problem 3 The following data were obtained for bio-fuel from pine and gasoline. Normalization and weighting factors are also included. Perform normalization and weighting to arrive at a single score for bio-fuel from pine and from gasoline. Comment on the results of the single score.

Indicator	Units	Bio-fuel – Loblolly Pine		Gasoline	
		Mean	SD	Mean	SD
Acidification	H+ moles eq	2.38E−03	1.61E−04	1.17E−02	8.43E−04
Carcinogens	kg benzene eq	3.04E−06	1.85E−06	3.57E−04	2.84E−04
Ecotoxicity	kg 2,4-D eq	1.23E−03	6.72E−04	8.00E−02	1.84E−02
Eutrophication	kg N eq	2.90E−06	6.43E−07	2.19E−05	5.63E−06
Global warming	kg CO_2 eq	2.17E−02	1.37E−02	1.00E−01	4.55E−03
Noncarcinogens	kg toluene eq	3.30E−02	2.39E−02	1.07E+01	9.17E+00
Ozone depletion	kg CFC-11 eq	9.30E−11	2.24E−11	1.93E−10	4.05E−11
Respiratory effects	kg PM2.5 eq	8.01E−06	4.06E−07	3.00E−05	1.93E−06
Smog	kg NO_x eq	2.77E−04	1.01E−05	1.46E−04	8.77E−06

Normalization factors

Impact category	Per year per capita	Total normalized value per capita
Global warming	kg CO_2 eq	24,500
Acidification	H+ moles eq	7440
Carcinogens	kg benzene eq	0.26
Noncarcinogens	kg toluene eq	1470
Respiratory effects	kg PM2.5 eq	76
Eutrophication	kg N eq	18
Ozone depletion	kg CFC-11 eq	0.31
Ecotoxicity	kg 2,4-D eq	74
Smog	kg NO_x eq	121

Weighting factors

Impact category	Weighting factor (%)
Global warming	29
Respiratory effects	9
Cancerous	8
Ecological toxicity	8
Eutrophication	6

Impact category	Weighting factor (%)
Noncancerous	5
Smog formation	4
Acidification	3
Ozone depletion	2

5.4 LCA Tools for Forest Biomaterials

There are many tools developed in various software languages that are aimed at specific aspects of LCA or examining an area of products. The focus of this book and chapter is aimed at biomaterials; we review two tools that are commonly used to perform LCAs of wood-based products and bio-fuels. These tools are thought to be robust, have ample documentation, and a wide breadth of analysis scope that is useful to learning and performing analyses in these areas.

5.4.1 FICAT

The Forest Industry Carbon Accounting Tool (FICAT) produced by the National Council for Air and Stream Improvement (NCASI) is "a [free] user-friendly tool that enables users to model the greenhouse gas 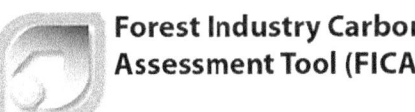 and carbon impacts of forest products sector projects, and to identify potential opportunities for improvements." (NCASI). The FICAT model programed by NCASI has several tabs, as listed in the following that walk the user through performing an analysis.

- Welcome
- 1. Carbon in forest ecosystems
- 2. Carbon in forest products
- 3. Greenhouse gas emissions from forest product manufacturing facilities
- 4. Greenhouse gas emissions associated with forestry operations, fiber recovery operations and non-wood fiber production
- 5. Greenhouse gas emissions associated with producing other raw materials/fuels
- 6. Greenhouse gas emissions associated with purchased electricity, steam, and heat
- 7. Transport-related greenhouse gas emissions
- 8. Emissions associated with product use
- 9. Emissions associated with product end-of-life
- 10. Avoided emissions
- Summary
- Uncertainty
- Benchmarking

The FICAT model has many default values included that can be used; however, it is encouraged that user- and project-specific data be used whenever possible. With some primary data, assumptions, and default data in FICAT, a carbon footprint of paper and wood products can be calculated relativity simply compared to modeling the same system in many other LCA software packages. The FICAT model is also useful when to calculate direct land-use change, carbon in products, energy emissions, other aspects of LCA pertaining to forest-based products. Emission factors, LUC results, and other aspect can be calculated using FICAT then used in other software or for other aspects of analysis outside of the software. Utilizing this tool specifically designed for forest-based products can reduce the effort required to do an LCA; however, the results are limited to GHG emissions, which may not be sufficient in all studies.

5.4.2 GREET Model

The Greenhouse Gases, Regulated Emissions, and Energy Use in Transportation (GREET) model created by Argonne National Laboratory is a tool used to calculate the environmental parameters surrounding  transportation fuels. There is an online version that is relatively simple to use as well as a more transparent Microsoft Excel-based model that has all the background data and calculations. This model useful not only for determining the environmental impacts of bio-fuels but also can be used in part for LCAs of other bio-based products. Many of the bio-fuel unit processes (e.g., feedstock handling, pretreatment, and hydrolysis) are common to other products besides bio-fuels, and the data in the Excel-based model can be useful to those studies as well. The documentation for this model is robust and provides in detail the methods used to perform the analysis. It is suggested that the readers explore this free model and read the documentation as it will develop understanding and skills surrounding bio-fuels LCA.

References

Bare, J., Gloria, T., and Norris, G. (2006). Development of the method and U.S. normalization database for life cycle impact assessment and sustainability metrics. *Environmental Science and Technology*, **40**(16), 5108–5115.

Curran, M. A. (Ed.). (2012). *Life cycle assessment handbook: a guide for environmentally sustainable products*. John Wiley & Sons.

Daystar, J., Venditti, R., and Kelley, S. S. (2016). Dynamic greenhouse gas accounting for cellulosic biofuels: implications of time based methodology decisions. *The International Journal of Life Cycle Assessment*, 1–15.

Gloria, T. P., Lippiatt, B. C., and Cooper, J. (2007). Life cycle impact assessment weights to support environmentally preferable purchasing in the United States. *Environmental Science & Technology*, **41**(21), 7551–7557.

Goedkoop, M., and Spriensma, R. (2001). The Eco-indicator99: A Damage Oriented Method for Life Cycle Impact Assessment: Methodology Report.

ISO 14040. (2006). *Environmental management – Life cycle assessment – Principles and framework*. International Organisation for Standardisation (ISO), Geneva.

ISO 14044. (2006). *Environmental management – Life cycle assessment – Requirements and guidelines.* International Organisation for Standardisation (ISO), Geneva. https://www.iso.org/obp/ui/#iso:std:iso:14044:ed-1:v1:en

ISO 14046. (2014). *Environmental management – Life cycle assessment – Requirements and guidelines.* International Organisation for Standardisation (ISO), Geneva.

Levasseur, A., Lesage, P., Margni, M., Deschênes, L., and Samson, R. (2010). Considering time in LCA: dynamic LCA and its application to global warming impact assessments. *Environmental science & technology*, **44**(8), 3169–3174.

Ministry of Housing. (2000). Spatial Planning and the Environment, https://www.pre-sustainability.com/download/EI99_Manual.pdf

Myhre, G., Shindell, D., Bréon, F.-M., Collins, W., Fuglestvedt, J., Huang, J., Koch, D., Lamarque, J.-F., Lee, D., Mendoza, B., Nakajima, T., Robock, A., Stephens, G., Takemura, T., and Zhang, H. (2013). Anthropogenic and natural radiative forcing. In: *Climate change 2013: The physical science basis. Contribution of Working Group I to the Fifth Assessment Report of the Intergovernmental Panel on Climate Change.* Stocker, T. F., Qin, D., Plattner, G.-K., Tignor, M., Allen, S. K., Boschung, J., Nauels, A., Xia, Y., Bex, V., and Midgley, P. M. (eds.). Cambridge University Press, Cambridge, United Kingdom and New York, NY, USA. Anthropogenic and Natural Radiative Forcing.

Novick, D. (1959). *The federal budget as an indicator of government intentions and the implications of intentions.* The RAND Corporation, Santa Monica, CA, P-1803.

Prado-Lopez, V., Seager, T. P., Chester, M., Laurin, L., Bernardo, M., and Tylock, S. (2014). Stochastic multi-attribute analysis (SMAA) as an interpretation method for comparative life-cycle assessment (LCA). *The International Journal of Life Cycle Assessment*, **19**(2), 405–416.

Ryberg, M., Vieira, M. D., Zgola, M., Bare, J., Rosenbaum, R. K. (2014). Updated US and Canadian normalization factors for TRACI 2.1. *Clean Technologies and Environmental Policy*, **16**(2), 329–339.

UCLA. (2016). UCLA Sustainability, https://www.sustain.ucla.edu/about-us/what-is-sustainability/.

Wolf, M. A., Pant, R., Chomkhamsri, K., Sala, S., and Pennington, D. (2012). The International Reference Life Cycle Data System (ILCD) Handbook-JRC Reference Reports.

6

First Principles of Pretreatment and Cracking Biomass to Fundamental Building Blocks

Amir Daraei Garmakhany[1] and Somayeh Sheykhnazari[2]

[1] *Buali Sina University, Toyserkan Faculty of Industrial Engineering, Department of Food Science and Technology, Hamedan, Iran*
[2] *Gorgan University of Agricultural Sciences & Natural Resources, Department of Wood and Paper Technology, Gorgan, Iran*

There is a crack in everything. That's how the light gets in.

Leonard Cohen, 1956–1968

6.1 Introduction

The growth of population has increased the need for food and energy, but unfortunately a large number of people in the world encounter shortage of food and energy because of inadequacy of food sources and fossil fuels.

According to the aforementioned reason, nowadays, preparing food materials and energy are two principal challenges in the world. The human used biotechnology and nuclear sciences to supply these needs. He has increased agricultural productions by using racial modification and repellence of plant diseases. Production of fuel and energy using nuclear energy is not accessible to all countries. Biotechnology (fermentation technology) can prepare many needs of society especially energy. Fermentation process requires several factors, including special microorganisms, media, culture, and optimal process (time, temperature, and feeding rate) conditions. Since the basic part of fermentation process costs is media culture, one of the goals in fermentation process is using low-cost media. Carbon is the energy source of media culture, and so the cost of media is mostly related to it. For this reason, the cost of the fermentation process decreases by using inexpensive carbon sources. Lignocellulosic materials are most abundant renewable materials on the earth (1×10^{10} million tons per year). They broadly comprise cellulose and hemicelluloses, which are composed of sugar units. Consequently, they can be used as a low-cost substrates to prepare the required energy in fermentation processes. Lignin can be used to produce chemicals or directly to supply thermal energy via combustion. The main sources of lignocellulosic materials consist of tree wood (softwood and hardwood), agricultural wastes, algae, sulfite wastes of paper-making factories, and so on. To utilize lignocellulosic materials several factors, including the annual production rate, accessibility, the costs of transportation

Introduction to Renewable Biomaterials: First Principles and Concepts, First Edition.
Edited by Ali S. Ayoub and Lucian A. Lucia.
© 2018 John Wiley & Sons Ltd. Published 2018 by John Wiley & Sons Ltd.

and conversion procedure, the amount of production yield of final products (sugars, alcohol, organic acids, …), and fermentation inhibitors should be considered.

Forests and trees are one of the main resources of lignocellulosic materials. But their production is a time consuming process, and irregular cutting of trees is forbidden in some countries. Marine algae is another resource of cellulose, but the amount of their cellulose content and dry matter is low; therefore it is not an economic source. The agricultural plants are generated in large quantities each year, and so they are suitable resources to produce cellulose. Besides this, the agricultural wastes such as bark and pulp of fruits after processing can be used to generate fermentative product feedstock. Eighty years ago, extensive researches were conducted to produce fermentative products especially bioethanol. In advanced countries like America, Canada, Brazil, Cuba, and European countries, bioethanol is used as a part of (20%) vehicle fuel. Industrial processes for bioethanol production use sugar cane, corn, and cereal grain as feedstocks, since the price of feedstock contributes more than 55% of the production cost, free or low-cost feedstocks such as hemicellulosic biomass and agri-food wastes can be used to reduce the cost of fermentation process (Del Campo et al., 2006). Here, there are several questions related to lignocellulosic materials. First, how can the potetial of lignocellulosic materials be used? How can these materials be processed? Will this process be economic and efficient and will not produce fermentative inhibitors? To answer these questions, the structure and chemical compositions of lignocellulosic materials should be attended. As stated earlier, stem of plants and tree trunk are composed of lignocellulosic materials. Cellulose is the main component of plant cell wall. It is responsible for plant's chemical and mechanical resistance (Raven et al., 1992). Hemicellulose is a copolymer with 5- or 6-carbon sugars. Lignin is an aromatic polymer that forms a shield on plant cell wall. It is evaluated that 7.5×10^{10} tons of cellulose is consume and regenerated per year (Kirk-Otmer, 2001). Thus cellulose is the most abundant polymer on the earth. There is a little protein, mineral, and so on in cellulosic material structures. The variety of lignin, hemicelluloses, and cellulose depends on the production source (hardwoods, softwoods, or grasses). Cellulose is a polymer of sugar units. Hence, it is fermentable and can be converted by chemical treatment in the form of different products such as carboxymethyl cellulose (CMC), hydroxypropyl methyl cellulose (HPMC), and crystalline cellulose. These are hydrocolloid materials and act as coating agent, thickening agent, and so on in different industries especially in the food industry. Cellulose is found in crystalline and amorphous forms. Few chains of polymer are connected together and form microfibril, and so connection of different microfibrils together lead to the formation of fibers. By this way, crystalline structure of cellulose is generated. To separate cellulose from lignocellulosic materials, first it should be released because cellulose and hemicellulose connect together, and there are cross-links between them. After that, crystalline structure of cellulose must be converted to amorphous structure that is hydrolyzed better by enzymes (Mtui, 2009).

Before two decades, extensive researches have been conducted about the efficient conversion of lignocellulosic materials to fermentable sugars. Based on these researches, we can say the majority of the carbohydrates can be hydrolyzed to single sugars or monosaccharides and fermented to ethanol. However, in the case of cellulose and hemicelluloses, the necessary technology for this process differs from that which is employed by the conventional starch-to-ethanol industry because of the complex chemical structure of their compounds. An initial pretreatment stage (acid hydrolysis, steam explosion, wet oxidation, etc.) is needed to soften the material and break down its structure to make it more susceptible to enzymatic attack before fermentation (Mtui, 2009). All pretreatments have severe reaction condition and high equipment and processing cost, and special problems related to the recovery of expensive catalyzer. During the pretreatment process, degradation compounds of pentoses and hexoses primarily furfural and 5-hydroxymethyl furfural (5-HMF) are formed. These components are toxic and inhibit the subsequent enzymatic and fermentative processes. Therefore, they must be removed or neutralized before the fermentation process (Palmqvist and Hahn-Hägerdal, 2000). After initial pretreatments on lignocellulosic materials and obtaining access to cellulose, enzymatic hydrolysis should be conducted . It is executed by cellulose enzymes such as endoglucanase, exoglucanase, and β-glucosidase. Enzymatic hydrolysis improve by optimization of substrate concentration, amount of enzyme, using surfactant and enzymatic mixtures. In the following section, pretreatment is discussed, but at the first it is better to discuss about lignocellulosic materials.

6.1.1 What Is Lignocellulosic Material?

6.1.1.1 Lignocellulosic Materials

Lignocellulosic materials are most abundant biopolymers on the earth. They formed from some basic constituents that have in their turn, also complicated internal structure (Chunping et al., 2008). For a clear understanding about lignocellulosic materials, an analysis of the structure and properties of lignocellulosic materials and its components are explained in this section. The physical properties of each of the components of lignocellulosic materials and how each of these components contributes to the behavior of the complex structure as a whole are addressed. As previously mentioned, the principle function of pretreatment process is to break down the biomass structure and the use of its constituent for producing fermentable sugars. Figure 6.1 shows the general structure of lignocellulosic biomass.

6.1.1.2 Cellulose

Cellulose, the most abundant constituent of the plant wall is a homopolysaccharide composed entirely of D-glucose linked together with β-glycosidic bonds and with a degree of polymerization (DP) up to 10,000 or higher (Keshwani, 2009). Indeed cellulose is the β-1,4-polyacetal of cellobiose (4-O-β-D-glucopyranosyl-D-glucose). Cellulose is more commonly considered as a polymer of glucose because cellobiose consists of two molecules of glucose. The chemical formula of cellulose is $(C_6H_{10}O_5)_n$, and the schematic structure of one chain of the polymer is given in Figure 6.2.

Many characteristics of cellulose depend on its DP, that is, the number of glucose units that make up one polymer molecule. The DP of cellulose can extend to a value of 17,000, even though more normally a number of 800–10,000 units are encountered (Kirk-Otmer, 2001). For example, cellulose from wood pulp has a DP between 300 and 1700.

The nature of the bond between the glucose molecules (β-1,4-glucosidic) allows the polymer to be arranged in long straight chains. The latter arrangement of the molecule, together with the fact that the hydroxides are evenly scattered on both sides of the monomers, leads to the formation of hydrogen bonds between the molecules of cellulose. The hydrogen bonds in turn result in the formation of a compound that

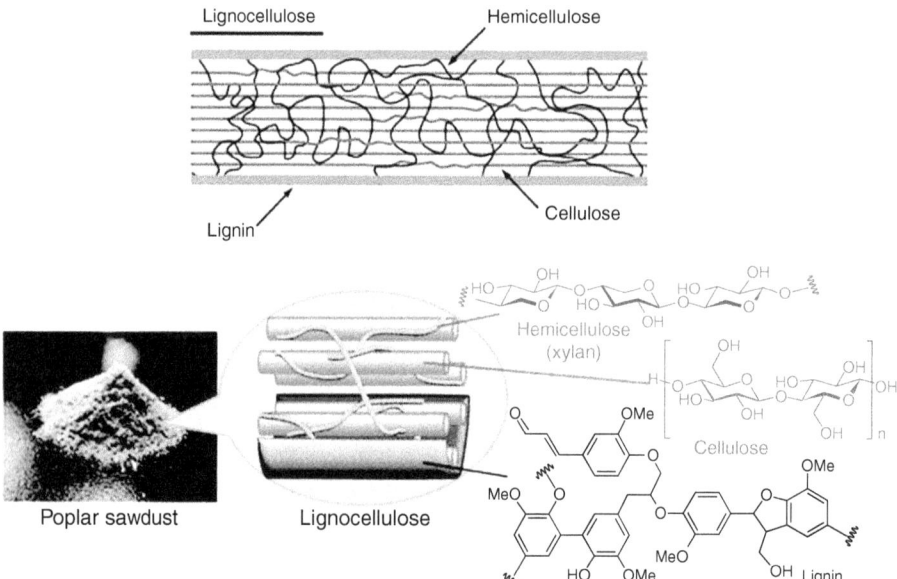

Figure 6.1 General structure of lignocellulosic materials.

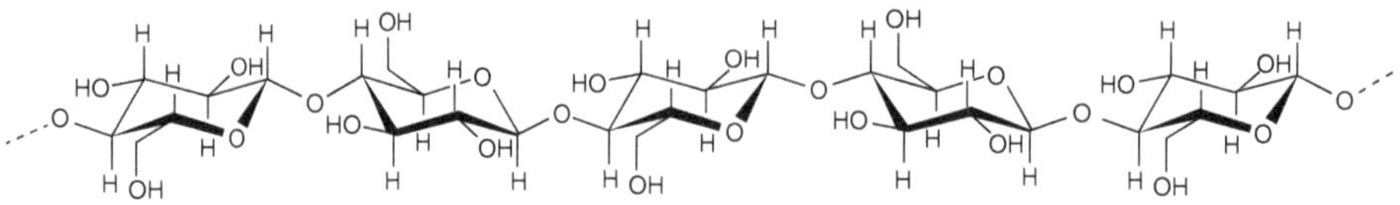

Figure 6.2 Schematic structure of cellulose molecule.

is comprised of several parallel chains attached to each other (Faulon et al., 1994). The schematic of the arrangement of the cellulose molecules in parallel chains and the associated hydrogen bonding is presented in Figure 6.3.

The linear structure of the cellulose chain enables the formation of both intra- and intermolecular hydrogen bonds resulting in the aggregation of chains into elementary crystalline. Fibrils of 36 cellulose chains form the structure of the elementary crystalline fibrils. The structure of cellulose along with the intermolecular hydrogen bonds increases cellulose tensile strength, makes it insoluble in most solvents, and is partly responsible for the resistance of cellulose against microbial degradation. The hydrophobic surface of cellulose leads to the formation of a dense layer of water that may hinder diffusion of enzymes and degradation products near the surface of cellulose (Kirk-Otmer, 2001; Jorgensen et al., 2007).

Cellulose is found in crystalline and the noncrystalline or amorphous structures. The coalescence of several polymer chains leads to the formation of microfibrils, which in turn are united to form fibers. By this way, cellulose gets a crystalline structure. Figure 6.4 shows the structure and the placement of cellulose in the plant cell wall.

Cellulose absorbs 8–14% water under normal atmospheric conditions (20 °C, 60% relative humidity), and so it is a relatively hygroscopic material. On the other hand, it is insoluble in water, where it swells, and it is also insoluble in dilute-acid solutions

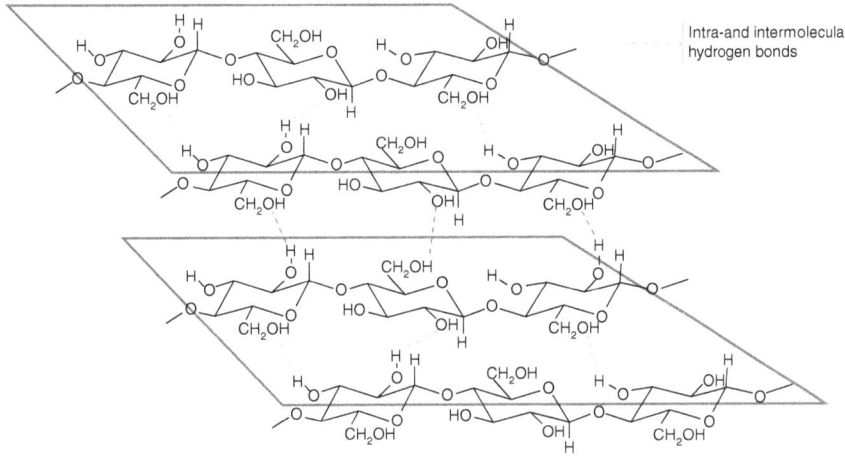

Figure 6.3 Demonstration of the hydrogen bonding that allows the parallel arrangement of the cellulose polymer chains.

at low temperature. The solubility of the polymer depends strongly on the degree of hydrolysis achieved. As a result, factors affecting the hydrolysis rate of cellulose also affect its solubility, however, with the molecule being in a different form than the native one. It becomes soluble at higher temperatures, since the energy provided is adequate to break the hydrogen bonds holding the crystalline structure of the molecule. Cellulose is also soluble in concentrated acids, but severe degradation of the polymer by hydrolysis is turning. In alkaline solutions, extensive swelling of cellulose takes place in addition to dissolution of the low-molecular-weight fractions of the polymer (DP < 200) (Krassig and Schurz, 2002). Different solvents of cellulose that have been applied in industrial or laboratory practice consist of uncommon and complex systems, such as cupriethylenediamine hydroxide or the cadmium complex cadoxen. Furthermore, aqueous salt solutions, such as zinc chloride, dissolve limited amounts of cellulose (Kirk-Otmer, 2001). Cellulose does not melt with temperature, but its decomposition starts at 180 °C.

6.1.1.3 Hemicellulose

Hemicelluloses are complex heterogeneous polysaccharides composed of monomeric residues including D-glucose, D-galactose, D-mannose, D-xylose, L-arabinose, D-glucuronic acid, and 4-O-methyl-D-glucuronic acid that are found in the plant cell wall and have different composition and structure depending on their source and the extraction method. Hemicelluloses have a DP lower than 200, side chains, and can be acetylated. Hemicelluloses are classified according to main sugar that exists in the backbone of the polymer, for example, xylan (β-1,4-linked xylose) or mannan (β-1,4-linked mannose). Plants belonging to the grass family (Poaceae), for example, rice, wheat, and switchgrass have hemicelluloses that are composed of mainly glucuronoarabinoxylans. In softwoods such as fir, pine, and spruce, galactoglucomannans are the principal hemicelluloses, while arabinoglucuronoxylans are second most abundant. In hardwoods species, such as birch, poplar, and so on, 4-O-methyl-glucuronoxylans are the most abundant hemicellulose units. Hemicelluloses with glucomannans are the second most

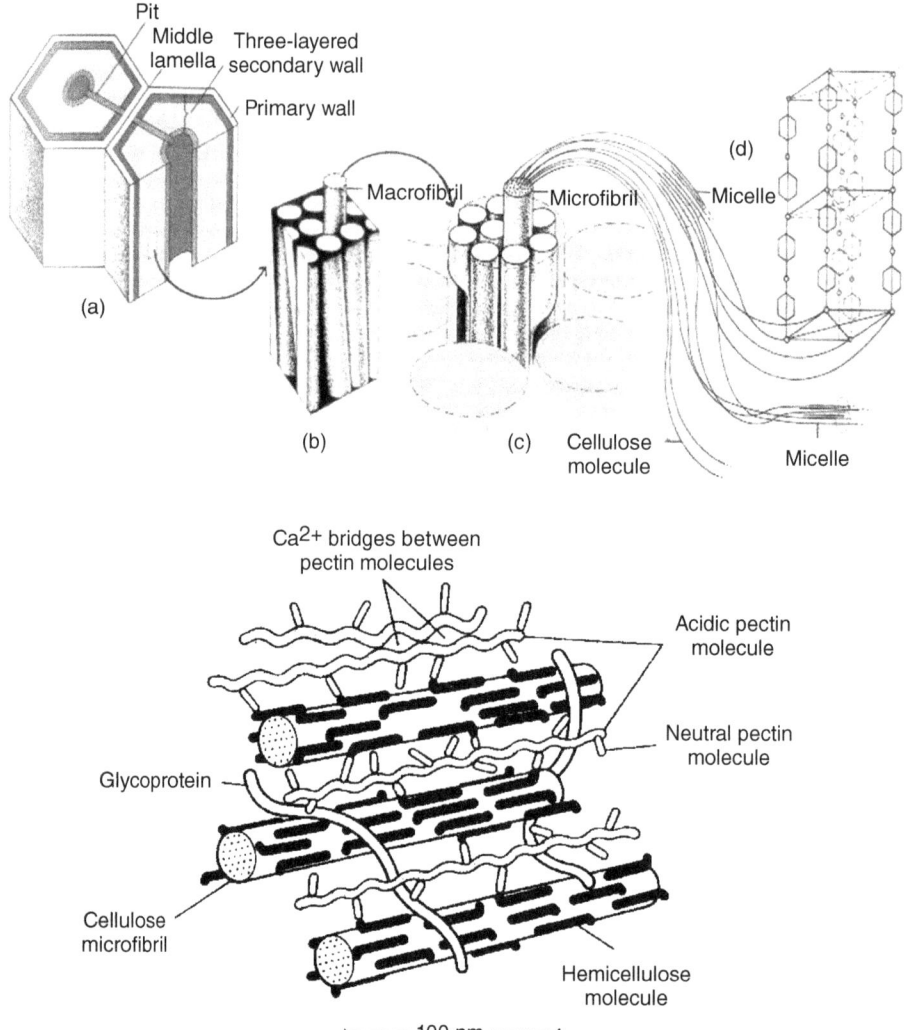

The relative arrangements of molecule types in a primary cell wall.

Figure 6.4 (a) Formation of micro- and macrofibrils (fibers) of cellulose and their position in the wall. (b) Magnified view of cellulose microfibril.

abundant. Due to these differences in hemicellulose compositions, agricultural waste products as well as hardwoods are rich in pentose sugars, such as xylose, whereas softwoods are rich in hexose sugars (Kirk-Otmer, 2001; Jorgensen et al., 2007).

As shown in Figure 6.5, the molecule of a xylan involves 1 → 4 linkages of xylopyranosyl units with α-(4-O)-methyl-D-glucuronopyranosyl units attached to anhydroxylose units. The result is a branched polymer chain that is largely composed of 5-carbon sugar monomers, xylose, and to a minor extent 6-carbon sugar monomer such as glucose.

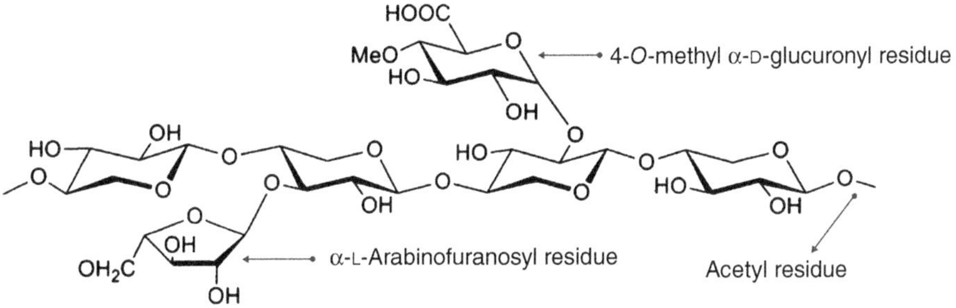

Figure 6.5 A schematic structure of the hemicellulose backbone.

The most important aspects of the structure and composition of hemicellulose are the lack of crystalline structure, mostly because of the highly branched structure, and the existence of acetyl groups linked to the polymer chain (Kirk-Otmer, 2001).

Hemicellulose produced from plants have a high degree of polydispersity, polydiversity, and polymolecularity (a wide range of size, shape, and mass characteristics). However, the DP does not top to the 200 units, while the minimum limit can be approximately 150 monomers.

Hemicellulose is insoluble in water at low temperature, while its hydrolysis begins at a lower temperature than cellulose, which renders it soluble at elevated temperatures. Its solubility in water improves in the presence of acid.

6.1.1.4 Lignin

Lignin is an aromatic and natural polymer with very complicated structure and nearly forms about 1/4 to 1/3 dry weights of trees. Lignin is composed of phenyl propane units and constitutes the most abundant nonpolysaccharide fraction in lignocellulosic biomass. The three monomers in lignin are *p*-coumaryl alcohol, coniferyl alcohol, and sinapyl alcohols (Figure 6.6) that are joined together through alkyl–aryl, alkyl–alkyl, and aryl–aryl other bonds and form amorphous three-dimensional polymer. Lignin embeds the cellulose thereby offering protection against microbial and chemical degradation. Furthermore, lignin is able to form different covalent bonds with some hemicelluloses, for example, benzyl ester bonds with the carboxyl group of 4-*O*-methyl-D-glucuronic acid in xylan. More stable ether bonds, also known as lignin carbohydrate complexes (LCCs) can be formed between lignin and arabinose or galactose side groups in xylans and mannans.

Figure 6.6 *p*-Coumaryl-, coniferyl-, and sinapyl alcohol: major building components of the three-dimensional polymer lignin.

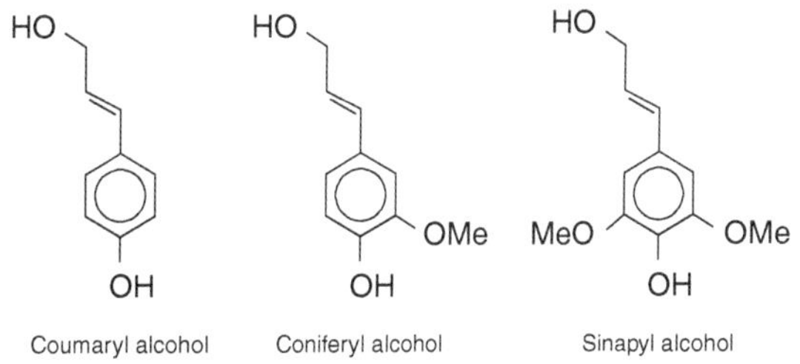

In general, herbaceous plants, such as grasses, have the lowest content of lignin, whereas softwoods have the highest lignin content (Jorgensen et al., 2007).

Higher plants can be divided into two classes: hardwood and softwood. It has been recognized that lignin from softwood is made up of more than 90% of coniferyl alcohol with the remaining being largely *p*-coumaryl alcohol units. In contrast to softwoods, lignin contained in hardwood is composed of varying ratios of coniferyl and sinapyl alcohol units (Kirk-Otmer, 2001). Figure 6.7 shows a model structure of lignin from softwoods. The solvents of lignin consist of low-molecular-weight alcohols, dioxane, acetone, pyridine, and dimethyl sulfoxide. Lignin softens and depolymerizes with increasing temperature, which allows the acceleration of depolymerization reactions of acidic or alkaline nature (Kirk-Otmer, 2001; O'Connor et al., 2007).

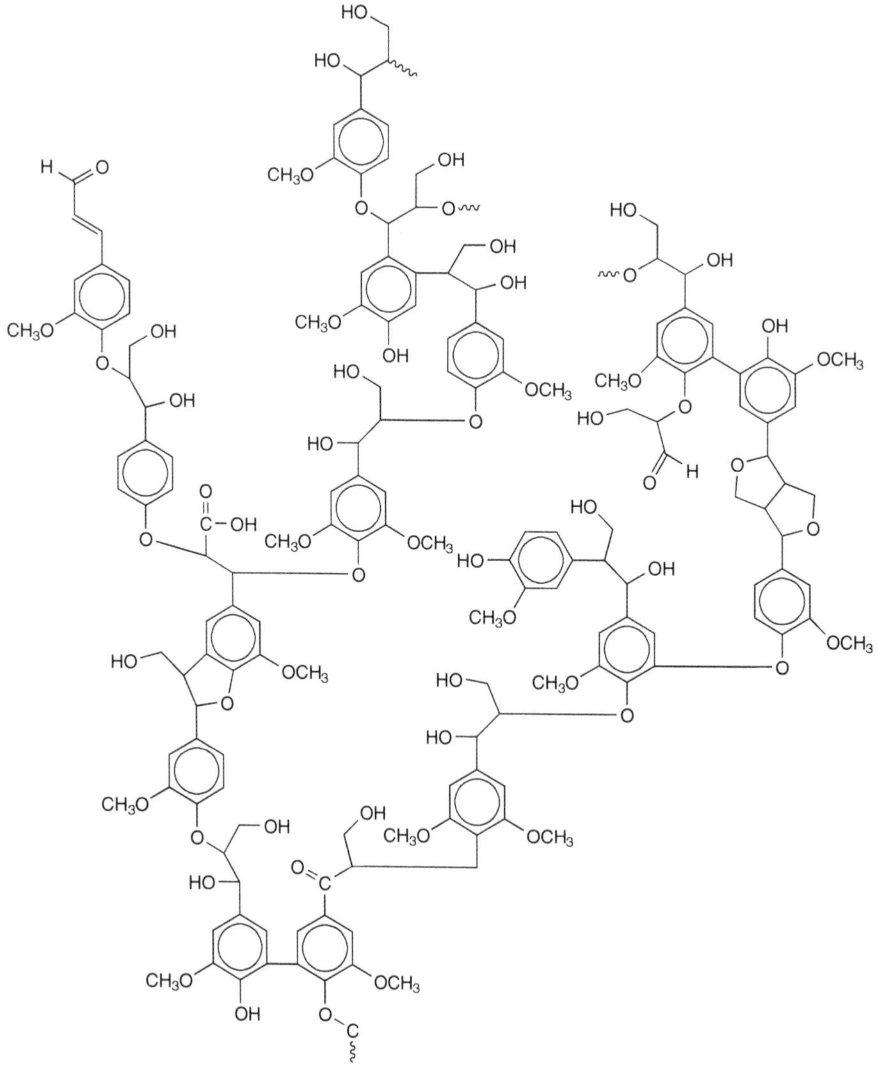

Figure 6.7 Schematic structure of softwood lignin.

6.2 What Difference Should Be Considered Between Wood and Agricultural Biomass?

Plant biomass is composed primarily of cellulose, hemicelluloses, lignin, and smaller amounts of water, pectin, protein, extractive materials, and ash, which do not contribute considerably in forming the structure of the material (Raven *et al.*, 1992). Cellulose, hemicelluloses, and lignin are present in various amounts in the different parts of the plant and they are intimately associated with each other to form the structural framework of the plant cell wall.

Cellulose preserves the crystalline structure, and it appears to be the core of the complex matrix. Hemicellulose is located both between the micro- and the macrofibrils of cellulose. Lignin provides a structural role of the matrix in which cellulose and hemicellulose are entrenched (Faulon *et al.*, 1994).

Since cellulose is the major material of the plant cell walls, most of the lignin is found in the interfibrous area, while a smaller part can also be found on the cell surface (Kirk-Otmer, 2001).

The composition of lignocellulosic materials depends on plant species, age and growth conditions. Distribution of cellulose, hemicellulose, and lignin as well as the content of the different sugars of the hemicellulose, varies significantly between different plants (Table 6.1).

There are four principal bonds recognized in lignocellulosic materials, including ether bonds, ester bonds, carbon—carbon (C—C) bonds, and hydrogen bonds.

These are primary bonds that form the intrapolymer linkages of lignocellulosic material components and connect the different units of lignocellulosic materials (interpolymer linkages). The situation and function of bonds are shown in Table 6.2 (Faulon *et al.*, 1994).

Table 6.1 Composition of different lignocellulosic materials (Jorgensen *et al.*, 2007).

Materials	Glucose[a]	Xylose[b]	Arabinose[b]	Mannose[b]	Lignin	References
Hardwood						
Birch	38.2	18.5	0[c]	1.2	22.8	Jørgensen *et al.* (2007)
Willow	43	24.9	1.2	3.2	24.2	Jørgensen *et al.* (2007)
Softwood						
Spruce	43.4	4.9	1.1	12	28.1	Jørgensen *et al.* (2007)
Pine	46.4	8.8	2.4	11.7	29.4	Jørgensen *et al.* (2007)
Grasses						
Wheat straw	38.2	21.2	2.5	0.3	23.4	Jørgensen *et al.* (2007)
Rice straw	34.2	24.5	n.d[d]	n.d[d]	11.9	Jørgensen *et al.* (2007)
Corn stover	35.6	18.9	2.9	0.3	12.3	Jørgensen *et al.* (2007)

a) Glucose is mainly obtained from cellulose.
b) Xylose, arabinose, and mannose make up hemicellulose.
c) Below detection limit.
d) Not determined.

Table 6.2 Summary of linkages between the monomer units that form the individual polymer lignin, cellulose, and hemicellulose, and between the polymers to form lignocellulosic biomass.

Polymeric intra bonds (bonds within different components)	
Ether bond	Lignin, (hemi)cellulose
Carbon to carbon bond	Lignin
Hydrogen bond	Cellulose
Ester bond	Hemicellulose
Polymeric inter bonds (bonds connecting different components)	
Ether bond	Cellulose–lignin, hemicellulose–lignin
Ester bond	Hemicellulose–lignin
Hydrogen bond	Cellulose–hemicellulose, cellulose–lignin, hemicellulose–lignin

6.2.1 Intrapolymeric Bonds

The most important bonds of lignin are C—C, ether, ester, and hydrogen bonds (Table 6.2). Ether bonds may be formed between allylic carbons, aryl, or carbon atoms of aryl–aryl and between two carbon atoms of allylic. Ether bonds formed about 70% of total bonds between monomeric units in lignin. The C—C bonds form 30% of remaining linkages between units (Kirk-Otmer, 2001).

Cellulose polymer is formed based on two main bonds:

1. Glucosidic bonds constitute the initial chain of a polymer. It is β-1-4-D-glucosidic bond that links glucose units together. Glucosidic bond is also considered as an ether bond. Since it is in fact the connection of two carbon atoms with an elementary oxygen interfering (Solomon, 1988)
2. Hydrogen bond is responsible for the crystalline structure of cellulose. Cellulosic chains are long, straight, and parallel; as a result, they can connect together with hydrogen bond that formed between two hydroxyl groups of different polymer chains (Faulon et al., 1994). Indeed hydroxyl groups are evenly distributed on both sides of the glucose monomer, and this leads to the long, straight, and parallel structure of cellulose polymer.

It has been distinguished that there are carboxyl groups in cellulose (one carboxyl per 100 or 1000 glucose units) (Krassig and Schurz, 2002). In hemicellulose, most principal bonds are ether bonds (as glucosidic and fructosidic bonds) and there are also carboxyl groups. Carboxyl groups can be found as carboxyl, ester, or salt in the molecule (Kirk-Otmer, 2001). The main difference between cellulose and hemicellulose is that the hydrogen bonds are absent in hemicellulose while there is a significant amount of carboxyl groups.

6.2.2 Polymeric Inter Bonds

In order to distinguish and determine the connections and bonds that link different polymers of lignocellulosic materials, first its structures should be broken and separated. The separation of polymers leads to alternation of their original structure, and consequently linkages are not determined clearly. Three types of hydrogen bonds connecting lignin to cellulose and hemicellulose have been found. In addition, the existence of covalence bonds between lignin and polysaccharides has been confirmed. It is also known that hemicellulose connects to lignin by ester bond, and there are ether bonds between lignin and cellulose or between lignin and hemicellulose. The hydrogen bonds between hemicellulose and cellulose are also identified, but these bonds are not strong enough due to the fact that hemicellulose does not have a primary alcohol functional group outside of the pyranoside ring.

6.2.3 Functional Groups and Chemical Characteristics of Lignocellulosic Biomass Components

The important ingredients to produce sugar monomers and finally alcohol or other compositions of lignocellulosic materials are

- functional groups and the ingredient that are involved in the hydrolysis of polysaccharides to their monomers also the consequent degradation reactions of these monomers to furfural
- functional groups and the ingredient that are involved in lignin de polymerization in order to increase cellulose accessibility for hydrolyzing enzyme.

The functional groups of all three components of lignocellulosic biomass are listed in Table 6.3.

The hydrogen bond is not a functional group, since its reaction does not lead to chemical change of the molecule. However, it changes the solubility of the molecule, and it is therefore important for the breakdown of lignocellulosic materials. Concerning the cellulose polymer, the principal goal of pretreatment is to break down the glucosidic (ester) bond that produces sugar monomers. The reactions that occur using functional groups of Table 6.3 are discussed in the following.

6.2.4 Aromatic Ring

Chlorination and nitration reactions are carried out via the electrophilic substitution mechanism. Finally, substitution of the aromatic ring of lignin is performed with chlorine or nitrate groups. This type of substitution is not achieved in the same way. Aromatic ring can be converted to cyclic structures and finally smaller molecules

Table 6.3 Functional groups of lignocellulosic biomass.

Functional group	Lignin	Cellulose	Hemicellulose
Aromatic ring	*		
Hydroxyl group	*		
Carbon to carbon bond	*		
Ether bond	*	*	*
Ester bond			*
Hydrogen bond		*	*

such as mono- and dicarboxylic acids by means of oxidation using different oxidants including chlorine, chlorine dioxide, and oxygen. Oxidative materials can also break the side chain of lignin monomeric units and generate components with three, two, or one carbon atoms (Kirk-Otmer, 2001).

6.2.5 Hydroxyl Group

The hydroxyl group initiates substitution reactions. Under acidic conditions hydroxyl group converted to aryl or allylic ether. Finally, ether substituted by an acid group (e.g., sulfonic acid). The advantage of latter reaction is that the presence of the acid group in the lignin polymer leads the polymer soluble in water and so it is named lignosulfonates (Kirk-Otmer, 2001).

6.2.6 Ether Bond

The ether bonds are most interesting functional groups in lignocellulosic materials because

1- thery are located between glucose monomers (glucosidic bond) and hold them in the polymer chain;
2- they are the primary bonds in the lignin polymer.

Thus, by the break down of the ether bond, lignin separates from polysaccharide matrix and then degradation of lignin and polysaccharides to lignin fragments and monomer sugars takes place.

The cleavage of the ether bond is carried out by solvolytic reactions that take place under acidic or alkaline conditions. The ether bond in lignin is converted to hydroxyl and then converted to carbonyl or carboxyl (before it is finally fragmented to C_3 or C_2 molecules), under acidic conditions. Under alkaline conditions, the mechanisms are different, and the aromatic ring is separated. Figure 6.8 shows an example of ether bond cleavage of the lignin polymer under alkaline condition. Decomposition of the ether bond increases with the addition of hydrosulfide. The ether bond of cellulose polymer can cleave under both acidic and alkaline conditions. In acidic conditions, acid acts as a catalyst protonating the oxygen atom. The charged group (ions) leaves the polymer chain and is substitute by the hydroxyl group of water (Figure 6.9).

In alkaline conditions, the mechanism involves the formation of 1–2 anhydro configuration (Figure 6.10). The intermediate form is a type of epoxide and due to the ring that is formed between the two carbon atoms and oxygen, allows via the S_N2 mechanism the nucleophilic substitution of hydrogen (Solomon, 1988). The use of concentrated alkali and a minimum temperature of 150 °C are required to reach a satisfactory reaction rate.

6.2.7 Ester Bond

Ester bonds are recognized between lignin and polysaccharides as well as hemicellulose polymer. The group of acetyl in hemicellulose is linked to the hydroxyl group of the main chain of the polysaccharide by an ester bond. With respect to LCC, it is not clear whether the ester bond is between lignin and cellulose or between lignin and hemicellulose or between lignin and both polysaccharides (Faulon et al., 1994). In general, hydrolysis is carried out to break the ester bond and results in carboxyl

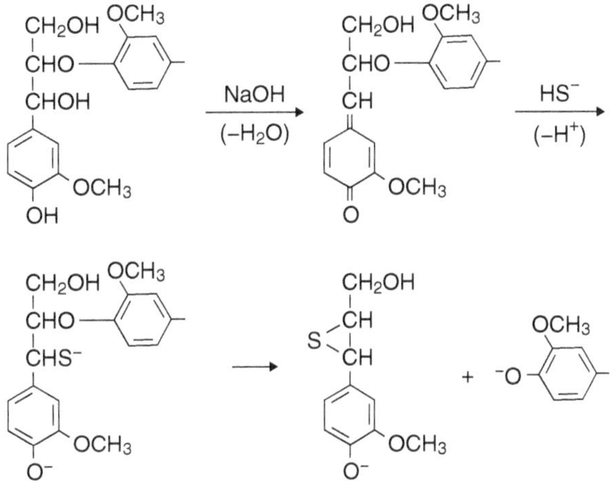

Figure 6.8 Cleavage of the ether bond of lignin in alkaline solution (Lin and Lin, 2002).

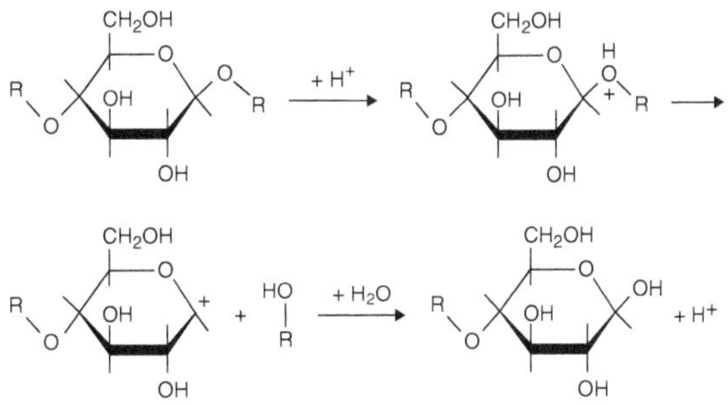

Figure 6.9 Hydrolysis of cellulose in acidic media (Krassig and Schurz, 2002).

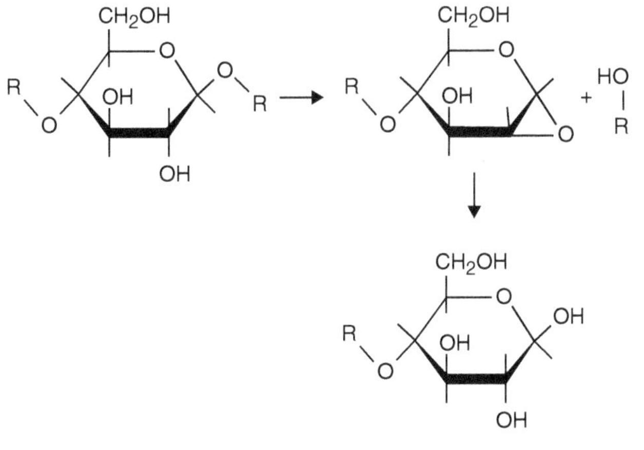

Figure 6.10 Hydrolysis of cellulose in alkaline conditions (Krassig and Schurz, 2002).

and hydroxyl groups. This reaction is reversible and endothermic. The balance is established by surplus water and high temperature. Acidic and alkaline catalysts are used in this procedure. Under alkaline conditions, the reaction is named saponification or esterification. The main difference between these two reactions is that alkaline reaction results in irreversible hydrolysis of ester (Solomon, 1988).

6.2.8 Hydrogen Bond

The existence of hydrogen bonds is distinguished between cellulose chains. The hydrogen bond is formed between a hydrogen atom of one hydroxyl group of a glucose monomer and the oxygen atom of a hydroxyl group of another glucose monomer in the parallel polymer chain of cellulose. Cellulose fiber formation and its insolubility in water is the result of hydrogen bonds. Also, hydrogen bond has been found in hemicellulose polymers. The connection of hemicellulose to cellulose is not very strong, which is due to the absence of primary alcohol functional groups outside the pyranoside ring that restrict the capacity of hemicellulose to form hydrogen bonds.

Breakdown of hydrogen bonds can be done by applying high temperature or by substituting the forming hydrogen bond molecules. Generally, there are two procedures to cleave a hydrogen bond:

1- Recognizing groups that they can form hydrogen bonds of higher energy than the ones formed in cellulose.
2- Varying and modifying the cellulose structure. This action can be achieved by physical destruction of the cellulose or by chemically producing cellulose derivative compound such as cellulose acetate (Bochek, 2003).

To produce the necessary carbon source for preparing media culture for fermentation process, various resources, including simple sugars, starch, fats, and agricultural and food wastes were used. The source of carbon is a large portion of fermentation process costs. Thus, using lignocellulosic wastes is economical and environmentally friendly. Agricultural wastes and crops are an attractive feedstock for use in food and bioenergy production. However, in Asia, Canada, and Europe, rapeseed and other crops are cultivated for the purpose of obtaining oil and foods for food and biodiesel resources. In general, agricultural straw is composed of three main fractions, including about 34% cellulose 19% hemicellulose, and 6.9% lignin. Due to their high content of fermentable sugars (more than 60%), cereal straw has been suggested as a raw material for use as a biofuel feedstock (Kyeong et al., 2011). In order to use this potential sugar source, pretreatment must be used. In the following, a summary of researches done in the field of lignocellulosic pretreatment is discussed. An initial pretreatment stage (acid and alkali hydrolysis, steam explosion, wet oxidation, etc.) is needed to soften the material and break down the biomass structure to make it more susceptible to enzymatic attack before fermentation (Mtui, 2009). After initial pretreatments on lignocellulosic material and access to cellulose, enzymatic hydrolysis must be applied. Cellulose enzymes such as endoglucanase, exoglucanase, and β-glucosidase are employed. Enzymatic hydrolysis can be improved by optimization of substrate concentration, enzyme-to-substrate ratio, and using surfactant and enzyme mixtures (Champagne, 2008).

6.3 Define Pretreatment

6.3.1 What Is the Purpose of Pretreatment?

The purpose of pretreatment is removal of lignin and hemicelluloses, reduction of cellulose crystallinity, and increase of biomass porosity. The ideal pretreatment consists of

- avoiding the need for reducing the size of biomass particles;

- preserving the pentose (hemicelluloses) fractions;
- preventing sugar degradation during the pretreatment process;
- limiting the formation of degradation compounds that inhibit the growth of fermenting microorganisms;
- minimizing energy demands and reducing fermentation costs.

6.4 Steps of Production of Cellulosic Ethanol

The process of bioethanol production from biomass consists of

1) pretreatment of biomass
2) hydrolysis of biomass
3) fermentation of released sugars
4) separation of produced ethanol.

The generation of ethanol from biomass is a complicated procedure.
For efficient ethanol production, these problems must be answered.

6.4.1 Pretreatment

To hydrolyze biomass, it must be attended to inhibitors, especially in pretreatment phase. At this stage, the crystalline structure of cellulose changes, lignin is distilled, and more surface area is formed for enzyme action. The pretreatment stage consists of the softening of biomass, breaking down of cellular structure, and increasing enzymatic digestion of biomass. Pretreatment is effective to produce sugar from biomass and help with the optimization of this goal (Mosier *et al.*, 2005).

6.4.2 Hydrolysis

Fermentation of cellulose to produce ethanol is not possible directly because it is a polymer.

To produce glucose, long chains of cellulose must be a broken down. This process is named hydrolysis or conversion to sugar.

The cellulose fraction of biomass is converted to glucose by hydrolysis. Hydrolysis is carried out by two methods:

- Chemical (acid) hydrolysis
- Enzymatic hydrolysis.

Hydrolysis by concentrated acids like sulfuric acid has been done since many years ago. The time of this procedure is short and its yield is high. Enzymatic hydrolysis is a new method. In this method, enzymes such as cellulase is used to break down cellulose. In order to make bioethanol production more economic, this process must be done fast.

6.4.3 What Are the Inhibitors for Biomass Carbohydrate Hydrolysis?

- Crystalline structure of cellulose with difficult hydrolysis.
- The existence of lignin and hemicellulose complex around cellulose component.
- Lignin links cells to each other and reduces the surface area.
- Hydrolysis of pentoses is more complicated than hexoses (Laureano-perez *et al.*, 2005; Mosier *et al.*, 2005).

6.4.4 Fermentation

The monosaccharides produced by hydrolysis are used in the fermentation process to produce ethanol. A new method known as simultaneous saccharification and fermentation (SSF) is done at one stage, and great effort has been made to perform saccharification and fermentation simultaneously at one stage (Ballesteros et al., 2004).

A part of saccharification is done to hydrolyze cellulose to monosaccharides. A new method of SSF was studied. In this method, first, 6-carbon monosaccharides like glucose are fermented and then 5-carbon monosaccharides such as pentose and arabinose are fermented in the same way by means of enzymes. This procedure needs to be considered closely because the pentose sugars fermentation is done hardly.

6.4.5 Formation of Fermentation Inhibitors

Pretreatment of lignocellulosic materials leads to the formation of different compounds with inhibitory effect on the fermentation process. These inhibitors have toxic effects on the fermenting microorganisms and therefore lead to decreased ethanol yield and production. The toxicity level of these materials depends on fermentation variables such as physiological conditions of microorganisms, dissolved oxygen concentration, and the pH of the medium. Toxic compounds can affect fermenting organisms to a point beyond which the efficient utilization of sugars is reduced thereby leading to decreased product formation.

Furthermore, the fermenting organisms may be resistant to inhibitors slightly or may become gradually adapted to their presence. Conversely, the best approach is to prevent the formation of inhibitors as much as possible through (adaptation of) the pretreatment process conditions or other measures.

The main types of inhibitors are summarized in Table 6.4 and discussed here. The effect of inhibitors is higher when they are present together due to their synergistic effect (Mussatto and Roberto, 2004).

6.4.6 Sugars Degradation Products

During hydrolysis, pentose sugars can degrade to a toxic compound, namely furfural, with its toxicity depending on its concentration in the fermentation media.

Table 6.4 Main types of fermentation inhibitors and their chemical structures.

Phenols	Furfural	5-Hydroxymethylfurfural (HMF)
Acetic acid	Levulinic acid	Formic acid

The relative toxicity of the different inhibitors for ethanol fermentation can be summarized as phenolic compounds > furfural > HMF > acetic acid > extractives.

Hydroxy methyl furfural is a toxic compound originating from hexose degradation. Hydroxy methyl furfural is considered less toxic than furfural, and its concentration in hemicellulose hydrolyzates is normally lower than furfural. The decomposition of hemicellulose is responsible for the formation of inhibitors. Kinetic studies have shown that the production of furfural increases after pretreatment for longer time and at highertemperature. The temperature higher than 160 °C and residence time of acid pretreatment longer than 4 h have been reported to be adequate to produce furfural or hydroxymethyl furfural, and their formation is significant at higher temperature or longer residence terms (McKillip and Collin, 2002).

6.4.7 Lignin Degradation Products

A large variety of compounds (aromatic, polyaromatic, phenolic, and aldehydic) are released from lignin during hydrolysis of lignocellulosic biomass. Phenolic compounds have a considerable inhibitory effect on the fermentation of lignocellulosic hydrolyzates and their toxicity (even at low concentrations) is higher than furfural and HMF, especially those with low molecular weight are the most toxic ones. Phenolic compounds cause a partition and loss of integrity of biological membranes of the fermenting organisms, reducing cell growth and sugar assimilation (Mussatto and Roberto, 2004). The main influencing factors for phenolic compounds formation are process time and temperature. Lignin degradation is negligible at temperatures lower than 180 °C in the absence of strong acid or alkaline conditions.

6.4.8 Acetic Acid

Acetic acid is derived from acetyl groups present in the hemicellulose structure. When the pH of the medium is low, acetic acid ($pk_a = 4.7$) appears in the undissociated form that is liposoluble and diffuses across the plasma membrane. The toxicity of acetic acid varies according to the fermentation conditions. Since the formation of acetic acid begins with hemicellulose hydrolysis, its formation cannot be prevented. However, a higher fermentation pH can decrease this effect or the acid can be neutralized before the fermentation process.

6.4.9 Inhibitory Extractives

Extractive compounds derived from lignocellulosic materials are acidic resins, taninic acids, and terpene acids. The toxicity of these materials is lower than lignin degradation products and acetic acid.

6.4.10 Heavy Metal Ions

Heavy metal ions such as iron, chromium, nickel, and copper can originate from corrosion of hydrolysis equipment during pretreatment process, and their toxicity inhibits the enzymes in the fermenting microorganism's metabolic pathways.

6.4.11 Separation

The separation phase is done after the fermentation stage and produced ethanol is extracted from the fermentation medium by different methods.

6.5 What Are the Key Considerations for Making a Successful Pretreatment Technology?

Pretreatment is a key procedure to produce cellulosic ethanol and plays an important role in decreasing the cost and increasing the total yield during cellulosic ethanol production. Pretreatment is the most expensive step of this process, after cost of original materials (20% of the total cost); therefore, so many researches has been done on biomass pretreatment (Merino and Cherry, 2007; Mosier et al., 2005, Yang and Wyman, 2008).

Pretreatment affects the amount of the necessary enzyme to hydrolyze the substrates, and it is also effective in influencing the factors on substrate and its digestibility by enzyme (Chandra et al., 2007).

The profile of pretreatment can decrease the cost of cellulosic ethanol production. Hydrolysis of biomass without pretreatment has a total yield of sugar, lower than 20%, but after pretreatment, the theoretical yield of sugar production increased to 90% (Alizadeh et al., 2005).

Hence, optimal pretreatment should decrease the amount of lignin and crystalline structure of cellulose and increase the accessible surface for enzymatic hydrolysis. The methods for increasing pretreatment efficiency are too complicated. Therefore, the method of pretreatment should be selected based on the following factors:

- Reduction of chemical consumption

Since most of the chemicals are toxic for the environment and microorganisms, pretreatments should consume lower chemicals.

- Lower medium temperature

The temperature of pretreatment should be reduced. Hemicellulose and lignin are degraded at higher temperatures (about 200 °C) and produce undesirable products such as furfural, HMF, and phenolic compounds that reduce bioethanol production yield and efficiency.

- Biocompatibility

The pretreatment should not generate surplus materials that are dangerous for the environment.

- Recovery of by-products

By-products such as lignin and hemicellulose should be recovered by economical processes.

- Proportional cost of the reactor

The cost of the reactor must be reduced, as increased reactor costs increase the total costs of cellulosic ethanol production.

There are many methods for biomass pretreatment, which have typical benefits and defects. Due to difference in structure of plant cell wall, some ways of pretreatment are effective on wood, whereas some others are effective on grassy plants. Therefore, one way cannot be effective for the all kinds of lignocellulosic materials.

Suitable pretreatment must

- produce maximum fermentable sugar yield;
- produce minimum carbohydrate degradation;
- minimize the formation of microorganism growth inhibitors;
- lead to low energy usage and have economical performance;
- be less expensive and economical in practice.

Simply the goal of pretreatment is the breakdown of lignocellulosic structure and its conversion to monosaccharide components in order to be used as fermentation substrate. To break down lignocellulosic materials, three main factors, including size of pores (Grous et al., 1986), cellulose crystallinity (Goldstein, 1983), and delignification (Dekker, 1988) must be considered. Accessibility to cellulose can be enhanced by hemicellulose separation. Hydrolysis of hemicellulose is simpler than cellulose and results in the formation of bigger pores in microfibrils (McMillan, 1994). This researcher showed that the increase of enzymatic digestibility of cellulose depends directly on the hemicelluloses separation. Grous et al. (1986) showed that there is a positive correlation between accessible surface (pores volume) and produced glucose from enzymatic hydrolysis. Cellulose crystallinity is the second factor to determine the produced glucose efficiency. Higher degree of crystallinity leads to decrease of the hydrolysis rate (Goldstein, 1983). Weimer et al. (1995) showed that thermal and chemical treatments increase the relative crystallinity index (RCI) of amorphous portions. Similar research showed that no significant increase was seen in the rate of RCI in the crystalline portion of cellulose. Delignification can increase polysaccharides' accessibility. Several pretreatment methods including dilute acids and alkali pretreatment, acid hydrolysis, ammonia fiber explosive, steam explosion, and enzymatic hydrolysis can be used for lignocellulosic biomass conversion to fermentable sugars.

The principal goals of these methods are delignification and breakdown of cellulose crystallinity structure as shown in Figure 6.11 (Harmsen et al., 2010).

6.5.1 Effect of Pretreatment on Hydrolysis Process

Pretreatments improve enzymatic hydrolysis by

- increasing the porosity and surface area;
- modifying the lignin structure;
- delignifying and removing lignin from biomass;
- partially depolymerizing of hemicelluloses;
- separating and removing hemicelluloses from biomass.
- reducing cellulose crystallinity.

6.6 What Are the General Methods Used in Pretreatment?

The following are a few pretreatment methods used:

- mechanical pretreatments like particle size reduction (chipping, grinding, and milling and using high pressure (extrusion process);
- physical pretreatments (using irradiation and microwaves, ultrasound and infrared, thermal pretreatment like as steam explosion and compressed hot water);

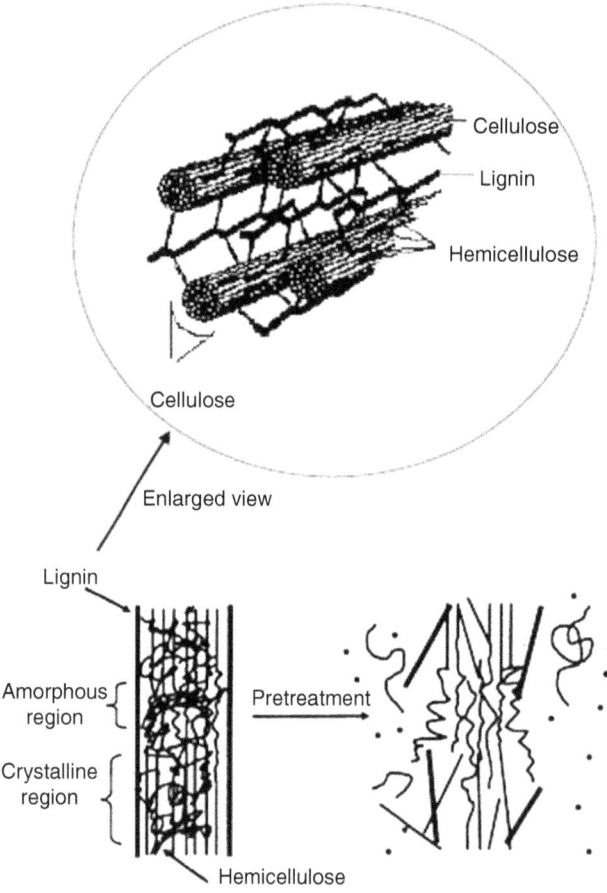

Figure 6.11 Schematic mechanism of effects of pretreatment on lignocellulosic biomass.

- chemical pretreatments such as using alkaline materials, dilute acid, oxidative agents, and organic solutions;
- physicochemical pretreatment (hydrothermal process, using steam autohydrolysis);
- biological and enzymatic pretreatments (using cellulase enzyme or microorganisms and fungus that produce pure cellulase enzyme);
- combined pretreatments in order to achieve maximum yield.

6.7 What Is Currently Being Done and What Are the Advances?

Currently, most researches focus on a combination of two or more pretreatment methods. Some of the pretreatment methods are listed in Table 6.5.

Physical pretreatment can be achieved by different methods, three types of which are discussed in the following.

Table 6.5 The different types of pretreatments and their effects on lignocellulosic biomass.

Mechanism	Example	Method
Particle size reduction Reduction of the cellulose crystallinity	Milling	Physical
Dissolving of hemicelluloses Transformation of lignin	Steam explosion	Physicochemical
Dissolving of hemicelluloses Reduction of the cellulose crystallinity	HCL	Chemical
Delignification Reduction of the cellulose crystallinity	NaOH	Chemical
Delignification	Fungi	Biological

This pretreatment consists of ball milling, chipping, and grinding. Ball milling is done on dry and wet biomass. Generally, ball mill size reduction increases the accessible surface area for enzymatic hydrolysis and reduces the crystalline structure of cellulose. This process is time consuming and not economic (Sun and Cheng, 2002). Lignocellulosic materials can be comminuted by a combination of chipping, grinding, and milling to reduce cellulose crystallinity. The size of the materials is usually 10–30 mm after chipping and 0.2–2 mm after milling or grinding (Sun and Cheng, 2002). The vibratory ball mill has been found to be more effective in breaking down the cellulose crystallinity of wood chips and improving the digestibility of the biomass than ordinary ball milling process (Millet et al., 1976). The power requirement for mechanical comminuting of lignocellulosic materials depends on the final particle size and lignocellulosic biomass characteristics (Cadoche and Lopez, 1989). The rate of energy consumption for material size reduction depends on the particle size and the amount of size reduction. These researchers showed that when the particle size is in the range of 3–6 mm, energy consumption will be less than 30 kWh per ton of biomass, but in practice consumption of energy was higher than this. Also the effect of γ-irradiation on cellulose showed that γ-irradiation breaks down β-1,4-glycosidic bonds, increases the surface area, and reduces the crystalline structure of cellulose, but it is a high-cost pretreatment and not economic. In conclusion, physical and mechanical pretreatments increase biomass hydrolysis efficiency by size reduction, increase accessible surface area for the pretreatment (chemical and enzymatic), and reduce the crystalline structure (Takacs et al., 2000; Sun and Cheng, 2002).

6.7.1 Steam Explosion

The steam-explosion process was first carried out in the year 1925 to produce fiberboard. Then, it was used to produce feed of livestock and pulp of wood. In the beginning 1980, steam explosion was used to pretreat biomass. Iotech Company did researches to investigate the effect of steam-explosion pretreatment on poplar. This company reported to the Department of Energy, United States, the effect of pressure, temperature, and residence time of this pretreatment on the rate of produced xylose and glucose. They showed the maximum of xylose and glucose generation different time and pressures and also xylose is released before glucose. They stated that the optimal condition to produce maximum holocellulose (xylose + glucose) is 500–550 psi pressure and 40 s process time. Later, many studies were conducted by other researchers about the effect of steam-explosion process on different biomass. Shultz et al. (1984) investigated the effect of steam-explosion pretreatment on different biomass such as pretreatment of mixture of wood, husk, corn stem, and bagasse. They showed that pretreatment (temperature 240–250 and time 1 min) increased the rate of enzymatic hydrolysis.

Steam explosion is one of the most common methods for the pretreatment of lignocellulosic materials. In this method, biomass is treated with high-pressure saturated steam, and then the pressure is suddenly reduced to air pressure, which makes the materials undergo an explosive decompression. Steam explosion is typically initiated at a temperature of 160–260 °C (corresponding pressure, 0.69–4.83 MPa) for several seconds to a few minutes before the material is exposed to atmospheric pressure (Sun and Cheng, 2002). In this method, no chemicals are used. During this pretreatment process, degradation compounds of pentoses and hexoses primarily furfural and 5-hydroxymethyl furfural (5-HMF) are formed (Del Campo et al., 2006).

The biomass/steam mixture is held for a period of time to promote hemicelluloses hydrolysis, and the process is terminated by an explosive decompression. The process causes hemicellulose degradation and lignin transformation due to high temperature, thus increasing the potential of cellulose hydrolysis. Hemicellulose is thought to be hydrolyzed by acids during steam-explosion pretreatment. Grous et al. (1986) reported that 90% efficiency of enzymatic hydrolysis was achieved in 24 h for wood chips pretreated by steam explosion, compared to only 15% hydrolysis yield of untreated wood chips. Removal of hemicelluloses from microfibrils is believed to expose the cellulose microfibrils to cellulose enzyme. Lignin is only to some extent removed from the material during the pretreatment but slightly redistributed on the cellulose fiber surfaces due to melting and depolymerization reaction that can be avoided by the quick release of biomass into atmospheric pressure. Hemicelluloses and lignin depolymerization increase the pore volume of the pretreated material (Li et al., 2007). Depending on the severity of the steam-explosion pretreatment, conversion of the cellulose to glucose will also take place. Addition or immersion of the material with H_2SO_4 or SO_2 (typically 0.3–3% w/w) prior to pretreatment can decrease the time and temperature of pretreatment and at the same time increase the recovery of sugars, reduce the formation of inhibitors, and improve the enzymatic hydrolysis (Ballesteros et al., 2006). For pretreatment of softwoods, the addition of an acid (H_2SO_4 or SO_2) is necessary to make the substrate accessible for enzymes (Jorgensen et al., 2007; Stenberg et al., 1998). Steam provides an effective vehicle to rapidly heat cellulose to the target temperature without excessive dilution of the resulting sugars. Rapid pressure release reduces the temperature and quenches the reaction at the end of the pretreatment. The rapid thermal expansion opens up the particulate structure of the biomass and enhances the digestibility of cellulose in the pretreated state (Jorgensen et al., 2007). The important variables in steam-explosion pretreatment are the time and temperature of pretreatment, particle size, and moisture content of biomass. Good hemicelluloses solubilization and hydrolysis can be achieved by either high temperature with short residence time process (27 °C for 1 min) or low temperature with long residence time process, for example, 190 °C for 10 min (Duff and Murray, 1996). The advantages of steam-explosion pretreatment include the low energy requirement compared to mechanical communition and no recycling or environmental costs. The conventional mechanical method requires 70% more energy than steam explosion to achieve the same particle size reduction (wright, 1998). Steam-explosion pretreatment with the addition of a catalyst is the technology that has been claimed to be the closest to commercialization (wright, 1998). This pretreatment has been tested extensively for a large number of different lignocellulosic feedstocks. The technology has been scaled-up and operated at the pilot-plant scale at the Iogen demonstration plant in Canada. The steam-explosion is recognized as one of the most cost-effective pretreatment processes for hardwoods and agricultural residues. But it is less effective for softwoods. Kobayashi et al. (2004) investigated methane production from bamboo wood using a steam-explosion pretreatment. Methane could not be produced from raw bamboo, but methane production is enhanced by steam-explosion pretreatment. The maximum amount of methane production of about 215 ml was obtained from 1 g of exploded bamboo wood at a steam pressure of 3.53 MPa and a steaming time of 5 min. Ballesteros et al. (2002) evaluated the effect of particle size on steam-explosion pretreatment of herbaceous lignocellulosic biomass. In this study, chipped biomass (5% moisture) with particle size (2–5, 5–8, 8–12 mm), temperature 190 and 210 °C, and residence time (4 and 8 min) was used. Larger size steam-exploded (8–12 mm) particles results in higher cellulose and enzymatic digestibility. After pretreatment, the water-soluble fiber was enzymatically hydrolyzed to determine the maximum accessible sugar yield. Cellulase enzyme loading was 15 filter paper unit (FPU) g^{-1} of substrate. Enzymatic hydrolysis was performed at 50 °C on a shaker incubator at 150 rpm for 72 h and at 2% (w/v) substrate concentration. Lower enzymatic hydrolysis yield (70%) was obtained at 190 °C for 4 and 8 min pretreatment. Higher enzymatic hydrolysis yields (about 99%) were obtained at 210 °C temperature. In a recent study, Viola et al. (2008) reported steam-explosion pretreatment of wheat, barley, and oat straw. The steam-explosion pretreatment was optimized at the batch scale on the basis of carbohydrate recovery.

Cara et al. (2008) investigated the production of ethanol fuel from olive tree. Olive tree pruning was subjected to steam-explosion pretreatment in the temperature range of 190–240 °C with or without previous impregnation with water or sulfuric acid solutions. The influence of both pretreatment temperature and impregnation conditions on sugar and ethanol yields was investigated by enzymatic hydrolysis and SSF on the pretreatment solids. Results showed that the maximum ethanol yield (7.2 g ethanol/100 g raw material) is obtained from water-impregnated, steam-pretreated residue at 240 °C. Nevertheless, if all sugars solubilized during pretreatment are taken into account, up to 15.99 g ethanol/100 g raw material may be obtained assuming theoretical conversion of these sugars to ethanol. Steam explosion has some limitations, including destruction of a portion of the xylan fraction, in complete disruption of the lignin–carbohydrate matrix, and generation of compounds that might have an inhibitory effect on microorganisms used in the downstream stage of the fermentation process. Due to the formation of the degradation products that may inhibit microbial growth, enzymatic hydrolysis and fermentation-pretreated biomass needs to be washed with water to remove the inhibitors along with water-soluble sugars obtained from hemicellulose hydrolysis (McMillan, 1994). The washing decreases the overall saccharification yield through the removal of soluble sugars, such as those generated by hydrolysis of hemicelluloses. Pretreatment using hot water is also used occasionally. In hot water pretreatment, pressure is used to maintain the water in the liquid state at elevated temperatures (Mosier et al., 2005). Flow-through processes pass water with high temperatures through lignocellulosic biomass. This type of pretreatment has been named hydrothermolysis, aqueous or steam/aqueous fractionation, uncatalyzed solvolysis, and aquasolv (Allen et al., 1996). The residence time for this process is usually around 15 min at temperatures in the range between 200 and 230 °C. Approximately 40–60% of the total biomass is dissolved. In this process, 4–22% of the cellulose, 35–60% of the lignin, and all of the hemicelluloses are removed. There are three types of liquid hot water reactor configurations, namely co-current, counter-current, and flow-through reactor. In co-current pretreatment conditions, water and lignocellulosic biomass move in the same direction. The biomass/water slurry is heated to a given temperature and maintained at pretreatment for a controlled residence time before being cooled. Counter-current pretreatment is designed to move water and lignin in opposite directions through the pretreatment reactor. In a flow-through reactor, hot water is passed over a stationary bed of lignocellulosic biomass. For liquid hot water pretreatment, size reduction of the biomass is not needed because the lignocellulosic biomass particles break apart when cooked in water (Weil et al., 1997).

Martinez et al. (1990) studied the pretreatment of few woods. The yield of sugar production after 48 h enzymatic hydrolysis of steam-explosion-pretreated samples at a temperature of 230 °C and residence time 1–2 min was higher than 90%. Others researches also showed similar results. Nunes and Pourquie (1996), Moniruzzaman (1996), and Daraei Garmakhany et al. (2014a) reported the effect of steam-explosion pretreatment on increasing the yield of saccharification of eucalyptus wood, rice straw, and canola straw, respectively.

6.7.2 Hydrothermolysis

In this way, biomass is exposed to high-pressure hot water at high temperatures (about 200 °C) for 30 min. Hydrothermolysis has been used for a long term, but it is expensive in large scale. Yu et al. (2010) investigated the effect of hot-compressed water (HCW) pretreatment on rice straw as a cost-effective pretreatment before enzymatic hydrolysis at a temperature ranging from 140 to 240 °C for 10 or 30 min. They evaluated the different characteristics of HCW-pretreated rice straw including sugar and inhibitor production as process yield. The maximum production of total glucose at 180 °C was 4.4–4.9% of glucan in raw material. Total maximum xylose production was obtained at 180 °C, accounting for 43.3% of xylan in raw material for 10 min pretreatment and 29.8% for 30 min pretreatment. The production of acetic acid increased at higher temperatures and longer pretreatment time that led to significant disruption of the lignocellulosic biomass structure. Maximum furfural formation (2.8 mg ml^{-1}) was achieved at 200 °C for both 10 and 30 min processing time. The glucose yield of enzymatic hydrolysis of pretreated rice straw was not lower than 85% for 30 min pretreatment at 180 °C were similar to 10 min pretreatment at 200 °C or higher temperature by considering sugar recovery, inhibitors formation and process severity. It is recommended that the temperature of 180 °C for 30 min can be the most efficient process for HCW pretreatment of rice straw. Jakobsson pretreated the wheat straw with steam at different temperatures (190, 200, and 210 °C) and residence times (2, 5, and 10 min). Sulfuric acid was used as a process catalyst. The straw was impregnated with sulfuric acid before pretreatment. For the evaluation of the pretreatment, enzymatic hydrolysis and fermentation used at 190 °C and 10 min resulted in the best overall yield for glucose as well as for xylose; 100 g wheat straw yielded 42.6 g glucose and 22 g xylose.

6.7.3 High-Energy Irradiations

The irradiation is a physical process. In this procedure, energy is transferred to material without needing molecules. It is essential that the energy does not cause radioactivity in the material. The irradiation dose is defined as the amount of energy absorbed per unit of mass. The dose of 1 Gy is the absorption of one joule of energy per gram of material. Irradiation process is selective and has high efficiency. The high yield is related to rotational electrons of atoms or food molecules or polluting compositions. When activated rotational electrons leave atoms, chemical changes are created in atoms or molecules, which is known as ionization. In this process, cation (an atom with a positive electrical charge) or an ion with negative electrical charge (anion) is formed. The ionization process constructs active atoms or molecules named free radicals. As the required energy for ionization is relatively low, foods do not change visibly by radiation process. A large portion of absorbed energy is used to produce free radicals and induce chemical reaction between radicals or between radicals and other molecules. A little part of absorbed energy is converted in the form of thermal energy. Consequently, food materials are processed with a little thermal energy and so sensory attributes and nutritional value of foods will conserved. The ionizing radiation is introduced as one of the suitable methods to conserve food and its safety. The application of this method for lignocellulosic biomass pretreatment has been small compared to chemical, thermal, and mechanical pretreatment. Recent methods are expensive and use high energy and produce fermentation inhibitors. To solve this problem, irradiation pretreatment can be used for its various advantages, for example, it does not produce inhibitors and is economic. High-energy ionizing radiation includes

- γ-irradiation produced from Co-60 or Cs-137;
- X-irradiation produced from machinery sources;
- electron beam irradiation.

All the abovementioned energy sources can have similar effects on food materials, but the main difference is their variable penetration rates. The γ- and X-irradiation can penetrate highly, and they can be used to treat foods as bulk, whereas electron beam irradiation is suitable for surface treatment or to treat thin boxes and packages.

The application of irradiation technology to pretreatment of lignocellulosic biomass was first described by Aoki *et al.* (1977). They showed that irradiation pretreatment by γ-irradiation resulted in the physical change of biomass. Other researchers reported the effect of irradiation degradation of various lignocellulosic materials such as sugarcane bagasse (Han *et al.*, 1981), rice straw (Kumakura and Kaetsu, 1979, 1984), corn cob, and peanut husk (Chosdu *et al.*, 1993) for increasing sugar yield. The irradiation-induced reactions in the macromolecules of the cellulose materials are known to be initiated through rapid localization of the absorbed energy within the molecules to produce long- and short-lived radicals, which caused the secondary degradation of materials through chemical reactions such as chain scission, cross-linking, and so on (Khan *et al.*, 2006). The efficiency of these two types of reactions depends mainly on the polymer structure and radiation dose (Charlesby, 1981). To reduce the required irradiation dose and acceleration of the degradation of lignocellulosic materials for economic reasons, irradiation pretreatment of rice straw with the addition of low concentrations of NaOH was studied (Lu and Kumakura, 1993).

High degradation of lignocellulosic biomass by a relatively small irradiation dose is attributed to the weakness of the chemical links between lignin, hemicelluloses, and cellulose units as well as the change of compositions through alkali swelling. After the termination of the irradiation process, produced radicals in amorphous regions will be extinct quickly, while others that are trapped in the crystalline and semicrystalline regions of the cellulose structure can decay at a certain period and cause a further degradation of lignocellulosic biomass. The latter phenomenon is called an after effect and was first described by Glegg and Kertesz (1956) in purified wood cellulose (the viscosity of cellulose described at least 2 weeks after irradiation). A similar result was reported by Gong et al. (1998) on cotton cellulose. Further researches on the postirradiation degradation were under taken by the analysis of radicals, and a significant decline of cellulosic radical was observed with extended storage time after irradiation and the rate was dependent on temperature, humidity, presence of oxygen, and so on (Polovka et al., 2006; Bayram and Delincée, 2004). The presence of lignin, which is a polyphenolic material, will affect the overall radiochemical events. Phenoxy radicals appeared to be important radical intermediates that ultimately transformed into o-quinonoid structures in lignin. The presence of quinine-type radicals in irradiated lignocellulosic materials also was reported by Simkovic et al. (1986). As the great influence of after effect on the irradiation degradation of lignocellulosic materials, some studies were carried out to accelerate the decaying rate of radicals after the termination of irradiation. The radiation degradation of straw was enhanced by heating after irradiation, which attributed to the energetic excitation of radicals (Kumakura and Kaetsu, 1984). The radiation pretreatment of the rice straw was also enhanced by heating after irradiation (Lu and Kumakura, 1993). However, the after effect of irradiation pretreatment on enzymatic hydrolysis of lignocellulosic biomass has not been reported. Chunping et al. (2008) evaluated the effect of irradiation pretreatment of wheat straw under different doses of CO-60 δ radiation. The weight loss and the fragility of wheat straw after irradiation, the combination effect of irradiation and mechanical crushing on enzymatic hydrolysis of wheat straw as well as the after effect of irradiation were examined. Results showed that irradiation pretreatment can cause significant breakdown of the structure of wheat straw. The weight loss of wheat straw increased, and the size distribution after crushing moves to fine particle at elevated irradiation doses. The glucose yield of enzymatic hydrolysis of wheat straw increased with increasing irradiation doses and achieved the maximum (13.40%) at 500 kGy. A synergistic effect was observed between doses of 500 kGy with powder size of 140 mesh number. The irradiation after effect had an important effect on enzymatic hydrolysis of wheat straw. The after effect of 400 kGy (at twenty-second day) was 20% of the original effect in glucose production while the after effect of 50, 100, 200 (at ninth day), and 300 kGy (at twentieth day) accounted for 12.9%, 14.9%, 8.9%, and 9.1%, respectively, for reducing sugar production. Stiku et al. (2008) investigated the effect of high dose irradiation as a pretreatment method on two common lignocellulosic materials (hardwood and softwood) by assessing the potential of cellulose enzyme derived from *Aspergillus flavus* Linn isolate NSPR 101 to hydrolyze the biomass materials. The irradiation strongly affected the materials, causing the enzymatic hydrolysis to increase by more than threefold. Maximum digestibility occurred in softwood at 40 kGy dosage of irradiation, while in hardwood it was at 90 kGy dosage. Several researches showed that irradiation pretreatment due to cellulose hydrolysis and lignin disruption lead to decreased mechanical properties of wood.

Charlesby (1995) showed that the molecular weight of cellulose decreases with increasing radiation dose. Many researchers have since studied the relationship between irradiation dose and cellulose fiber degradation (Dubey *et al.*, 2004). Takacs *et al.* (2000) reported that the DP of cotton cellulose was reduced from 1600 to 300 after 10 kGy of γ-irradiation. Other studies also showed that radiation pretreatment leads to a reduction of the cellulose crystallinity (Kasprzyk *et al.*, 2004; Alberti *et al.*, 2005). Kasprzyk *et al.* (2004) showed that γ-irradiation could reduce the amount of crystallinity in cellulose fibers which, is observed at dosage higher than 100 kGy. The crystallinity index of microcrystalline cellulose (MCC), flax, cotton, and viscose was reduced up to 12% with a irradiation dose of 200 kGy (Alberti *et al.*, 2005).

Mark *et al.* (2009) reported the reduction of crystallinity in biomass using electron beam pretreatment, and Bak *et al.* (2009) investigated the improvement of saccharification conversion using electron beam irradiation pretreatment. However, electron beam is a relatively low energy beam (usually 1–5 MeV). For pretreatment, it is necessary to use it at high dosage (Mark *et al.*, 2009; Bak *et al.*, 2009).

Proton beam irradiation (PBI) has been used in a variety of disciplines such as nanotechnology, medicine, information technology, and biotechnology due to its efficiency and benefits. It can be used in low doses for effective degradation of biomass. In addition, if PBI is used with liquid ammonia pretreatment, the yield of fermentable sugars increases due to the increasing accessibility of enzyme to cellulose (Kim *et al.*, 2008).

6.7.4 Acid Pretreatment

Acid pretreatment involves the use of concentrated and diluted acids to break down the rigid structure of the lignocellulosic material. The most commonly used acids are sulfuric acid, hydrochloric acid, nitric acid, and phosphoric acid. Due to the ability of the acid to remove hemicelluloses, acid pretreatments have been used as a part of overall processes in fractionating the components of hemicellulosic from lignocellulosic biomass. Acid pretreatment (removal of hemicelluloses) followed by alkali pretreatment (removal of lignin) results in relatively pure cellulose (Brodeur *et al.*, 2011). In acid pretreatment process, dilute or concentrated acid is added to the biomass and then allowed to stand for a given time and temperature.

Acid hydrolysis has been one of the traditional pretreatment methods to preterits lignocellulosic materials before the fermentation process. Bracont (1819) found that pretreatment of wood by concentrated sulfuric acid leads to production of glucose (Goldstein, 1983). Franzidis and Porteous (1981) investigated the first commercial acid hydrolysis process. The American process that is named as the sisman was used between 1910 and 1922. Wood particles were hydrolyzed under a batch process by using 0.5% sulfuric acid and high pressure steam at 912 kPa. The total ethanol yield from this process was a 22 Gal per ton that is uneconomical. A few years later, another process was invented by Heinrich Scholler in Germany that produced improved production yield of ethanol (55–58 Gal per ton of biomass) in 13–20 h. The Scholler method is a percolation method. In this method, sulfuric acid with a concentration of 0.8% and at temperature between 120 and 180 °C is passed through wood wastes. The maximum

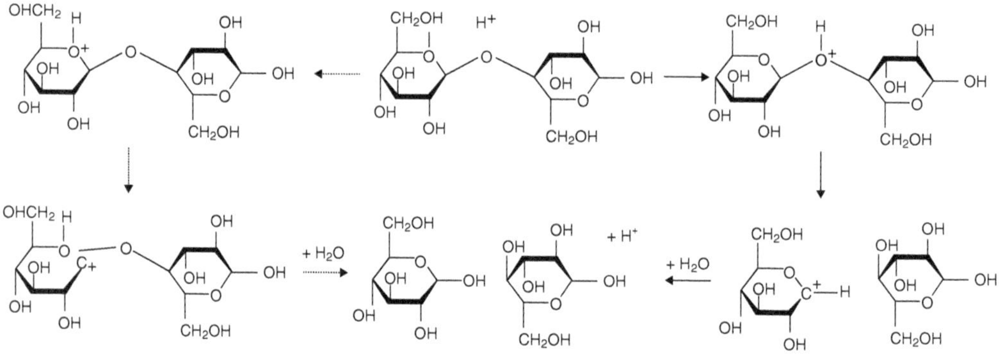

Figure 6.12 The major mechanism of acid hydrolysis of glycosidic bonds. Adapted from Fengel and Wegener (1984).

use of the Scholler method dates back to World War II in Germany. American forest products laboratory improved Scholler method and increased the production yield of ethanol to 64 Gal per ton of wastes in a mere 3 h hydrolysis time. This method is named Madison. This procedure was never truly established commercially because it could not compete effectively with ethanol derived from petroleum sources.

6.7.5 Mechanism of Acid Hydrolysis

At first, we should say, acid hydrolysis is a relatively efficient method to break cellulose component. Principal catalyst in this method is hydrogen aqueous ions (4 A°). Bigger particles (to 51 A°) can penetrate into pores of microfibrils; therefore hydrogen ions can penetrate more easily and not face the problem of accessibility compared to cellulase enzymes. The main mechanism of acid hydrolysis is relatively simple (Figure 6.12). It is similar to hydrolysis of other glycosidic bonds such as starch ($\alpha 1 \rightarrow 4$ linked glucose chains, with $\alpha 1 \rightarrow 6$ branches).

Step 3 (Figure 6.12) is the rate-limiting step of the process because of the formation of the high-energy half-chair configuration by the cyclic carbonium ion (Fengel and Wegener, 1984, Goldstein, 1983).

Initial hydrolysis rates are typically very rapid (Goldstein, 1983). Grethlein (1991) performed experiments to show that in the initial steps of the hydrolysis reaction, larger pore volumes do correspond to faster reaction rates. However, after limited hydrolysis, the reaction rate slows down considerably (Goldstein, 1983). The glycosidic bonds that are most susceptible to hydrolysis are those either at the surfaces or in the amorphous regions of cellulose. Rapid hydrolysis rates reflect hydrolysis activity in these regions and can be seen as a decrease in the DP from several thousand to about 200 (Ladisch, 1989). This point is referred to as the leveling of degree of polymerization (LODP). Further hydrolysis is much more difficult beyond the LODP because of the high crystallinity of the remaining cellulose molecules.

6.7.6 Alkaline Pretreatment

In the alkaline pretreatment of lignocellulosic biomass, different chemicals such as NaOH, Ca(OH)$_2$, and KOH are used (Figure 6.13). This procedure is used for delignification, and it is more effective on agricultural wastes. A large portion of lignin

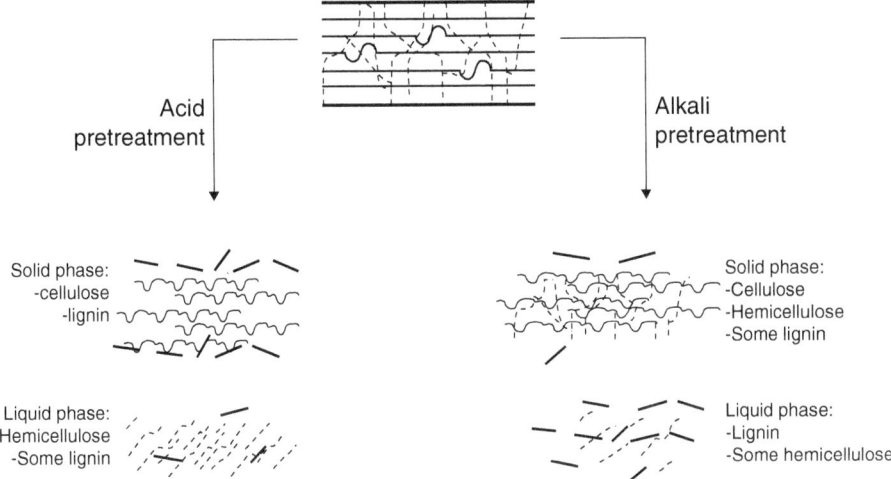

Figure 6.13 Separation of lignocellulosic biomass components in acidic and alkaline pretreatment conditions.

and hemicellulose is distilled during alkaline pretreatment (Hamelinck et al., 2005). Kyeong et al. (2011) has investigated the effect of sodium hydroxide pretreatment on canola straw. The results showed that in pretreatment condition with 7.9% sodium hydroxide concentration, 5.5 h of reaction time, and 68.4 °C of reaction temperature, the maximum glucose yield, which can be recovered by enzymatic hydrolysis at the optimum conditions was 95.7%. An increase of the surface area and pore size in pretreated canola straw by sodium hydroxide pretreatment was observed by scanning election microscope (Daraei Garmakhany et al., 2014a,b). Saha and Cotta (2006) studied the ethanol production from alkaline peroxide pretreated and enzymatically saccharified wheat straw. The results showed that maximum yield of monomeric sugars from wheat straw (8.6% w/v) was achieved by alkaline peroxide pretreatment (21.5% H_2O_2 v/v, pH 11.5, 35 °C, 24 h) and enzymatic saccharification (45 °C, pH 5, 120 h) by three commercial enzyme preparations (cellulase, β-glucosidase, and xylanase) and using 0.16 ml of each enzyme preparation per gram of straw was $672 \pm 4 \, mg\,g^{-1}$ (96.7% yield). During the pretreatment, no measurable quantities of furfural and hydroxymethyl furfural were produced. The concentration of ethanol (per liter) from alkaline peroxide pretreated and enzymatic saccharified wheat straw (66 g) by recombinant *Escherichia coli* strain FBR5 at pH 6.5 and 37 °C in 48 h was 18.9 ± 0.9 g with a yield of 0.46 g/g of available sugars (0.29 g/g straw). The ethanol concentration (per liter) was 15 ± 0.1 g with a yield of 0.23 g/g of straw in the case of SSF by the *E. coli* strain at pH 6 and 37 °C in 48 h. Alkaline pretreatment disrupted the cellulose–lignin bond (Sun and Cheng, 2002; Merino and Cherry, 2007).

Alkaline pretreatment has several defects such as

- high cost of chemicals (NaOH has a higher cost than H_2SO_4)
- higher concentration of alkaline materials than acids
- problems of recovery
- recovery of chemicals increases the cost of the process
- requires washing of biomass after pretreatment (extra cost of waste water filtering).

6.7.7 Ammonia Pretreatment

Ammonia pretreatment is used usually for delignification of biomass. It is performed through two methods, namely ammonia recycle percolation (ARP) and ammonia fiber expansion (AFEX).

6.7.8 Ammonia Recycle Percolation (ARP)

ARP is a process based on ammonia. Aqueous solution of ammonia (5–15% w/v) is passed through a reactor packed with biomass at elevated temperatures (100–180 °C), and then the ammonia in the effluent is separated and recycled. This method leads to biomass swelling and delignification. When incorporated into a biomass, saccharification and enzymatic hydrolysis increase. ARP technology almost completely fractionates biomass into the three major constituents. During the ARP process, all xylane dissolves in ammonia while more than 92% of cellulose remains in the biomass.

6.7.9 Ammonia Fiber Expansion (AFEX)

The AFEX process treats lignocellulosic materials with liquid ammonia under pressure and then rapidly releases the pressure, which leads to the following: (i) cellulose is decrystallized, (ii) hemicelluloses are prehydrolyzed, (iii) lignin in the pretreated material is altered, and (iv) the fiber structure is disrupted and accessible surface area for enzymatic hydrolysis is increased.

AFEX can achieve higher than 90% conversion of cellulose and hemicellulosic materials including wheat straw, rice straw, corn fiber, sugarcane, switchgrass, and so on (Sun and Cheng, 2002; Teymouri et al., 2005; Alizadeh et al., 2005).

6.7.10 Defects of AFEX Process

- The cost of ammonia.
  It is an expensive method, and for the commercialization of this process, ammonia must be recycled.
- High temperature leads to xylose degradation.
  AFEX method was developed by Dill et al. in Michigan State University for pretreatment of switchgrass and corn cob. It is more effective for the pretreatment of grassy plants such as switchgrass than woody plant with higher lignin content.
- Cellulase enzyme cannot function in the presence of ammonia; thus it should be extracted from biomass before enzymatic hydrolysis.
- Pollution of air due to release of ammonia is undesirable.

6.7.11 Enzymatic Pretreatment

Enzymatic pretreatment is used to break down celluloses and hemicelluloses to fermentable sugars like xylose and glucose. Enzymatic pretreatment is a biocompatible procedure and involves using carbohydrate enzymes (cellulase and hemicellulase) to hydrolyze lignocellulosic materials to fermentable sugars (Keshwani and Cheng, 2010; Hamelinck et al., 2005; Laureano-Perez et al., 2005). Enzymatic hydrolysis usually is conducted by cellulase enzyme. Cellulase enzyme is produced by some species of bacteria and fungi. Cellulase enzyme derived from fungi is the best enzyme in commercial scale (Keshwani and Cheng, 2010).

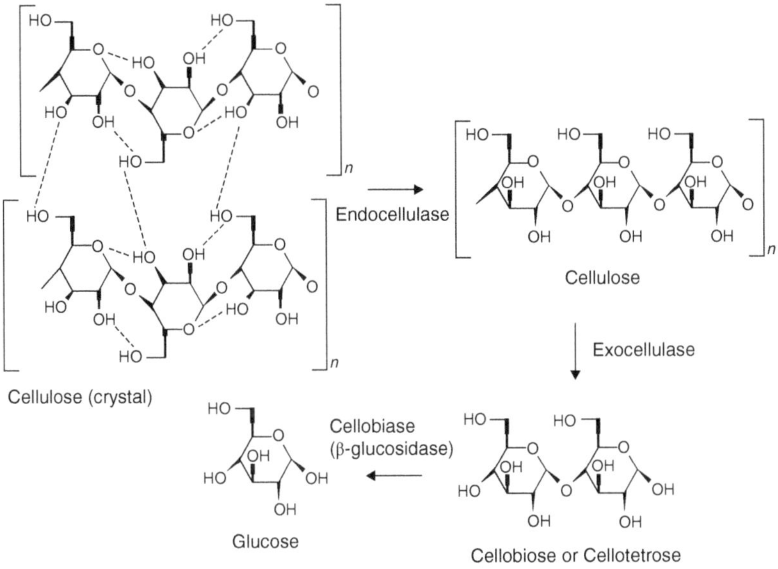

Figure 6.14 The mechanism of enzymatic hydrolysis of cellulose to glucose.

Cellulase enzymes consist of 1,4-β-D-glucanohydrolase, 4-β-D-glucanocellobiohydrolase, and β-glucosidase.

These enzymes belonged to three groups: endoglucanase, exoglucanase, and cellobiase, respectively. Endoglucanases break down cellulose chains randomly and produce cellubiose, glucose, and cellotriose. Exoglucanases attack the nonreducing end of cellulose and release cellobiose units at the end. Cellubiase enzyme converts cellubiose units to glucose (Figure 6.14).

6.7.12 Advantages of Biological Pretreatment

- Low chemical and lower energy usage
- Biocompatibility
- Formation of inhibitors in this method is lower than with other methods.

6.7.13 Defects of Biological Pretreatment

- The process is slow and complete hydrolysis is time consuming.
- There is enzymatic inhibition.
- Some of the sugars produced during hydrolysis can inhibit cellulose enzyme activity.
- Enzymes are not cost-effective.
- To maintian optimal conditions, more care must be taken.

6.8 Summary

An increased use of biofuels would contribute to sustainable development by reducing greenhouse gas emissions and the use of nonrenewable resources. Lignocellulosic biomass, including agricultural and forestry residues instead of traditional feedstocks (starch crops), could prove to be an ideally inexpensive and abundantly available source of sugar for fermentation into transportation fuels. Cellulose crystallinity, accessible surface area, protection by lignin, and sheathing by hemicelluloses all contribute to the resistance of cellulose in biomass to hydrolysis. The biomass pretreatment and the intrinsic structure of the biomass itself are primarily responsible for its subsequent hydrolysis. The conditions employed in the chosen pretreatment method will affect various substrate characteristics, which, in turn, govern the subsequent fermentation of the released sugars. Therefore, pretreatment of biomass is an extremely important step in the synthesis of biofuels from lignocellulosic biomass, and there is a critical need to understand the fundamentals of various processes, which can help in making a suitable choice depending on the structure of the biomass substrate and the hydrolysis agent. A vast array of materials are suitable for the production of biofuels. It must be emphasized that it is not always possible to transfer the results of pretreatment from one type of material to another. Furthermore, one technology that is efficient for a particular type of biomass material might not work for another material. Various pretreatment processes for lignocellulosic biomass have their specific advantages and disadvantages. The choice of the pretreatment technology used for a particular biomass depends on its composition and the by-products produced as a result of pretreatment. These factors significantly affect the costs associated with a pretreatment method.

References

Alberti, A., Bertini, S., Gastaldi, G., Iannaccone, N., Macciantelli, D., Torri, G., Vismara, E., 2005. Electron beam-irradiated textile cellulose fibers. *European Polymer Journal*. **41** (8), 1787–1797.

Alizadeh, H., Teymouri, F., Gilbert, T. I., and Dale, B. E. 2005. Pretreatment of switchgrass by ammonia fiber explosion (AFEX). *Applied Biochemistry and Biotechnology*, **124**, 1133–1141.

Allen, S. G., Kam, L. C., Zemann, A. J., Antal, M. J. 1996. Fractionation of sugar cane with hot compressed liquid water. *Industrial Engineering Chemistry Research*, **35**, 2709–2715.

Aoki, T., Norimoto, M., Yamada, T. 1977. Some physical properties of wood and cellulose irradiated with gamma rays. *Wood Research*, **62**, 19–28.

Bak, J. S., Ko, J. K., Han, Y. H., Lee, B. C., Choi, I. G., Kim, K. H. 2009. Improved γ beam irradiation pretreatment. *Bioresource Technology*, **100**, 1285–1290.

Ballesteros, I., Negro, M. J., Oliva, J. M., Cabanas, A., Manzanares, P., Ballesteros, M. 2006. Ethanol production from steam explosion pretreated wheat straw. *Applied Biochemistry and Biotechnology*, **70**, 3–15.

Ballesteros, M. J., Oliva, I., Negro, M. J., Manzanares, P., Ballesteros, M. 2002. Enzymic hydrolysis of steam exploded herbaceous agricultural waste (*Brassica carinata*) at different particle sizes. *Process Biochemistry*, **38**, 187–192.

Ballesteros, M., Oliva, J. M., Negro, M. J., Manzanres, P., Ballesteros, I. 2004. Ethanol from lignocellulosic materials by a simultaneous saccharification and fermentation process (SFS) with Kluyveromyces marxianus CECT 10875. *Process Biochemistry*, **39**, 1843–1848.

Bayram, G., Delincée, H. 2004. Identification of irradiated Turkish foodstuffs combining various physical detection methods, *Food Control*, **15**, 81–91.

Bochek, A. M. 2003. Effect of hydrogen bonding on cellulose solubility in aqueous and nonaqueous solvents. *Russian Journal of Applied Chemistry*, **76** (11), 1711–1719.

Brodeur, G., Yau, E., Badal, K., Collier, J., Ramachandran, K. B., Ramakrishnan, S. 2011. A review of chemical and physicochemical pretreatment of lignocellulosic biomass. *Enzyme Research*, **10**, 4061–4078.

Cadoche, L., Lopez, G. D. 1989. Assessment of size reduction as a preliminary step in the production of ethanol from lignocellulosic wastes. *Biological Wastes*, **30**, 153–157.

Cara, C., Ruiz, C., Ballesteros, M., Manzanares, P., Negro, M. J., Castro, E. 2008. Production of fuel ethanol from steam-explosion pretreated olive tree pruning. *Fuel*, **87**, 692–700.

Champagne, P. 2008. Bioethanol form agricultural waste residues. *Environmental Progress*, **27**, 51–57.

Chandra, R. P., Bura, R., Mabee, W. E., Berlin, A., Pan, X., and Saddler, J. N. 2007. Substrate pretreatment: The key to effective enzymatic hydrolysis of lignocellulosics. *Biofuels*, **108**, 67–93.

Charlesby, A., 1981. Crosslinking and degradation of polymers. *Radiation Physics and Chemistry*, **18**, 59–66.

Charlesby, A. 1995. Degradation of cellulose by ionizing radiation. *Journal of Polymer Science*, **15**, 263–270.

Chosdu, R., Hilmy, N., Erizal, E. T. B., Abbas, B. 1993. Radiation and chemical pretreatment of cellulosic waste. *Radiation Physics and Chemistry*, **42**, 695–698.

Chunping, Y., Zhiqiang, Sh., Guoce, Y., Jianlong, W. 2008. Effect and after effect of c radiation pretreatment on enzymatic hydrolysis of wheat straw. *Bioresource Technology*, **99**, 6240–6245.

Daraei Garmakhany, A., Kashaninejad, M., Aalami, M. 2014a. Physicochemical characterization of canola straw pretreated with steam explosion for enhancing fermentable sugar production. *Minerva Biotechnologica*, **26** (4), 241–246.

Daraei Garmakhany, A., Kashaninejad, M., Aalami, M., Maghsoudlou, Y., Khomieri, M., Tabil, L. G. 2014b. Enhanced biomass delignification and enzymatic saccharification of canola straw by steam-explosion pretreatment. *Journal of the Science of Food and Agriculture*, **94** (8), 1607–1613.

Dekker, R. F. H. 1988. Steam explosion: an effective pretreatment method for use in the bioconversion of lignocellulosic materials. In B. Focher, A. Marzetti and V. Crescenzi (Eds.) Proceedings of the International Workshop on Steam Explosion Techniques: Fundamentals and Industrial Applications, 277–305.

Del Campo, I., Alegrìa, I., Zazpe, M., Echeverrìa, M., Echeverrìa, I. 2006. Dilute acid hydrolysis pretreatment of agri-food wastes for ethanol production. *Industrial Crop Production*, **24**, 214–221.

Dubey, K. A., Pujari, P. K., Ramnani, S. P., Kadam, R. M., Sabharwal, S. 2004. Microstructural studies of electron beam irradiated cellulose pulp. *Radiation Physics and Chemistry*, **69** (5), 395–400.

Duff, S. J. B., Murray, W. D. 1996. Bioconversion of forest products industry waste cellulosics to fuel ethanol: A review. *Bioresource Technology*, **55**, 1–33.

Faulon, J. L., Carlson, G. A., Hatcher, P. G. 1994. A three-dimensional model for lignocellulose from gymnospermous wood. *Organic Geochemistry*, **21**, 1169–1179.

Fengel, D., Wegener, G. 1984. *Wood: Chemistry, Ultrastructure, Reactions*, Berlin, Walter de Gruyter.

Franzidis, J. P., Porteous A. 1981. Chapter 14: Review of recent research on the development of a continuous reactor for the acid hydrolysis of cellulose. In D. L. Klass and G. H. Emert (Eds.) *Fuels From Biomass and Wastes.* (pp. 267–296). Ann Arbor, Michigan: Ann Arbor Science Publishers, Inc.

Glegg, R. E., Kertesz, Z. I. 1956. After effect in the degradation of cellulose and pectin by gamma rays. *Science*, **124**, 893–894.

Goldstein, I. S. 1983. Acid processes for cellulose hydrolysis and their mechanisms. In E. J. Soltes (Ed.) *Wood and Agricultural Residues* (pp. 315–328). New York, NY: Academic Press, Inc.

Gong, N., Chang, D., Zhang, J., Liu, J. 1998. After effect in the radiation degradation of cotton cellulose. *Journal of Beijing Institute of Technology*, **18**, 647–650.

Grous, W. R., Converse, A. O., Grethlein, H. E. 1986. Effect of steam explosion pretreatment on pore size and enzymatic hydrolysis of poplar. *Enzyme and Microbial Technology*, **8**, 274–280.

Hamelinck, C. N., Hooijdonk, G., Faaij, A. P. C. 2005. Ethanol from lignocellulosic biomass: Techno-economic performance in short, middle and long term. *Biomass and Bioenergy*, **28**, 384–410.

Han, Y. W., Timpa, J., Clegler, A., Courtney, J., Curry, W. F., Lambremont, E. N. 1981. γ-Ray induced degradation of lignocellulosic materials. *Biotechnology and Bioengineering*, **23**, 2525–2535.

Harmsen, P., Huijgen, W., Bermudez, L., Bakker, R. 2010. A review of physical and chemical pretreatment processes for lignocellulosic biomass. *Food and Biobased Research*, **1184**, 1–54.

Jorgensen, H., Kristensen, J. B., Felby, C. 2007. Enzymatic conversion of lignocellulose into fermentable sugars: Challenges and opportunities. *Biofuels, Bioproducts and Biorefining*, **1**, 119–134.

Jørgensen, H., Vibe-Pedersen, J., Larsen, J., Felby, C. (2007) Liquefaction of lignocellulose at high-solids concentrations. *Biotechnology and Bioengineering*, **96** (5), 862–870

Kasprzyk, H., Wichlacz, K., Borysiak, S. 2004. The effect of gamma radiation on the supramolecular structure of pine wood cellulose in situ revealed by X-ray diffraction. *Electronic Journal of Polish Agricultural Universities*, **7** (1), 29.

Keshwani, D. R. 2009. Microwave pretreatment of switchgrass for bioethanol production. Dissertation. Graduate Faculty of North Carolina State University, Raleigh, North Carolina.

Keshwani, D. R., Cheng, J. J. 2010. Modeling changes in biomass composition during microwave based alkali pretreatment of switchgrass. *Biotechnology and Bioengineering*, **105** (1), 88–97.

Khan, F., Ahmad, S. R., Kronfli, E. 2006. γ-Radiation induced changes in the physical and chemical properties of lignocellulose. *Biomacromolecules*, **7**, 2303–2309.

Kim, T. H., Taylor, F., Hicks, K. B. 2008. Bioethanol production from barley hull using SAA (soaking in aqueous ammonia) pretreatment. *Bioresource Technology*, **99**, 5694–5702.

Kirk-Otmer 2001 *Encyclopedia of Chemical Technology*, 4th edition, Vol. 5, Wiley.

Kobayashi, F., Take, H., Asada, C., Nakamura, Y. 2004. Methane production from steam exploded bamboo. *Journal of Bioscience Bioengineering*, **97**, 426–428.

Krassig, H., Schurz, J. 2002. *Ullmann's Encyclopedia of Industrial Chemistry*, Sixth edition, Weinheim, Germany, Wiley-VCH.

Kumakura, M., Kaetsu, I. 1979. Radiation induced decomposition and enzymatic hydrolysis of cellulose. *International Journal of Applied Radiation and Isotopes*, **30**, 139–141.

Kumakura, M., Kaetsu, I. 1984. Effect of electron beam current on radiation pretreatment of cellulosic wastes with electron beam accelerato. *Radiation Physics and Chemistry*, **23**, 523–527.

Kyeong, K., Jeong, G., Park, D. 2011. Pretreatment of rapeseed straw by sodium hydroxide. *Biosystems Engineering*, **45**, 101–109.

Ladisch, M. R. 1989. Hydrolysis. In O. Kitani and C. W. Hall (Eds.) *Biomass Handbook*: 434–451. New York, NY: Gordon and Breach Science Publishers.

Laureano-Perez, L., Teymouri, F., Alizadeh, H., Dale, B. 2005. Understanding factors that limit enzymatic hydrolysis of biomass. *Applied Biochemistry and Biotechnology*, **121**, 1081–1099.

Li, J., Henriksson, G., Gellerstedt, G. 2007. Lignin depolymerization/repolymerization and its critical role for delignification of aspen wood by steam explosion. *Bioresource Technology*, **98**, 3061–3068.

Lin, S. Y., Lin, I. S. 2002. *Ullmann's Encyclopedia of Industrial Chemistry*, Sixth edition, Weinheim, Germany, Wiley-VCH.

Lu, Z., Kumakura, M. 1993. Effect of radiation pretreatment on enzymatic hydrolysis of rice straw with low concentrations of alkali solution, *Bioresource Technology*, **43**, 13–17.

Mark, D., Arthur, S., William, W., Kun, C., Mellony, M., Jessica, S. 2009. Electron beam irradiation of cellulose. *Radiation Physics and Chemistry*, **78**, 539–542.

Martinez, J., Negro, M. J., Saez, F., Manero, J., Saez, R., Martin, C. 1990. Effect of acid steam explosion on enzymatic hydrolysis of O. nervosum and C. cardunculus. *Applied Biochemistry and Biotechnology*, **24/25**, 127–134.

McKillip, W. J., G. Collin. 2002. *Ullmann's Encyclopedia of Industrial Chemistry*, Sixth edition, Weinheim, Germany, Wiley-VCH.

McMillan, J. D. 1994. Pretreatment of lignocellulosic biomass. In M. E. Himmel, J. O. Baker, R. P. Overend (Eds.) *Enzymatic Conversion of Biomass for Fuels Production*. (pp. 292–324). Washington, DC: American Chemical Society.

Merino, S. T., Cherry, J. 2007. Progress and challenges in enzyme development for biomass utilization, *Biofuels*, **108**, 95–120.

Millet, M. A., Baker, A. J., Scatter, L. D. 1976. Physical and chemical pretreatment for enhancing cellulose saccharification. *Biotechnology and Bioengineering Symposium*, **6**, 125–153.

Moniruzzaman, M. 1996. Saccharification and alcohol fermentation of steam exploded rice straw. *Bioresource Technology*, **55**, 111–117.

Mosier, N., Wyman, C., Dale, B., Elander, R., Lee, Y. Y., Holtzapple, M., Ladisch, M. 2005. Features of promising technologies for pretreatment of lignocellulosic biomass. *Bioresource Technology*, **96**, 673–686.

Mtui, G. Y. S. 2009. Recent advances in pretreatment of lignocellulosic wastes and production of value added products. *African Journal of Biotechnology*, **8**, 1398–1415.

Mussatto, S. I., Roberto, I. C. 2004. Alternatives for detoxification of diluted-acid lignocellulosic hydrolyzates for use in fermentative processes: A review. *Bioresource Technology*, **93** (1), 1–10.

Nunes, A P., Pourquie, J. 1996. Steam explosion pretreatment and enzymatic hydrolysis of eucalyptus wood. *Bioresource Technology*, **57**, 107–110.

O'Connor, R. P., Woodley, R., Kolstad, J. J., Kean, R., Glassner, D. A., Mastel, B., Ritzenthaler, J. M., John, H., Warwick, J., Hettenhaus, J. R., Brooks, R. K. 2007. Process for fractionating lignocellulosic biomass into liquid and solid products, assignee U. S. A. Nature-works LLC, patent number WO 2007120210.

Palmqvist, E., Hahn-Hägerdal, B. 2000. Fermentation of lignocellulosic hydrolyzates. I: Inhibition and detoxification. *Bioresource Technology*, **74**, 17–24.

Polovka, M., Brezová, V., Staško, A., Mazúr, M., Suhaj, M., Simko, P. 2006. EPR investigations of gamma-irradiated ground black pepper. *Radiation Physics and Chemistry*, **75**, 309–321.

Raven, P. H., R. F. Evert, Susan, EE. 1992. *Biology of Plants* (sixth edition), W. H. Freeman and company/Worth Publishers.

Saha, B. C., Cotta, M. A. 2006. Ethanol production from alkaline peroxide pretreated enzymatically saccharified wheat straw. *Biotechnology Progress*, **22** (2), 449–453.

Shultz, T. P., Templeton, M. C., Biermann, C. J., Mc Ginnis, G. D. 1984. Steam explosion of mixed hardwood chips, rice hulls, corn stalks, and sugar cane bagasse. *Journal of Agricultural and Food Chemistry*, **32**, 1166–1172.

Simkovic, I., Ebringerova, A., Tino, J., Placek, J., Manasck, Z., Zilka, L. 1986. ESR study of soda waste liquors, *Holzforschung*, **40**, 15–18.

Solomon, T. W. G. 1988. *Organic Chemistry* (fourth edition), John Wiley & Sons.

Stenberg, K., Tengborg, C., Galbe, M., Zacchi, G. 1998. Optimization of steam pretreatment of SO_2 impregnated mixed softwoods for ethanol production. *Journal of Chemistry Technology and Biotechnology*, **71**, 299–308.

Sun, Y., Cheng, J. Y. 2002. Hydrolysis of lignocellulosic materials for ethanol production: A review. *Bioresource Technology*, **83**, 1–11.

Takacs, E., Wojnarovits, L., Foldvary, C., Hargittai, P., Borsa, J., Sajo, I. 2000. Effect of combined gamma-irradiation and alkali treatment on cotton cellulose. *Radiation Physics and Chemistry*, **57**, 399–403.

Teymouri, F., Laureano-Perez, L., Alizadeh, H., Dale, B. E. 2005. Optimization of the ammonia fiber explosion (AFEX) treatment parameters for enzymatic hydrolysis of corn stover. *Bioresource Technology*, **96**, 2014–2018.

Viola, E., Zimabardi, F., Cardinale, M., Cardinale, G., Braccio, G., Gamabacorta, E. 2008. Processing cereal straws by steam explosion in a pilot plant to enhance digestibility in ruminants. *Bioresource Technology*, **99**, 681–689.

Weil, J. R., Sarikaya, A., Rau, S. L., Goetz, J., Ladisch, C. M., Brewer, M., Hendrickson, R., Ladisch, M. R. 1997. Pretreatment of yellow poplar sawdust by pressure cooking in water. *Applied Biochemistry and Biotechnology*, **681**, 21–40.

Weimer, P. J., J. M. Hackney, A. D. French. 1995. Effects of chemical treatments and heating on the crystallinity of celluloses and their implications for evaluating the effect of crystallinity on cellulose biodegradation. *Biotechnology and Bioengineering* **48**, 169–178.

Wright, J. D. 1998. Ethanol from biomass by enzymatic hydrolysis. *Chemistry Engineering Progress*, **84**, 62–74.

Yang, B., Wyman, C. E. 2008. Pretreatment: The key to unlocking low-cost cellulosic ethanol. *Biofuels, Bioproducts and Biorefining*, **2**, 26–40.

Yu, G., Yano, Sh., Inoue, H., Inoue, S., Endo, T., Sawayama, Sh. 2010. Pretreatment of rice straw by a hot-compressed water process for enzymatic hydrolysis. *Applied Biochemistry and Biotechnology*, **160**, 539–551.

7

Green Route to Prepare Renewable Polyesters from Monomers: Enzymatic Polymerization

Toufik Naolou

Institute of Biomaterial Science and Berlin-Brandenburg Centre for Regenerative Therapies, Helmholtz-Zentrum Geesthacht, Teltow, Germany

7.1 Philosophic Statement

Enzymatic polymerization has emerged in the last two decades due to their advantages for polymer synthesis compared to conventional catalysts. Besides the nontoxicity, enzymes can also support sustainably the synthesis of polymer by allowing the polymerization process to be conducted under mild conditions and in a shorter time. Additionally, enzymes allow clean synthesis of linear and comb-like polyesters with free reactive groups, such as double bond, hydroxyl, thiol, or epoxy groups, in one step without the need for protection/deprotection steps, which otherwise would be necessary when conventional catalysts are used for such synthesis. These unique features of enzymes make them suitable candidate for the green synthesis of polyesters.

7.2 Introduction

The last century had witnessed a revolutionary advance toward the development of novel materials that could meet the needs of our modern life. Polymers, indeed, occupy a prominent place among these materials, with a worldwide demand in excess of 393 million tons in 2013 [1]. Polymers have become indispensable for our today's life, due to their superior advantages and properties such as attractive cost/performance ratio, rapid production with low cost, ease of processing, low weight, high mechanical performance, good barrier properties, good heat stability, and excellent corrosion resistance [2–4]. Albeit, since the late 1980s of the last century, the concept of "green chemistry" or sustainable chemistry started to emerge mostly driven by dwindling of fossil resources, global warming, and environmental concerns caused by hazardous chemical processes, nondegradable waste, harmful emissions, and so on. The number of researches in the field of green chemistry increase steadily every year [5]. Thus it is not surprising to recognize the efforts made in the field of polymer science during the last two decades toward optimized synthetic pathways to meet the requirements of green chemistry concept and produce novel sustainable polymers that are able to be alternatives to the existing "petro-polymers" with competitive performance properties and price but

Introduction to Renewable Biomaterials: First Principles and Concepts, First Edition.
Edited by Ali S. Ayoub and Lucian A. Lucia.
© 2018 John Wiley & Sons Ltd. Published 2018 by John Wiley & Sons Ltd.

mostly to be also safely recycled or degraded [6]. In that regard, polyesters represent an important class of polymers that can meet a lot of green chemistry requirements as they can be mostly synthesized starting from monomers obtained from biomass feedstocks [2]. Additionally, their biodegradability causes no environmental pollution compared to that of nondegradable plastic and makes them suitable for biomedical applications [7–10]. The wide spectrum of monomers, which can be used for polyester synthesis, with their various chemical structures allows for the synthesis of polyester with tailored chemical structure able to meet specific properties needed for different applications. Some of the polyesters synthesized completely or partially from renewable feedstock have already found their way to the market, that is, poly(lactic acid) (PLA), poly(butylene succinate) (PBS), poly(butylene adipate-co-terephthalate), and poly(hydroxyalkanoates) with worldwide production of 195,000, 100,000, 75,000, 32,000 tons in 2013, respectively [1, 11]. Polymer synthesis starting from sustainable monomers, however, is not enough to meet the demands of green chemistry, as industrial synthetic processes involve utilization of toxic catalyst as well as using high reaction temperature. Utilization of enzymes in organic synthesis has emerged as an attractive green synthetic pathway alternative to conventional chemical catalyst [12]. Scientists have investigated the potential application of enzymes as catalyst for polymers synthesis. In fact, utilization of enzymes as catalyst for polyester synthesis supports the requirement of green principles for polymer production from many aspects [2, 13] including the following:

- Enable the synthesis of polyesters under mild conditions, which minimize the amount of energy needed for polymer production and avoid undesirable side products.
- Allow the synthesis of functional polyester in one step, without the need for protection–deprotection steps, due to the chemo-, enantio-, and regioselectivity of enzymes.
- Contrary to most catalysts used currently for polyester synthesis, enzymes are classified as nontoxic.
- Enzymes are derived from renewable resources (sustainability).
- Enzymes are recyclable when they are immobilized, which reduces the catalyst costs.

This chapter presents an overview of the application of enzymatic polymerization to catalyze the synthesis of polyesters using different types of monomers prepared from biomass, where some recently and interesting published reports are highlighted. A deeper insight, however, into enzymatic polymerization topic is available from other published excellent reviews [13–17].

7.3 Lipase-Catalyzed Ring-Opening Polymerizations of Cyclic Monomeric Esters (Lactones and Lactides)

Lactones are an important class of monomers for the synthesis of aliphatic polyesters via chain-growth mechanism, which offers a lot of attractive advantages over monomers that react via step-growth mechanism [18]; for example, the polymerization reaction proceeds in one direction without generating leaving groups [19], synthesizing high molar mass and low polydispersity polyesters within shorter reaction time and the capability to control end groups of resulting polyester [20]. Many reports have appeared recently on the synthesis of useful lactones from biomass (Figure 7.1). γ-Butyrolactone

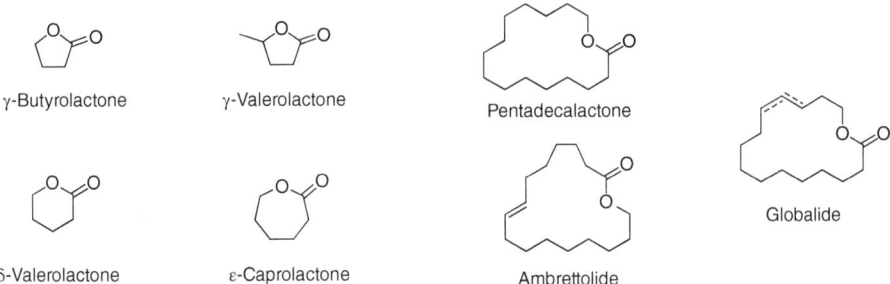

Figure 7.1 The chemical structure of some lactones derived from renewable resources.

(γ-BL), for instance, can be produced from succinic acid, which is in turn produced from glucose via fermentation [21–23]. Furthermore, γ-valerolactone (γ-VL) can be prepared directly from levulinic acid (LA), which is another platform of chemicals prepared from chemical transformation of cellulose and hemicelluloses, by hydrogenation process [24]. δ-Valerolactone (DVL) can be produced in two synthesis steps starting from furfural [25]. Interestingly, it has been recently reported that ε-caprolactone (ε-CP) can also be obtained using bio-based synthetic route in four steps starting from hydroxymethylfurfural (HMF) [26].

Utilization of lipase as catalyst for ring-opening polymerization (ROP) of lactones was discovered by two independent research groups, that is, Kobayashi and Knani, in 1993 [27, 28]. Since then, enzyme-catalyzed ROP of lactones have been extensively investigated to find the effect of different reaction conditions, for example, enzyme origin [29–31], water content [32, 33], solvent [34, 35], temperature [36, 37], and concentration [34], on polymerization kinetic and the final product,

The overall synthesis is shown in Figure 7.2.

It is well accepted that the lipase-catalyzed ROP of lactones proceeds via several steps including first activation of the carbonyl ester by formation of an acyl-enzyme intermediate between the lactone and the OH (hydroxyl groups) of serine residue in the active center of the enzyme. This is followed by the attack of the initiator, for example, water or alcohol, to deacylate the acyl-enzyme intermediate and form the corresponding ω-hydroxycarboxylic acid/ester. The later step is known as the initiation step where the type of initiator will determine the end groups of resulting polyester chains. In the propagation step, the deacylation of acyl-enzyme intermediate is carried out usually by the terminal hydroxyl group of the growing chains to increase the length of resulting chain by one monomer unit. The polymerization and copolymerization of substituted and unsubstituted lactones with different ring sizes have been widely investigated [16].

Figure 7.2 Lipase-catalyzed ROP of lactones.

R: H, alkyl…
n: 2, 3, 4, …, 15

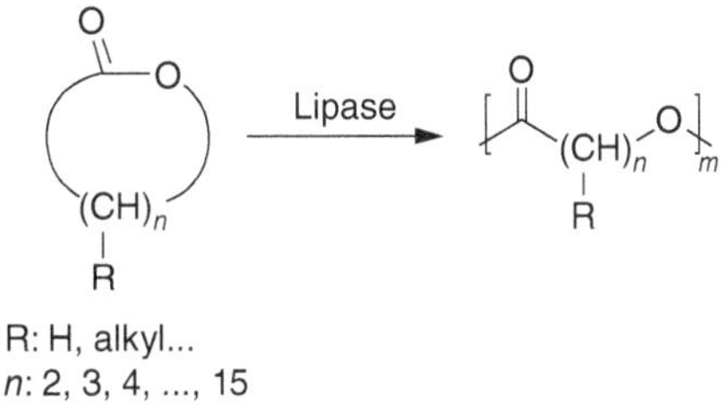

Among them, lipase-catalyzed ROP of seven-membered lactone (ε-CP) is the most investigated one due to the importance of its polymer in many applications [35, 38–40]. The best results were observed when immobilized lipase from *Candida antarctica* lipase B (N435) was used as a catalyst. In this case, poly(ε-caprolactone) (poly(ε-CP)) of molecular weight 25 kDa was obtained in toluene at 70 °C [41]. Thurecht *et al.* [42] were able to prepare poly(ε-CP) of molecular weight up to 50 kDa when the polymerization process was carried out in supercritical carbon dioxide (scCO$_2$). The polymer in this case has, however, a polydispersity of about 2 due to enzyme-catalyzed transesterification reaction. γ-BL showed only low activity toward lipase-catalyzed ROP, as only polymer with molar mass of 888 Da was obtained after 430 h of polymerization in *n*-hexane at a temperature of 60 °C using lipase from *Burkholderia cepacia* (lipase PC) as a catalyst [43]. The copolymerization of γ-BL with ε-CP was carried out in toluene at 70 °C using N435 as catalyst for 4–48 h, which resulted in copolymers of M_n, based on the γ-BL/ε-CP feed ratio, in the range between 9600 and 16,100 Da. Many attempts have been made to enzymatically polymerize δ-VL to produce poly(δ-valerolactone) using wide varieties of enzymes as catalyst [40, 44]. The highest molar mass ($M_n = 3200$ Da) was obtained at 45 °C using isooctane as solvent and lipase CC as catalyst [45].

Interestingly, the kinetic investigation of lipase-catalyzed ROP of unsubstituted lactones of various ring sizes revealed that the reactivity of monomer and the achievable molecular weight increases by increasing the ring size, which contrast to the trend observed for the metal–organic catalyzed ROP of lactones due to the remarkably decrease of ring strain in large lactones [46, 47]. Consequently, increasing attention has been paid toward lipase-catalyzed ROP of macrolactones that can be derived from biomass. Of particular interest is the lipase-catalyze ROP of pentadecalactone (PDL), a naturally produced material with worldwide consumption in the range of 100–1000 metric tons per year, mainly used as fragrance ingredient of many finished consumer product categories [48]. Poly(pentadecalactone) (PPDL) has been obtained with M_w up to 143 kDa at reaction temperature of 85 °C for 72 h using N435 as catalyst [49]. Taden *et al.* [50] reported the synthesis of PPDL with molecular weight of 200 kDa at 45 °C in miniemulsion using lipase PC as a catalyst. In fact, PPDL showed mechanical and crystallization properties similar to those of polyethylene (PE) [51–54]. The potential applications of this polymer were investigated by preparation fibers from high-molecular-weight PPDL, which revealed a tensile strength up to 0.74 GPa [49].

Another interesting example is the lipase-catalyzed ROP of unsaturated macrolactones, namely, ambrettolide and globalide, to investigate their potential application as biomaterials. The resulting polymers revealed high crystallinity, nontoxicity, and were able to cross-link in the melt to yield fully amorphous materials [55].

Lactide is a cyclic diester of lactic acid, which is produced through bacterial fermentation of glucose and other biomass sources [56]. The monomer is used mainly to produce PLA, the second largest produced biopolymer, through Sn(II)-catalyzed ROP. Utilization of lipase for the ROP of lactide resulted in PLA with low molecular weights and/or low reaction yield. While lipase PS was able to catalyze the ROP of lactide at a temperature of 80–130 °C to give PLA of M_w up to 12,600 Da, the yield of resulting polyester was significantly low (3–16%) [57]. Under these conditions, DLLA (DL-Lactide) resulted in higher molecular weight compared to LLA or DLA. When N435, however, was used as catalyst instead of lipase PS, only DLA could participate in the polymerization reaction in toluene under mild conditions to give PLA with M_n value 3300 Da, while

LLA showed now reactivity [58]. However, N435 was able to catalyze the ROP of LLA when the polymerization reaction was carried out in supercritical CO_2, which resulted in PLLA (Poly-L-Lactide) with good M_w of 12,900 Da but with low yield (≤12%) [59].

7.4 Lipase-Catalyzed Polycondensation

Despite the many advantages of using ROP to synthesize polyesters, polycondensation is considered as the main synthetic route to prepare commercial polyesters [13]. In fact, this polymerization technique allows direct synthesis of wide spectrum of polyesters form naturally derived monomers. Polyesters can be synthesized by condensation polymerization using either AA- and BB-type monomers or AB-type monomers by either esterification or transesterification reaction route (Figure 7.3). In both cases, however, small molecular weight compound, for example, water, hydrochloric acid, alcohol, and so on, is usually produced besides the main product during the condensation reaction. Elimination of this by-product during the polymerization process is necessary in order to shift the equilibrium reaction toward products, which can be achieved by carrying out polymerization reaction (i) in bulk under driven conditions, for example, high temperature, vacuum or gas stream [60], (ii) in the presence of solvent that is able to form an azeotropic mixture with the resulting by-product, which is then constantly eliminated by passing the mixture through Soxhlet apparatus packed with molecular sieve of suitable pore size [61].

Monomers with activated acyl donors, such as thioester, oxime ester, and anhydrides, have been used for lipase-catalyzed polycondensation of polyesters as they own higher reactivity toward polymerization process and due to ease of elimination the resulting by-products. However, utilization of monomers with enol ester, such as vinyl esters, seems to be the most effecting synthetic strategy as the resulting enol is not stable and thus tautomerizes readily to give the corresponding aldehydes or ketones, which makes the polymerization process irreversible [12, 62]. A lot of commercial available monomers obtained from renewable resources have been tested for lipase-catalyzed polycondensation. While the lipase-catalyzed polymerization of lactic acid gave a low-molecular-weight PLA [63], polyesters with higher molecular weight have been enzymatically produced by utilization of another monomer.

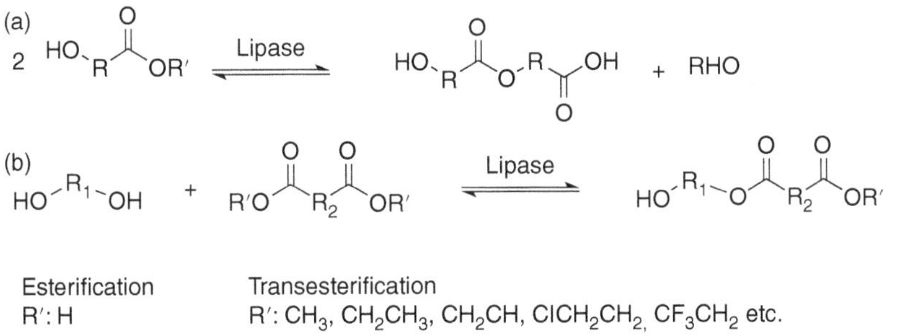

Esterification
R': H

Transesterification
R': CH_3, CH_2CH_3, CH_2CH, $ClCH_2CH_2$, CF_3CH_2 etc.

Figure 7.3 Lipase-catalyzed polycondensation to obtain polyester via (a) AB-type monomers, (b) AA-, BB-type monomers.

7.4.1 Dicarboxylic Acid or Its Esters with Diols

The first publication reported lipase synthesis of polyesters from various dicarboxylic acids and diols appeared in 1984 where only oligomers were obtained using lipase from *Aspergillus niger* as a catalyst [64]. Binns *et al.* reported lipase-catalyzed synthesis of polyesters using adipic acid with 1,4-butanediol – both monomers are derived from glucose [65, 66] – in diisopropyl ether, which resulted in a polyester with degree of polymerization (DP) 20 [67]. Utilization of divinyl adipate instead of adipic acid together with 1,4-butanediol in diisopropyl ether at 45 °C for 48 h in the presence of lipase PC as a catalyst afforded polyester with molecular weight of 2.1×10^4 Da [68]. Interestingly, similar polymerization was conducted in $scCO_2$, environmentally friendly solvent, instead of diisopropyl ether, which afforded the same polyester with M_n 3900 Da [69]. Poly(1,4-butyl sebacate) was produced by lipase-catalyzed polycondensation of sebacic acid with 1,5-butanediol; both monomers can be synthesized from renewable resource, in bulk under reduced pressure yield polyester of M_n 14 kDa [70]. The authors also demonstrated increases in polyester molecular weight by increasing the length of methylene chain of monomers. Linko *et al.* [71] was able, however, to produce poly(1,4-butyl sebacate) with molecular weight up to 42 kDa using *Mucor miehei* lipase as catalyst under high vacuum. PBS is an important biodegradable polyester due to its excellent biodegradability, thermal processability, and balanced mechanical properties [72]. PBS can be produced by polymerization of 1,4-butanediol with succinic acid, which is in turn synthesized by microbial fermentation of renewable feedstock such as glucose, starch, xylose, and so on [73]. In 2006, a relatively high-molecular-weight PBS of $M_w = 38$ kDa was produced by polymerization of diethyl succinate and 1,4-butanediol using N435 as a catalyst in a temperature-varied, two-stage polymerization process [74]. In this report, the reaction temperature was increased from 80 to 95 °C after 21 h in order to keep the resulting polymer soluble in the reaction medium.

Itaconic acid is an unsaturated carboxylic acid and considered as an important renewable monomer for the production of polyesters. It can be obtained via fermentation of carbohydrates, for example, glucose, and has a current worldwide production of about 15,000 tons per year [75]. In fact, introducing itaconic acid as building block yields polyester containing double bond within the main chain, which can potentially be used later for further chemical modification, for example, crosslinking reaction. Fully bio-based and cross-linkable polyester was synthesized by lipase-catalyzed polycondensation of 1,4-butanediol, succinic acid, and itaconic acid using lipase CA (lipase from *Candida antarctica*) as a catalyst [76].

Only oligomers can be obtained when succinic acid and itaconic acid used in carboxylic acid form to produce polyester, while alkyl diesters form of these monomers resulted polyester with molecular weight and molar compositions affected mainly by the applied polymerization method. The highest molecular weight, $M_w = 23$ kDa, was obtained by carrying out the enzymatic polymerization reaction under azeotropic condition where a mixture of cyclohexane and toluene was used as solvent. Loos *et al.* investigated the N435-catalyzed polycondensation of bio-based dimethyl itaconate, 1,4-butanediol, and various diacid ethyl esters of methylene with a chain length of 2–10 (Figure 7.4) [77]. The authors could synthesize polyesters of molecular weight up to 94 kDa by utilizing the two-stage method in diphenyl ether. The authors demonstrated also that N435 prefers diacid ethyl ester monomers with methylene chain length ($n > 2$),

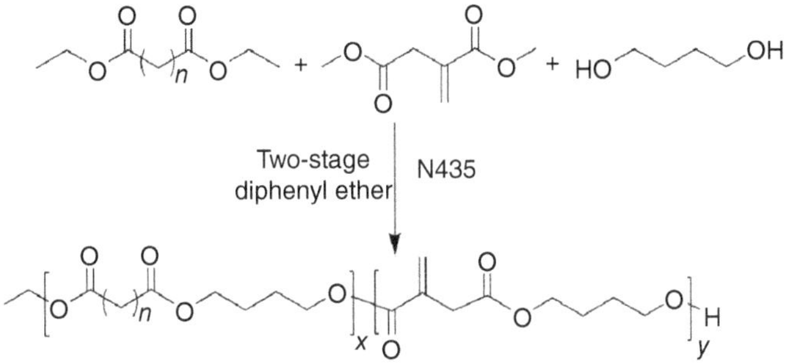

$n = 2$, succinate; $n = 3$, glutarate; $n = 2$, adipate; $n = 6$, subarate; $n = 8$, sebacate; $n = 10$, dodecanedioate;

Figure 7.4 Lipase-catalyzed synthesis of bio-based polyester via a two-stage method in diphenyl ether.

with highest specificity for diethyl adipate. Deterioration of the double bond is the main side reaction observed when itaconic acid is used as a monomer to synthesize polyester using conventional polymerization techniques. No deterioration of the double bond has been observed due to the chemoselectivity of lipase and the ability to achieve the polymerization process under mild reaction conditions when enzyme is used as catalyst.

7.4.2 Dicarboxylic Acid or Its Esters with Polyols

Presence of pendant functional groups, for example, hydroxyl, carboxylic acid, or mercapto, within the structure of polyesters is of great interest as they can affect the physical and mechanical properties of the resulting polymer, for example, increase solubility of polymer in water, or can be potentially used later to attach some interesting molecules onto the polymer chain, such as drug or fluoresce dye, or allow to attach further polymer chains in order to modify the mechanical, biological, and physical properties to meet the technical needs of many applications. The regioselectivity of enzymes enables the syntheses of linear or nearly linear polyesters in one step starting from multifunctional monomers (functionality ≥ 3). Thus, many interesting bio-based polyols such as glycerol, sucrose, have been investigated for potential use as a building block to prepare linear polyesters with free hydroxyl (OH) pendant groups. The first report appeared in 1991, in which water-soluble sucrose oligoesters with DP of 11 were successfully synthesized with high regioselectivity by reacting sucrose and bis(2,2,2-trifluoroethyl) sebacate in pyridine at 45 °C for 5 days using the protease Proleather as a catalyst. Russell and his coworkers reported lipase-catalyzed synthesis of polyester with free pendant OH by the reaction of divinyl adipate with triols of various lengths at 50 °C for 24 h [78]. The resulting polyesters have M_w ranging from 3 to 14 kDa based on the kind of triol utilized.

The authors demonstrated that the utilized reaction conditions resulted in polyester with free OH pendant groups that are 90–95% secondary and 5–10% primary without evidence for network formation. Deeper investigation of previous reaction was carried out later by Kobayashi and his coworkers where they studied the factors, that is, reaction temperature, lipase origin, feed ratio of monomers, and influencing stereoselectivity of lipase-catalyzed polycondensation reaction between divinyl sebacate and various diols (Figure 7.5) [79]. The report showed a perfect control of polymerization regiospecific at

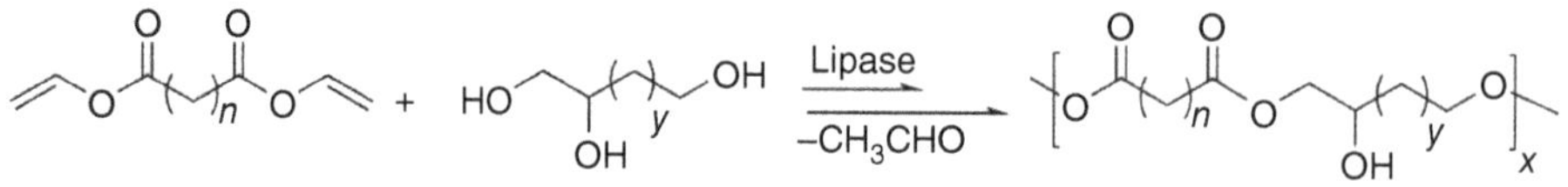

Figure 7.5 Lipase-catalyzed polycondensation to prepare linear polyester with free hydroxyl pendant groups [78].

n = 2, divinyl adiapte; n = 8 divinyl sebacat
y = 0, glycerol; y = 1, 1,2,4-butanetriol; y = 3, 1,2,6-hexanetriol

the α,ω-position of the utilized triol when the polymerization was carried out at 45 °C. A sugar containing polyester was prepared by lipase-catalyzed polycondensation of divinyl sebacate and sorbitol in acetonitrile at 60 °C for 72 h [80]. The resulting polymer has an average molecular weight of 10 kDa with exclusive acylation at α,ω-position of sorbitol. Carrying out the polymerization reaction at a lower temperature of 20 °C increased the yield of polymerization process but resulted in polymers with smaller molecular weight. Potential application of renewable alditol polyols to prepare polyester with free pendant group was carried out later using N435-catalyzed polycondensation between adipic acid and various alditol polyols, that is, erythritol, xylitol, ribitol, D-glucitol, D-mannitol, and D-galactitol [81]. The polymerization process was carried out in bulk for up to 46 h at temperature of 90 °C. The resulting polyesters had a M_w ranging from 11 kDa (galactitol) to 73 kDa (D-mannitol) without correlation between sugar reactivity and corresponding chain length.

7.4.3 Polyesters from Fatty Acid-Based Monomers

Vegetable oils are considered as one of the most important renewable platform chemicals because of their abundant availability in low price [82]. The two main components that can be derived from vegetable oils are glycerol and fatty acids. Therefore, utilization of fatty acids to prepare polymers has been shown great interest over the past two decades [83]. In the field of lipase-catalyzed synthesis of polyesters, fatty acids were used either as a building block to prepare the polyester main chain or were introduced to polymer structure as side chains [84].

7.4.3.1 Lipase-Catalyzed Polycondensation of α,ω-Dicarboxylic Acids and Diols

A series of linear unsaturated or epoxidized thermoplastic polyesters were prepared by enzymatic polymerization of unsaturated or epoxidized α,ω-dicarboxylic acid methyl esters of chain lengths (C_{18}, C_{20}, C_{26}) with 1,3-propanediol or 1,4-butanediol using N435 as a catalyst [85]. The utilized α,ω-dicarboxylic acid methyl esters were obtained from vegetable oils by transesterification reaction of their triglycerides with methanol, followed by metathetical cleavage of the resulting methyl esters using Grubbs' catalysts [86]. The resulting terminally unsaturated fatty acid part was then participated in a metathetical condensation reaction using Grubbs' catalysts to yield finally long-chain symmetrically unsaturated α,ω-dicarboxylic acid dimethyl esters. The double bond of the resulting products was then epoxidized using hydrogen peroxides in methyl acetate and in the presence of N435 as a catalyst. The enzymatic polymerization of unsaturated or epoxidized α,ω-dicarboxylic acid methyl esters with 1,3-propanediol resulted in polyesters with molecular weight in the range of 1950–3300 Da and had melting points in the range of 47–75 °C, while with 1,4-butanediol polyesters with higher molecular weights (7900–11,600 Da) were produced. Both double bonds and epoxy groups remained stable during the polymerization process due to the chemoselectivity of the enzyme and the ability, which was given by them, to carry out the polymerization process under mild conditions. Completely green synthetic strategy to synthesize both monomers and corresponding polyesters has been suggested later by Gross and his coworkers [87], where they prepared the α,ω-dicarboxylic acid monomers from fatty acids using *Candida tropicalis* ATCC20962 or related engineered strains. Accordingly, three bio-based monomers, 18-*cis*-9-octadecenedioic, 1,22-*cis*-9-docosenedioic, and 1,18-*cis*-9,10-epoxy-octadecanedioic acids, were synthesized from oleic, erucic, and epoxy stearic acids, respectively. The resulting unsaturated and epoxidized α,ω-dicarboxylic acid were then used as a building block besides various diols,1,3-propanediol, 1,8-octanediol, and 1,16-hexadecanediol, to produce aliphatic polyesters containing double bond- or epoxy-functionalized polyesters using N435 as a catalyst. The resulting polyesters had a molecular weight in the range from 25 to 57 kDa based on the utilized monomers and/or reaction conditions.

7.4.3.2 Lipase-Catalyzed Polycondensation of Hydroxy Fatty Acids

Ricinoleic acid (RA), unsaturated hydroxyl fatty acid, is the main component of the seed oil obtained from castor plant. In 2005, RA was used as a monomer to synthesize lipophilic star-shaped polyester in two steps [88]. First, poly(ricinoleic acid) was synthesized in bulk using N435 as a catalyst at 70 °C for 10 days. Only oligomers of molecular weight 1040 Da was obtained due to the high viscosity of RA and its chemical structure, which contains only secondary hydroxyl group accessible for the polymerization reaction. In the next step, the branched star-shaped polyester was enzymatically prepared by the reaction of resulting poly(ricinoleic acid) and polyols, for example, trimethylolpropane or pentaerythritol, using immobilized lipase from *C. antarctica* or *Rhizomucor miehei* at 70 °C for 14 days. The resulting polymer had an average molecular weight up to 4850 Da and owned high viscosity and high viscosity index, which makes it a good candidate to apply in the field of environmentally friendly lubricant materials. Using enzymes to produce the previous polymer enabled its production under mild conditions in order to avoid the discoloration, odor, dehydration. Further attempts to enzymatically synthesize poly(RA) were carried out later by Matsumura and his coworkers (Figure 7.6) [89]. The authors screened the activity of many enzymes to test their reactivity toward enzymatic polycondensation of RA or methyl ricinoleate. Among the tested enzymes, immobilized lipase PC showed the best activity and was able to produce poly(RA) with M_w 5600 Da.

Poly(RA) with molecular weight of 100,600 Da could be achieved by utilization of methyl ricinoleate instead of the RA, placing 4 Å molecular sieves in the vapor phase

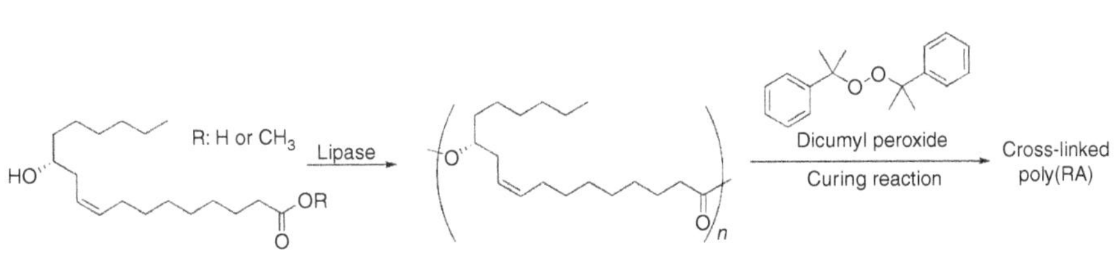

Figure 7.6 Lipase-catalyzed polycondensation of ricinoleic acid or methyl ricinoleate followed by cross-linking reaction using dicumyl peroxide [89].

to capture the resulting methanol and thus shifting the equilibrium reaction toward the product, and increasing the amount of utilized lipase to a concentration of 150 wt%. The final polymer was a viscous liquid with a glass transition temperature of −74.8 °C.

Cis-9,10-epoxy-18-hydroxyoctadecanoic acid (EHC) is another interesting renewable monomer, which forms about 100 g kg^{-1} of dry outer bark of *Betula verrucosa* [90]. In contrast to RA, the chemical structure of EHC is too suitable for the enzymatic polymerization as it owns primary hydroxyl group, which should be more accessible to enzyme than the secondary one. EHC was used as a monomer to prepare linear epoxy-functionalized polyesters using N435 as a catalyst [91]. The effect of solvent on polymerization process was investigated first using three solvents: acetonitrile, dioxane, and toluene. The highest molecular weight, 20 kDa, was obtained when toluene was used as a solvent for 68 h. Similar molecular weight, 16 kDa, in a much shorter time, 6 h, could be achieved when the polymerization process was carried out in bulk.

7.4.3.3 Fatty Acids as Side Chains to Modify Functional Polyesters

Utilization of sustainable saturated and unsaturated fatty acids as side chains to modify linear functional polyester, that is, prepared enzymatically by copolymerization of dicarboxylic acids or their esters with polyols, have attracted increasing attention due to the potential applications of the resulting polymers in many interesting fields. Linear reactive polyester grafted with unsaturated fatty acid chains was prepared by lipase-catalyzed polycondensation of divinyl sebacate and glycerol in the presence of unsaturated higher fatty acids, for example, oleic, linoleic, and linolenic acids [79, 92]. The reaction was carried out in bulk for 24 h using equivalent amount of each component. The polymer yield and molecular weight increased by increasing the reaction temperature, while the polymer composition was nearly the same. The currying process of the resulting polyester was induced either by oxidation with cobalt naphthenate catalyst or by thermal treatment to yield finally cross-linked transparent film. Comb-like, epoxide-containing polyesters were prepared later via two synthetic routes [93]. In the first one, aliphatic polyester owing unsaturated groups in the side chain was prepared in one step by lipase-catalyzed polycondensation of divinyl sebacate, glycerol, and the unsaturated fatty acids. The double bond in the side chains were then epoxidized using hydrogen peroxide in the presence of lipase. In the second route, unsaturated fatty acid was first epoxidized, using hydrogen peroxide and lipase, subsequently used in the enzymatic reaction to prepare the final epoxide comb-like polyester. Currying of the resulting polymer was thermally achieved, which resulted in transparent polymeric film owing glassy surface and pencil hardness higher than the cured film obtained from polyester own unsaturated fatty acids. In fact, both utilized routes are considered as green synthetic strategy as no toxic catalyst was needed to prepare the polyester main chain and to epoxidize the fatty acid side chains. Poly(glycerol adipate) (PGA) is a linear aliphatic polyester with free pendant OH groups and can be obtained by lipase-catalyzed polycondensation of divinyl/or dimethyl adipate with glycerol under mild conditions. A library of comb-like polyesters has been prepared by acylation of PGA with various fatty acids, for example, laurate, stearate, and behenate, with various degrees of substitution. The properties of the resulting polymers were investigated in bulk, water, and at air/water interface. In water, the resulting polymers were able to form well-defined particles of small size and high homogeneity [94]. By using interfacial deposition method, Mäder *et al.* were able to prepare various self-stabilizing nanoparticles with well-defined sizes and narrow size distributions [95]. The shape of resulting nanoparticles depends strongly on the length of side chains and on the degree of substitution [61]. Furthermore, the resulting polymers showed no toxicity to HepG2 cells [96]. The ease of synthesis besides the unique characteristic of resulting polymers strongly suggests that these polymers can be used for nano-drug delivery or as injectable or implantable carrier for the controlled release of active ingredient [97].

7.4.4 Polyester Using Furan as Building Block

The importance of incorporating furan moieties within the structure of polyester stems not only from their wide availability starting from renewable resources but also from the possibility of developing novel chemical structure with original properties simulating the properties and applications of currently available polymers derived from fossil resources [83]. Furan dicarboxylic acid (FDCA), for instance, is a member of the furan family, obtained by oxidation of HMF, which is in turn derived from dehydration of certain sugars [56, 98]. The monomer has been widely studied as a potential bio-based alternative to a petroleum-based monomer terephthalic acid (TA), which is used to prepare the most important commercial polyester poly(ethylene terephthalate) (PET) [99]. Interestingly, recent studies showed superior improved barrier properties for poly(ethylene furanoate) (PEF), which has a water diffusion coefficient 5 times lower than that of PET besides a 10 times lower oxygen permeability [100, 101]. It is expected, therefore, that by 2020 over 60% of global production of FDCA will be used to produce bio-based alternatives to PET. 2,5-Bis(hydroxymethyl)furan (BHMF) is another interesting bio-based furan monomer, which can be obtained by the reduction of HMF [102]. Despite the great importance of furan monomers as building blocks in polymer science, investigation of the possibility of using lipase as catalyst for their polymerization has been reported just recently. Boeriu et al. reported lipase-catalyzed polycondensation of dimethyl furan-2,5-dicarboxylate and linear α, ω-aliphatic diols with chain length in the range from C2 to C12, under anhydrous conditions provided by a mixture of anhydrous toluene and tert-butanol of (70:30 wt%), using a one-stage method [103]. Only a mixture of linear and cyclic furan oligomers, however, could be obtained under the investigated reaction conditions. Loos and her coworker could, however, synthesize similar polymer with molecular weight up to 100 kDa by optimizing the reaction conditions [104]. The authors investigated the N435-catalyzed polymerization of dimethyl 2,5-furandicarboxylate (DMFDCA) with various renewable aliphatic diols using two-stage method in diphenyl ether (Figure 7.7).

In another recent study, the authors investigated also the N435-catalyzed polymerization of BHMF with various diacid ethyl esters, that is, diethyl succinate, diethyl

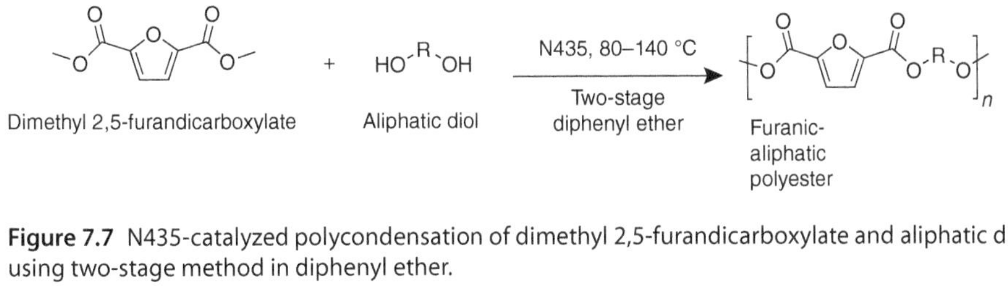

Figure 7.7 N435-catalyzed polycondensation of dimethyl 2,5-furandicarboxylate and aliphatic diol using two-stage method in diphenyl ether.

glutarate, diethyl adipate, and diethyl sebacate (98+%), using three-stage method [105]. Only polyester with M_w in the range of 1800–2900 could be obtained, which was explained by the observed etherification side reaction.

7.4.5 Conclusions and Remarks

Enzyme-catalyzed polymerization has been revealed to be a green metal-free synthetic route to prepare polyesters from renewable materials under mild conditions in a relatively short time. Enzymes also allow the production of clean and attractive reactive polyesters, with double bond, OH, epoxy groups, in one step, which can be used later in many interesting applications.

It should be stressed finally that, while the sustainability is a mankind's dream, the current achievements in polymer field still fall below the expectations of this dream. Although the current statistics indicates clearly that the bio-based share grows in a faster rate than the growth of global polymer market [1], the recent drop in oil price could, however, reduce this growth. On the other hand, despite the aforementioned advantages of using lipase for polyester synthesis, the enzymes are used yet for industrial production of polyesters due to some obstacles such as (i) the large quantity of enzymes required for the polymerization process, which increases significantly the production cost, (ii) relatively limit the number of polymers with high molar mass, which can be obtained using this polymerization technique, (iii) and the low reactivity of lipases toward some important commercially available monomers, for example, lactides [13, 106]. Therefore, significant scientific efforts should be made in order to overcome these critical difficulties to make our planet livable for our children and grandchildren.

7.4.6 Questions for Further Consideration

1. How can we increase the efficiency of lipases toward enzymatic polymerization?
2. How can we increase the temperature tolerance of lipases?
3. What should be done in order to increase the regioselectivity of enzymes at temperatures higher than 50 °C?
4. Which strategies should be followed to increase the molar mass of polyesters produced via enzymatic polymerization?
5. How can we reduce the cost of polymer production when enzymes are used as catalyst?

List of Abbreviations

PLA	poly(lactic acid)
PBS	poly(butylene succinate)
γ-BL	γ-butyrolactone
γ-VL	γ-valerolactone
LA	levulinic acid
DVL	δ-valerolactone
ε-CP	ε-caprolactone
HMF	hydroxymethylfurfural/(hydroxymethyl) furfural
ROP	ring-opening polymerization

N435	immobilized lipase from *Candida antarctica* lipase B
Poly(ε-CP)	poly(ε-caprolactone)
scCO$_2$	supercritical carbon dioxide
Lipase PC	lipase from *Burkholderia cepacia*
M_n	number average molecular weight
M_w	weight average molecular weights
Da	dalton
PDL	pentadecalactone
PE	polyethylene
PPDL	poly(pentadecalactone)
DP	degree of polymerization
lipase CA	lipase from *Candida antarctica*
OH	hydroxyl groups
RA	ricinoleic acid
EHC	*cis*-9,10-epoxy-18-hydroxyoctadecanoic acid
PGA	poly(glycerol adipate)
FDCA	furan dicarboxylic acid
TA	terephthalic acid
PET	poly(ethylene terephthalate)
PEF	poly(ethylene furanoate)
BHMF	2,5-bis(hydroxymethyl)furan
DMFDCA	dimethyl 2,5-furandicarboxylate

References

1 F. Aeschelmann, and M. Carus, Biobased building blocks and polymers in the world: capacities, production, and applications–status quo and trends towards 2020, *Ind. Biotechnol*, **11**(3), 154–159 (2015).

2 R. Mülhaupt, Green polymer chemistry and bio-based plastics: dreams and reality, *Macromol. Chem. Phys*, **214**(2), 159–174 (2013).

3 C.K.S. Pillai, Challenges for natural monomers and polymers: novel design strategies and engineering to develop advanced polymers, *Des. Monomers Polym*, **13**(2), 87–121 (2010).

4 M. Jamshidian, E.A. Tehrany, M. Imran, *et al.*, Poly-lactic acid: production, applications, nanocomposites, and release studies, *Compr. Rev. Food Sci. Food Saf*, **9**(5), 552–571 (2010).

5 J.A. Linthorst, An overview: origins and development of green chemistry, *Found. Chem*, **12**(1), 55–68 (2010).

6 M.A. Hillmyer, and W.B. Tolman, Aliphatic polyester block polymers: renewable, degradable, and sustainable, *Acc. Chem. Res*, **47**(8), 2390–2396 (2014).

7 A. Díaz, R. Katsarava, and J. Puiggalí, Synthesis, properties and applications of biodegradable polymers derived from diols and dicarboxylic Acids: From Polyesters to poly(ester amide)s, *Int. J. Mol. Sci*, **15**(5), 7064–7123 (2014).

8 M. Vert, Aliphatic polyesters: Great degradable polymers that cannot do everything, *Biomacromolecules*, **6**(2), 538–546 (2005).

9 A.L. Andrady, Microplastics in the marine environment, *Mar. Pollut. Bull*, **62**(8), 1596–1605 (2011).

10 A. Sivan, New perspectives in plastic biodegradation, *Curr. Opin. Biotechnol*, **22**(3), 422–426 (2011).

11 G.Q. Chen, A microbial polyhydroxyalkanoates (PHA) based bio- and materials industry, *Chem. Soc. Rev*, **38**(8), 2434–2446 (2009).

12 J.E. Puskas, K.S. Seo, and M.Y. Sen, Green polymer chemistry: Precision synthesis of novel multifunctional poly(ethylene glycol)s using enzymatic catalysis, *Eur. Polym. J*, **47**(4), 524–534 (2011).

13 R.A. Gross, M. Ganesh, and W. Lu, Enzyme-catalysis breathes new life into polyester condensation polymerizations, *Trends Biotechnol*, **28**(8), 435–443 (2010).

14 S. Kobayashi, Lipase-catalyzed polyester synthesis – A green polymer chemistry, *Proc. Japan Acad. Ser B*, **86**(4), 338–365 (2010).

15 S. Kobayashi, and A. Makino, Enzymatic polymer synthesis: an opportunity for green polymer chemistry, *Chem. Rev*, **109**(11), 5288–5353 (2009).

16 S. Kobayashi, Enzymatic ring-opening polymerization and polycondensation for the green synthesis of polyesters, *Polym. Adv. Technol*, **26**(7), 677–686 (2015).

17 B. Yeniad, H. Naik, and A. Heise, Lipases in polymer chemistry, *Adv. Biochem. Eng. Biotechnol*, **125**, 69–95 (2011).

18 A. Löfgren, A.C. Albertsson, P. Dubois, and R. Jérôme, Recent advances in ring-opening polymerization of lactones and related compounds, *J. Macromol. Sci. Part C Polym. Rev*, **35**(3), 379–418 (1995).

19 N. Miletić, K. Loos, R.A. Gross, Enzymatic Polymerization of Polyester, in K. Loos (Ed.), *Biocatalysis in Polymer Chemistry*, Wiley-VCH Verlag GmbH & Co. KGaA, Weinheim, Germany, pp. 83–129 (2010).

20 X. Lou, C. Detrembleur, and R. Jérôme, Living cationic polymerization of δ-valerolactone and synthesis of high molecular weight homopolymer and asymmetric telechelic and block copolymer, *Macromolecules*, **35**(4), 1190–1195 (2002).

21 D.W. Hwang, P. Kashinathan, J.M. Lee, *et al.*, Production of γ-butyrolactone from biomass-derived 1,4-butanediol over novel copper-silica nanocomposite, *Green Chem*, **13**(7), 1672–1675 (2011).

22 G. Budroni, and A. Corma, Gold and gold–platinum as active and selective catalyst for biomass conversion: Synthesis of γ-butyrolactone and one-pot synthesis of pyrrolidone. *J. Catal*, **257**(2), 403–408 (2008).

23 I. Meynial-Salles, S. Dorotyn, and P. Soucaille, A new process for the continuous production of succinic acid from glucose at high yield, titer, and productivity, *Biotechnol. Bioeng*, **99**(1), 129–135 (2008).

24 A. Corma, S. Iborra, and A. Velty, Chemical routes for the transformation of biomass into chemicals. *Chem. Rev*, **107**(6), 2411–2502 (2007).

25 L. Wang, K. Thai, and M. Gravel, NHC-catalyzed ring expansion of oxacycloalkane-2-carboxaldehydes: A versatile synthesis of lactones, *Org. Lett*, **11**(4), 891–893 (2009).

26 T. Buntara, S. Noel, and P.H. Phua, *et al.*, Caprolactam from renewable resources: catalytic conversion of 5-Hydroxymethylfurfural into caprolactone, *Angew. Chemie Int. Ed*, **50**(31), 7083–7087 (2011).

27 H. Uyama, and S. Kobayashi, Enzymatic ring-opening polymerization of lactones catalyzed by lipase. *Chem. Lett*, **22**(7), 1149–1150 (1993).

28 D. Knani, A.L. Gutman, and D.H. Kohn, Enzymatic polyesterification in organic media. Enzyme-catalyzed synthesis of linear polyesters. I. Condensation polymerization of linear hydroxyesters. II. Ring-opening polymerization of ε-caprolactone, *J. Polym. Sci. Part A Polym. Chem*, **31**(5), 1221–1232 (1993).

29 S. Namekawa, S. Suda, H. Uyama, and S. Kobayashi, Lipase-catalyzed ring-opening polymerization of lactones to polyesters and its mechanistic aspects, *Int. J. Biol. Macromol*, **25**(1–3), 145–151 (1999).

30 S. Noda, N. Kamiya, M. Goto, and F. Nakashio, Enzymatic polymerization catalyzed by surfactant-coated lipases in organic media, *Biotechnol. Lett*, **19**(4), 307–310 (1997).

31 T. Nakaoki, Y. Mei, L.M. Miller, *et al.*, Lipase B catalyzed polymerization of lactones: effects of immobilization matrices on polymerization kinetics & molecular weight, *Ind. Biotechnol*, **1**, 126–134 (2005).

32 M. Matsumoto, D. Odachi, and K. Kondo, Kinetics of ring-opening polymerization of lactones by lipase, *Biochem. Eng. J*, **4**(1), 73–76 (1999).

33 M. De Geus, J. Peeters, M. Wolffs, *et al.*, Investigation of factors influencing the chemoenzymatic synthesis of block copolymers, *Macromolecules*, **38**(10), 4220–4225 (2005).

34 A. Córdova, T. Iversen, K. Hult, and M. Martinelle, Lipase-catalysed formation of macrocycles by ring-opening polymerisation of ε-caprolactone, *Polymer*, **39**(25), 6519–6524 (1998).

35 R.T. MacDonald, S.K. Pulapura, Y.Y. Svirkin, *et al.*, Enzyme-catalyzed epsilon-Caprolactone ring-opening polymerization, *Macromolecules*, **28**(1), 73–78 (1995).

36 Y. Mei, A. Kumar, and R.A. Gross, Probing water-temperature relationships for lipase-catalyzed lactone ring-opening polymerizations, *Macromolecules*, **35**(14), 5444–5448 (2002).

37 H. Uyama, K. Takeya, and S. Kobayashi, Synthesis of polyesters by enzymatic ring-opening copolymerization using lipase catalyst, *Proc. Japan Acad. Ser. B Phys. Biol. Sci*, **69**(8), 203–207 (1993).

38 H. Uyama, K. Takeya, and S. Kobayashi, Synthesis of polyesters by enzymatic ring-opening copolymerization using lipase catalyst, *Proc. Japan Acad. Ser. B Phys. Biol. Sci*, **69**(8), 203–207 (1993).

39 H. Uyama, S. Namekawa, and S. Kobayash, Mechanistic studies on the lipase-catalyzed ring-opening polymerization of lactones, *Polym. J*, **29**(3), 299–301 (1997).

40 H. Kikuchi, H. Uyama, and S. Kobayashi, Lipase-catalyzed ring-opening polymerization of substituted lactones, *Polym. J*, **34**(11), 835–840 (2002).

41 H. Uyama, S. Suda, H. Kikuchi, and S. Kobayashi, Extremely efficient catalysis of immobilized lipase in ring-opening polymerization of lactones, *Chem. Lett*, **11**, 1109–1110 (1997).

42 K.J. Thurecht, A. Heise, M. Degeus, *et al.*, Kinetics of enzymatic ring-opening polymerization of ε-caprolactone in supercritical carbon dioxide, *Macromolecules*, **39**(23), 7967–7972 (2006).

43 G.A.R. Nobes, R.J. Kazlauskas, and R.H. Marchessault, Lipase-catalyzed ring-opening polymerization of lactones: a novel route to poly(hydroxyalkanoate)s. *Macromolecules*, **29**(14), 4829–4833 (1996).

44 Y. Suzuki, S. Taguchi, T. Hisano, *et al.*, Correlation between structure of the lactones and substrate specificity in enzyme-catalyzed polymerization for the synthesis of polyesters, *Biomacromolecules*, **4**(3), 537–543 (2003).

45 S. Kobayashi, K. Takeya, S. Suda, and H. Uyama, Lipase-catalyzed ring-opening polymerization of medium-size lactones to polyesters, *Macromol. Chem. Phys*, **199**(8), 1729–1736 (1998).

46 A. Duda, A. Kowalski, S. Penczek, *et al.*, Kinetics of the ring-opening polymerization of 6-, 7-, 9-, 12-, 13-, 16-, and 17-membered lactones. Comparison of chemical and enzymatic polymerizations. *Macromolecules*, **35**(11), 4266–4270 (2002).

47 L. van der Mee, F. Helmich, R. de Bruijn, *et al.*, Investigation of lipase-catalyzed ring-opening polymerizations of lactones with various ring sizes: kinetic evaluation, *Macromolecules*, **39**(15), 5021–5027 (2006).

48 D. McGinty, C.S. Letizia, and A.M. Api, Fragrance material review on ω-pentadecalactone, *Food Chem. Toxicol*, **49**, S193–S201 (2011).

49 M. de Geus, I. van der Meulen, B. Goderis, *et al.*, Performance polymers from renewable monomers: high molecular weight poly(pentadecalactone) for fiber applications, *Polym. Chem*, **1**(4), 525–533 (2010).

50 A. Taden, M. Antonietti, and K. Landfester, Enzymatic polymerization towards biodegradable polyester nanoparticles, *Macromol. Rapid Commun*, **24**(8), 512–516 (2003).

51 B. Lebedev, and A. Yevstropov, Thermodynamic properties of polylactones, *Die Makromol. Chemie*, **185**(6), 1235–1253 (1984).

52 P. Skoglund, and Å. Fransson, Crystallization kinetics of polytridecanolactone and polypentadecanolactone, *Polymer*, **39**(14), 3143–3146 (1998).

53 M. Gazzano, V. Malta, M.L. Focarete, *et al.*, Crystal structure of poly(ω-pentadecalactone), *J. Polym. Sci. Part B Polym. Phys*, **41**(10), 1009–1013 (2003).

54 J. Cai, C. Liu, M. Cai, *et al.*, Effects of molecular weight on poly(ω-pentadecalactone) mechanical and thermal properties, *Polymer*, **51**(5), 1088–1099 (2010).

55 I. van der Meulen, M. de Geus, H. Antheunis, *et al.*, Polymers from functional macrolactones as potential biomaterials: enzymatic ring opening polymerization, biodegradation, and biocompatibility, *Biomacromolecules*, **9**(12), 3404–3410 (2008).

56 J.J. Bozell, and G.R. Petersen, Technology development for the production of biobased products from biorefinery carbohydrates—the US Department of Energy's "Top 10" revisited *Green Chem*, **12**(4), 539–554 (2010).

57 S. Matsumura, K. Mabuchi, and K. Toshima, Lipase-catalyzed ring-opening polymerization of lactide, *Macromol. Rapid Commun*, **18**(6), 477–482 (1997).

58 M. Hans, H. Keul, and M. Moeller, Ring-opening polymerization of dd-lactide catalyzed by novozyme 435, *Macromol. Biosci*, **9**(3), 239–247 (2009).

59 R. García-Arrazola, D.A. López-Guerrero, M. Gimeno, and E. Bárzana, Lipase-catalyzed synthesis of poly-l-lactide using supercritical carbon dioxide. *J. Supercrit, Fluids*, **51**(2), 197–201 (2009).

60 C.K. Williams, Synthesis of functionalized biodegradable polyesters, *Chem. Soc. Rev*, **36**(10), 1573–1580 (2007).

61 T. Naolou, V.M. Weiss, D. Conrad, et al., Fatty Acid Modified Poly(glycerol adipate) - Polymeric Analogues of Glycerides, in *Tailored Polymer Architectures for Pharmaceutical and Biomedical Applications*, vol. **1135**, American Chemical Society, Washington, DC, pp. 39–52 (2013).

62 Y.F. Wang, J.J. Lalonde, M. Momongan, et al., Lipase-catalyzed irreversible transesterifications using enol esters as acylating reagents: preparative enantio- and regioselective syntheses of alcohols, glycerol derivatives, sugars and organometallics, *J. Am. Chem. Soc*, **110**(21), 7200–7205 (1988).

63 K.R. Kiran, and S. Divakar, Lipase-catalysed polymerization of lactic acid and its film forming properties, *World J. Microbiol. Biotechnol*, **19**(8), 859–865 (2003).

64 S. Okumura, M. Iwai, and Y. Tominaga, Synthesis of ester oligomer by *Aspergillus niger* lipase, *Agric. Biol. Chem*, **48**(11), 2805–2808 (1984).

65 K.M. Draths, and J.W. Frost, Environmentally compatible synthesis of adipic acid from D-glucose, *J. Am. Chem. Soc*, **116**(1), 399–400 (1994).

66 H. Yim, R. Haselbeck, W. Niu, et al., Metabolic engineering of Escherichia coli for direct production of 1,4-butanediol, *Nat. Chem. Biol*, **7**(7), 445–452 (2011).

67 F. Binns, S.M. Roberts, A. Taylor, and C.F. Williams, Enzymic polymerisation of an unactivated diol/diacid system, *J. Chem. Soc. Perkin Trans*, **1**(8), 899–904 (1993).

68 H. Uyama, S. Yaguchi, and S. Kobayashi, Lipase-catalyzed polycondensation of dicarboxylic acid-divinyl esters and glycols to aliphatic polyesters, *J. Polym. Sci. Part A Polym. Chem*, **37**(15), 2737–2745 (1999).

69 T. Takamoto, H. Uyama, and S. Kobayashi, Lipase-catalyzed synthesis of aliphatic polyesters in supercritical carbon dioxide, *e-Polymers*, **1**(1) (2001).

70 H. Uyama, K. Inada, and S. Kobayashi, Lipase-catalyzed synthesis of aliphatic polyesters by polycondensation of dicarboxylic acids and glycols in solvent-free system. *Polym. J*, **32**(5), 440–443 (2000).

71 Y.Y. Linko, Z.L. Wang, and J. Seppälä, Lipase-catalyzed synthesis of poly(1,4-butyl sebacate) from sebacic acid or its derivatives with 1,4-butanediol, *J. Biotechnol*, **40**(2), 133–138 (1995).

72 J. Xu, and B.H. Guo, Poly(butylene succinate) and its copolymers: Research, development and industrialization, *Biotechnol. J*, **5**(11), 1149–1163 (2010).

73 J.M. Pinazo, M.E. Domine, V. Parvulescu, and F. Petru, Sustainability metrics for succinic acid production: A comparison between biomass-based and petrochemical routes, *Catal. Today*, **239**, 17–24 (2015).

74 H. Azim, A. Dekhterman, Z. Jiang, and R.A. Gross, *Candida antarctica* lipase B-catalyzed synthesis of poly(butylene succinate): Shorter chain building blocks also work, *Biomacromolecules*, **7**(11), 3093–3097 (2006).

75 M. Rose, and R. Palkovits, Cellulose-based sustainable polymers: State of the art and future trends, *Macromol. Rapid Commun*, **32**(17), 1299–1311 (2011).

76 Y. Jiang, G.O.R.A. van Ekenstein, A.J.J. Woortman, and K. Loos, Fully biobased unsaturated aliphatic polyesters from renewable resources: enzymatic synthesis, characterization, and properties, *Macromol. Chem. Phys*, **215**(22), 2185–2197 (2014).

77 Y. Jiang, A. Woortman, G. Alberda van Ekenstein, and K. Loos, Environmentally benign synthesis of saturated and unsaturated aliphatic polyesters via enzymatic

polymerization of biobased monomers derived from renewable resources, *Polym. Chem*, **6**, 1–9 (2015).

78 B.J. Kline, E.J. Beckman, and A.J. Russell, One-step biocatalytic synthesis of linear polyesters with pendant hydroxyl groups, *J. Am. Chem. Soc*, **120**(37), 9475–9480 (1998).

79 H. Uyama, K. Inada, and S. Kobayashi, Regioselectivity control in lipase-catalyzed polymerization of divinyl sebacate and triols, *Macromol. Biosci*, **1**(1), 40–44 (2001).

80 H. Uyama, E. Klegraf, S. Wada, and S. Kobayashi, Regioselective polymerization of sorbitol and divinyl sebacate using lipase catalyst, *Chem. Lett*, **174**(7), 800–801 (2000).

81 J. Hu, W. Gao, A. Kulshrestha, and R.A. Gross, "Sweet polyesters"s: Lipase-catalyzed condensation - polymerizations of alditols, *Macromolecules*, **39**(20), 6789–6792 (2006).

82 G. Lligadas, J.C. Ronda, M. Galià, and V. Cádiz, Renewable polymeric materials from vegetable oils: a perspective, *Mater. Today*, **16**(9), 337–343 (2013).

83 A. Gandini, T.M. Lacerda, A.J.F. Carvalho, and E. Trovatti, Progress of polymers from renewable resources: Furans, vegetable oils, and polysaccharides, *Chem. Rev*, **116**(3), 1637–1669 (2016).

84 Y. Tai, and K. Zhang, Enzyme-Assisted Synthesis of Plant Oil-Based Polymers, in M.R. Kessler, C. Zhang, and S. Madbouly (Eds.), *Bio-Based Plant Oil Polymers and Composites*, Elsevier Inc., pp. 127–148 (2016).

85 S. Warwel, C. Demes, and G. Steinke, Polyesters by lipase-catalyzed polycondensation of unsaturated and epoxidized long-chain α,ω-dicarboxylic acid methyl esters with diols, *J. Polym. Sci. Part A Polym. Chem*, **39**(10), 1601–1609 (2001).

86 P. Schwab, M.B. France, J.W. Ziller, and R.H. Grubbs, A series of well-defined metathesis catalysts–synthesis of [RuCl2(=CHR′)(PR3)2] and its reactions, *Angew. Chemie Int. Ed. English*, **34**(18), 2039–2041 (1995).

87 Y. Yang, W. Lu, X. Zhang, *et al*., Two-step biocatalytic route to biobased functional polyesters from ω-carboxy fatty acids and diols, *Biomacromolecules*, **11**(1), 259–268 (2010).

88 A.R. Kelly, and D.G. Hayes, Lipase-catalyzed synthesis of polyhydric alcohol-poly(ricinoleic acid) ester star polymers, *J. Appl. Polym. Sci*, **101**(3), 1646–1656 (2006).

89 I. Ebata, K. Toshima, and S. Matsumura, Lipase-catalyzed synthesis and curing of high-molecular-weight polyricinoleate, *Macromol. Biosci*, **7**(6), 798–803 (2007).

90 R. Ekman, The suberin monomers and triterpenoids from the outer bark of *Betula verrucosa*, Ehrh. *Holzforschung*, **37**(4), 205–211 (1983).

91 A. Olsson, M. Lindström, and T. Iversen, Lipase-catalyzed synthesis of an epoxy-functionalized polyester from the suberin monomer cis-9,10-epoxy-18-hydroxyoctadecanoic Acid, *Biomacromolecules*, **8**(2), 757–760 (2007).

92 T. Tsujimoto, H. Uyama, and S. Kobayashi, Enzymatic synthesis of cross-linkable polyesters from renewable resources. *Biomacromolecules*, **2**(1), 29–31 (2001).

93 H. Uyama, M. Kuwabara, T. Tsujimoto, and S. Kobayashi, Enzymatic synthesis and curing of biodegradable epoxide-containing polyesters from renewable resources, *Biomacromolecules*, **4**(2), 211–215 (2003).

94 P. Kallinteri, S. Higgins, G.A. Hutcheon, et al., Novel functionalized biodegradable polymers for nanoparticle drug delivery systems, *Biomacromolecules*, **6**(4), 1885–1894 (2005).

95 aV.M. Weiss, T. Naolou, G. Hause, et al., Poly(glycerol adipate)-fatty acid esters as versatile nanocarriers: From nanocubes over ellipsoids to nanospheres, *J. Control. Release*, **158**(1), 156–164 (2012) bV.M. Weiss, T. Naolou, T. Groth, et al., In vitro toxicity of stearoyl-poly(glycerol adipate) nanoparticles, *J. Appl. Biomater. Funct. Mater*, **10**(3), 163–169 (2012).

96 K. Mäder, V. Weiss, J. Kressler, and T. Naolou, Injectable and implantable carrier systems based on modified poly(dicarboxylic acid multi-oil esters) for the controlled release of active ingredient. WO2015161841 A2, issued (2015).

97 R.J. Van Putten, J.C. Van Der Waal, E. De Jong, et al., Hydroxymethylfurfural, a versatile platform chemical made from renewable resources, *Chem. Rev*, **113**(3), 1499–1597 (2013).

98 A. Pellis, E. Herrero Acero, L. Gardossi, et al., Renewable building blocks for sustainable polyesters: new biotechnological routes for greener plastics, *Polym. Int*, **65**(8), 861–871 (2016).

99 S.K. Burgess, O. Karvan, J.R. Johnson, et al., Oxygen sorption and transport in amorphous poly(ethylene furanoate), *Polymer*, **55**(18), 4748–4756 (2014).

100 S.K. Burgess, J.E. Leisen, B.E. Kraftschik, et al., Chain mobility, thermal, and mechanical properties of poly(ethylene furanoate) compared to poly(ethylene terephthalate), *Macromolecules*, **47**(4), 1383–1391 (2014).

101 X. Tong, Y. Ma, and Y. Li, Biomass into chemicals: Conversion of sugars to furan derivatives by catalytic processes, *Appl. Catal. A Gen*, **385**(1–2), 1–13 (2010).

102 Á. Cruz-Izquierdo, L.A.M. van den Broek, J.L. Serra, et al., Lipase-catalyzed synthesis of oligoesters of 2,5-furandicarboxylic acid with aliphatic diols, *Pure Appl. Chem*, **87**(1), 59–69 (2015).

103 Y. Jiang, A.J.J. Woortman, G.O.R. Alberda van Ekenstein, and K. Loos, A biocatalytic approach towards sustainable furanic–aliphatic polyesters, *Polym. Chem*, **6**(29), 5198–5211 (2015).

104 Y. Jiang, A.J.J. Woortman, G.O.R. Alberda van Ekenstein, et al., Enzymatic synthesis of biobased polyesters using 2,5-Bis(hydroxymethyl)furan as the building block, *Biomacromolecules*, **15**(7), 2482–2493 (2014).

105 A. Albertsson, and R. Srivastava, Recent developments in enzyme-catalyzed ring-opening polymerization, *Adv. Drug Deliv. Rev*, **60**(9), 1077–1093 (2008).

106 S. Kobayashi, Lipase-catalyzed polyester synthesis – A green polymer chemistry, *Proc. Japan Acad. Ser. B*, **86**(4), 338–365 (2010).

8

Oil-Based and Bio-Derived Thermoplastic Polymer Blends and Composites

Alessia Quitadamo[1], Valerie Massardier[1] and Marco Valente[2]

[1] *INSA de Lyon, IMP/CNRS 5223, Lyon, France*
[2] *University of Rome La Sapienza, Department of Chemical and Material Engineering, Rome, Italy*

> There must *be a* better way *to make the* things we want, *a* way *that doesn't spoil the* sky, *or the* rain *or the* land.
>
> Paul McCartney

8.1 Introduction

The importance assumed by polymer materials in our world is evident and is more ancient than we can imagine.

Polymers like wood, silk, wool and cotton are among the first materials known by the men, though their deep nature has been known only in the twentieth century.

Important discoveries have been done thanks to observation of nature, trying to learn from its structures and organizations, as a consequence many synthetic polymers often hydrophobic have been produced, revolutionizing our routine.

Great importance is related to thermoplastic polymers, thanks to their mechanical properties, production processes, cheapness and versatile applications. Moreover, natural polymers are mainly thermoplastics after plasticization process, with properties rather different from fuel-based thermoplastics such as their hydrophilicity. Hence, this chapter focuses on both oil-based and bio-derived thermoplastic polymers.

Environmental impact is one of the biggest problems related to synthetic polymers: end of life of fuel-based polymers can be a problem, especially when they are disposed off in oceans or on earth. Although many progresses have been done, there are still problems related to the disposal of non-biodegradable polymer wastes.

The increasing consumption of polymer materials and the increasing importance of environmental protection directed researches towards new material classes, such as bio-derived polymers. Many bio-derived polymers are biodegradable, helping reducing environmental impacts caused by fuel-based polymers. In order to obtain polymers with new properties and to face the problem of solid polymer wastes, thermoplastic blends have been developed.

Introduction to Renewable Biomaterials: First Principles and Concepts, First Edition.
Edited by Ali S. Ayoub and Lucian A. Lucia.
© 2018 John Wiley & Sons Ltd. Published 2018 by John Wiley & Sons Ltd.

During last decades, many researches have also studied composites with natural fillers. These works allowed obtaining composites with reduced environmental impact, using renewable sources as fillers, and renewable sources or not as matrices.

The first part of this chapter focuses on the main feature of oil-based and bio-derived thermoplastic polymers and their blends, while the second part covers composites with natural fillers.

8.2 Oil-Based and Bio-Derived Thermoplastic Polymer Blends

8.2.1 Comparison Between Oil-Based and Bio-Derived Thermoplastic Polymers

Bio-derived thermoplastic polymers were developed during last decades in order to potentially substitute traditional polyolefins in their common applications such as

- packaging and food (films, containers, food wrapper, composting bags, plates, cutlery)
- agricultural (planting containers, controlled release of chemicals)
- textile (wipes, filters, geotextiles, diapers)
- medicine (release of drugs, suture, screw, orthopaedic products).

The importance of bio-based polymers in these fields concerns the reduction of environmental impacts and the improved compatibility with organic material. Nevertheless, hydrophilicity of bio-based polymers can be a problem, concerning the compatibility with traditional hydrophobic polyolefins.

Polylactic acid (PLA) is one of the most known bio-derived polymer first produced through polymerization by Carothers in 1932 [1] and lately patented by DuPont in 1954. PLA is an aliphatic thermoplastic polyester generally produced through ring-opening polymerization of lactides.

Mechanical properties of PLA are good considering the tensile modulus, tensile strength and flexural strength, comparable or higher than traditional polyolefins such as polyethylene, polypropylene and polyethylene terephthalate. Table 8.1 reports the main mechanical features of PLA and traditional polyolefin.

Otherwise, one of the main drawbacks of PLA is its brittleness; in fact tensile modulus and tensile strength of PLA are comparable to PET, but PLA elongation at break limited its applications in PET fields [2].

In order to improve PLA applications, brittleness has to be decreased. During the last years, researchers proposed different ways to face this problem, such as biocompatible

Table 8.1 Some PLA mechanical features in comparison to traditional polyolefin.

	PLA	PET	HDPE	PP
Tensile modulus (GPa)	3.4	3.3	1.3	1.8
Tensile strength (MPa)	53	50–70	15–20	40–50
Flexural strength (MPa)	80	80	20	40
Elongation at break (%)	6	50–150	700–800	100–600
Cost ($/lb)	1–1.5	0.70–0.72	0.27	0.28

plasticizers, blending with other polymers (developed in the paragraph thermoplastic blends) and recent study also proposed the use of ionic liquids.

Baiardo *et al.* [3] analysed the influence of two different plasticizers, in different percentages: acetyl tri-*n*-butyl citrate (ATBC) and PEGs with different molecular weights. Both acted as plasticizers, significantly increasing the elongation at break of PLA. As reported in the literature, the main problem of plasticizers is the decrease of tensile strength and tensile modulus; this problem was confirmed by this study and Table 8.2 displays these results.

Martin and Avérous [4] analysed the effect of different plasticizers such as glycerol, PEG and oligomeric lactic acid (OLA). Glycerol was the less effective plasticizer, while

Table 8.2 Tensile test results of pure PLA and PLA modified with polyethylene glycol (PEG) and acetyl tri-*n*-butyl citrate (ATBC) in different percentages.

Material	σ_b (MPa)	E (GPa)	ε_b (%)
Pure PLA	66	3.3	1.8
PEG 400 (wt%)			
5	41.6	2.5	1.6
10	32.5	1.2	140
12.5	18.7	0.5	115
15	19.1	0.6	88
20	15.6	0.5	71
PEG 1.5 K (wt%)			
5	52.3	2.9	3.5
10	46.6	2.8	5
12.5	18.5	0.7	194
15	23.6	0.8	216
20	21.8	0.6	235
PEG 10 K (wt%)			
5	53.9	2.8	2.4
10	48.5	2.8	2.8
15	42.3	2.5	3.5
20	22.1	0.7	130
ATBC (wt%)			
5	53.4	3.2	5.1
10	50.1	2.9	7
12.5	17.7	0.1	218
15	21.3	0.1	299
20	23.1	0.1	298

Baiardo *et al.* 2003 [3]. Reproduced with permission of John Wiley and Sons.

Table 8.3 Influence of molecular weight and content of triethyl citrate (TEC) and acetyl tributyl citrate (ATBC) on polylactic acid thermal properties.

Sample (numbers represent ratios)	Name	Crystallinity (%)	T_c (°C)	ΔH_c (J g^{-1})	T_m (°C)	ΔH_m (J g^{-1})
Treated PLA	TrtPLA	0.45	131.81	4.937	152.03	4.517
PLA 95/TEC 5	PLA-TEC5	2.61	120.87	16.12	148.92	18.43
PLA 90/TEC 10	PLA-TEC10	3.32	119.67	9.640	147.97	12.42
PLA 85/TEC 15	PLA-TEC15	3.84	104.42	20.19	145.94	23.23
PLA 80/TEC 20	PLA-TEC20	5.88	91.63	18.52	143.52	22.90
PLA 70/TEC 30	PLA-TEC30	7.49	70.56	10.95	138.55	15.83
PLA 95/ATBC 5	PLA-ATBC5	1.35	115.73	20.07	146.96	21.27
PLA 90/ATBC 10	PLA-ATBC10	2.68	111.56	20.77	150.08	23.02
PLA 85/ATBC 15	PLA-ATBC15	3.42	101.06	23.68	148.07	20.97
PLA 80/ATBC 20	PLA-ATBC20	4.36	92.01	19.61	146.74	22.86
PLA 70/ATBC 30	PLA-ATBC30	9.73	71.75	13.90	143.20	20.24

Adapted from ref. 5.

PEG and OLA acted as best plasticizers, with decreasing glass transition temperature, melting point and increasing up to 200% the elongation at break. As always, the plasticizer effect was also evident in a decrease in elastic modulus.

The influence of molecular weight and content of plasticizers introduction was analysed by Maiza et al. [5]. They evaluate the effect of different percentages of triethyl citrate (TEC) and acetyl tributyl citrate (ATBC) on properties of films for food packaging made of PLA.

Increasing the plasticizer contents leads to lower T_g, melting and crystallization temperature and melting and crystallization heat. Meanwhile, an increased crystallinity was experienced with the rise of the plasticizer. Some results are shown in Table 8.3 and Figure 8.1.

During last years, a new method of improving PLA toughness has been developed through addition of ionic liquids, which are molten salt, achieving great attention because of their high thermal stability, low vapour pressure, recyclability and non-flammability; moreover, ionic liquids are considered as green solvents.

Ionic liquids are characterized by liquid in which the ions are slightly coordinated, reaching low melting points (below 100 °C). Their structure is generally composed of organic cations and inorganic anions, influencing their properties. In fact, the low melting point is a direct consequence of organic cation asymmetry compared to inorganic part: in this way, lactic energy is low and so the melting point.

They are usually categorized into four different families on their cations:

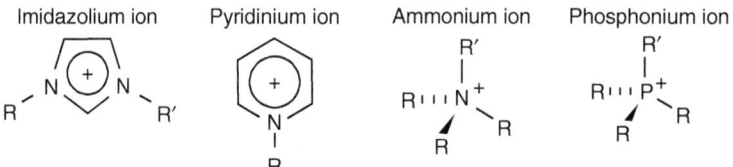

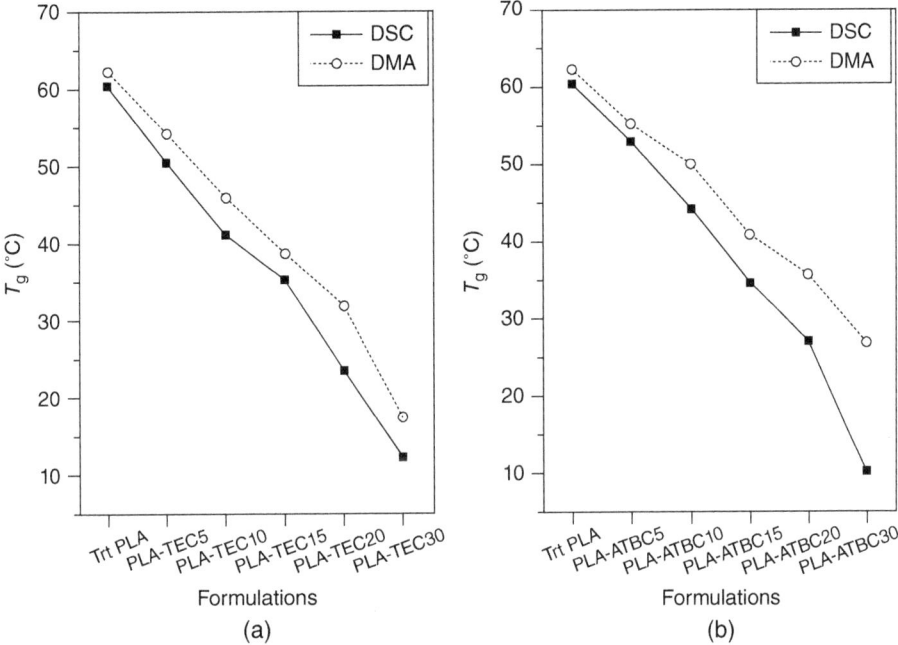

Figure 8.1 Influence of TEC (a) and ATBC (b) on glass transition temperature of PLA ref. [5].

Chen *et al.* [6] studied the effect of four different ionic liquids:

1-methyl-3-pentylimidazolium tetrafluoroborate ([MPI][BF$_4$]),
1-methyl-3-pentylimidazolium hexafluorophosphate ([MPI][PF$_6$]),
1-methyl-3-pentylimidazolium bis(trifluoromethanesulfonyl)imide ([MPI][TFSI]),
1,2-dimethyl-3-pentyl imidazolium bis(trifluoromethanesulfonyl)imide ([DMPI][TFSI]).

Their main results are reported in Tables 8.4 and 8.5 and Figure 8.2.

Table 8.4 The effect of four different ionic liquids on PLA mechanical properties.

Sample	Tensile strength (MPa)	Elongation at break (%)	Young modulus (GPa)
PLA	49.2 ± 2.6	2.5 ± 0.5	3.29 ± 0.16
PLA + 10% [MPI][BF$_4$]	24.1 ± 1.1	26.9 ± 1.0	2.13 ± 0.23
PLA + 20% [MPI][BF$_4$]	16.4 ± 1.6	32.8 ± 3.8	1.96 ± 0.03
PLA + 10% [MPI][BF$_4$]	26.8 ± 1.5	29.7 ± 1.6	2.54 ± 0.16
PLA + 20% [MPI][BF$_4$]	20.6 ± 1.3	61.6 ± 5.8	1.84 ± 0.03
PLA + 10% [MPI][TFSI]	30.9 ± 1.7	15.9 ± 1.8	1.92 ± 0.28
PLA + 20% [MPI][TFSI]	26.0 ± 1.3	28.9 ± 2.2	1.48 ± 0.06
PLA + 10% [DMPI][TFSI]	27.4 ± 1.4	16.1 ± 2.9	2.27 ± 0.04
PLA + 20% [DMPI][TFSI]	13.7 ± 1.6	34.1 ± 3.8	1.25 ± 0.08

Chen *et al.* 2013 [6]. Reproduced with permission of Elsevier.

Table 8.5 Effect of four different ionic liquids on PLA thermal properties.

Sample	Decomposition temperature (°C)	Glass transision temperature (°C)	Melting peak (°C)
PLA	357	60.6	164.2
PLA + 10% [MPI][BF$_4$]	364	62.8	174.1
PLA + 20% [MPI][BF$_4$]	368	62.3	174.3
PLA + 10% [MPI][BF$_4$]	366	63.6	174.6
PLA + 20% [MPI][BF$_4$]	370	62.9	175.6
PLA + 10% [MPI][TFSI]	362	53.8	172.6
PLA + 20% [MPI][TFSI]	364	51.4	172.9
PLA + 10% [DMPI][TFSI]	363	59.6	173.8
PLA + 20% [DMPI][TFSI]	365	57.9	174.2

Chen *et al*. 2013 [6]. Reproduced with permission of Elsevier.

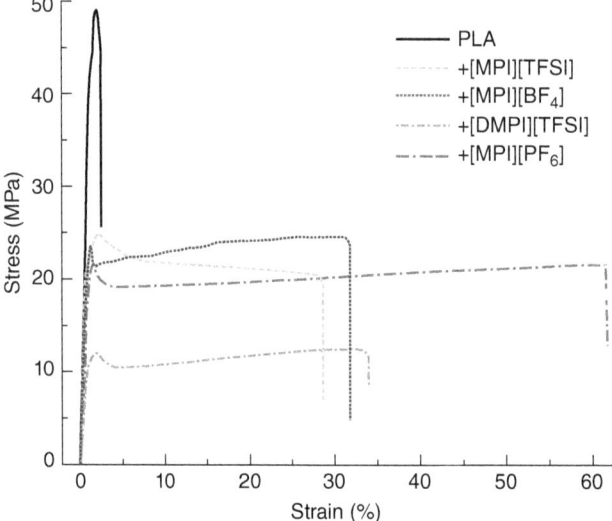

Figure 8.2 Stress–strain curves of neat PLA and PLA modified with four different ionic liquids. Chen *et al*. 2013 [6]. Reproduced with permission of Elsevier.

They demonstrated that ionic liquids can improve ductility of PLA and slightly improve its thermal stability but with a considerable reduction in both tensile strength and modulus.

Another interesting and fairly new thermoplastic bio-derived polymer is thermoplastic starch (TPS). Starch is a high available renewable source, present in many different plants such as potato, corn, wheat or waxy corn [7]. Granules of starch are composed of amylose and amylopectin in different percentages, affecting their mechanical properties. In fact, amylose is a linear polymer of D-glucose units attached at α-1,4, and amylopectin is a highly branched polymer of D-glucose units attached at α-1,4 and α-1,6. The amylose/amylopectin ratio depends on plant source, and this effect is evident in Figure 8.3.

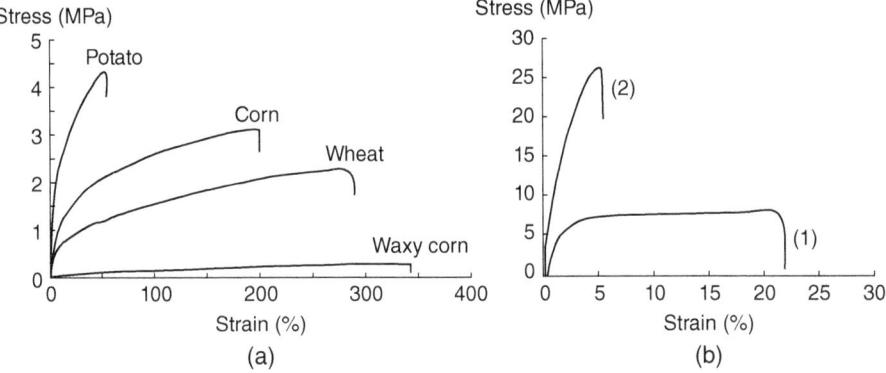

Figure 8.3 (a) The influence of plant sources in derived starch products. (b) The different stress and strain behaviour of pure amylopectin (1) and pure amylose (2). Saiah *et al.* 2012 [7]. *Source:* http://www.intechopen.com/books/thermoplastic-elastomers/properties-and-biodegradation-nature-of-thermoplastic-starch-Used under CC BY 3.0, https://creativecommons.org/licenses/by/3.0/.

In order to be processed, starch needs a plasticizing process reachable with water, additives (glycerol is one of the widely used), heat and shear. During this process, named gelatinization, there is the disruption of crystalline phase in the granules to obtain an amorphous TPS [8]. The plasticizers act by breaking the starch–starch bonds, replacing them with starch–plasticizer interactions. The main features of a plasticizer for starch are compatibility with starch macromolecules, polarity, hydrophilicity, smallness and boiling point high enough to avoid evaporation during the process [9].

Usually, plasticizers used with TPS are glycerol, sorbitol, glycols and urea [10–15]. TPS plasticized with glycerol showed, after a stored period, retrogradation behaviour, resulting in embrittlement.

Retrogradation, called also ageing, happened because of the non-thermodynamical stability of amorphous phase. This problem is strictly related to temperature; higher the difference between storage temperature and glass transition temperature, the higher the possibility of ageing phenomenon, especially in high humidity conditions or high amount of plasticizers [9].

Ageing problem has been overcome through plasticizers based on amide groups, which unfortunately showed healthy problems. Hence, Jiugao *et al.* used different plasticizers, such as citric acid (one of the same plasticizers of PLA) [16]. They found a positive effect of citric acid in preventing the retrogradation processes. In fact, citric acid showed stronger interactions with C–O group than glycerol, assisting to avoid the formation of hydrogen bonds between hydroxyl groups of TPS.

Recently, TPS has been obtained by adding ionic liquid. Sankri *et al.* [17] analysed the plasticization of starch by 1-butyl-3-methylimidazolium chloride [BMIM]Cl and water (30%), compared to glycerol performances and water (30%). Table 8.6 displays the main results of this study.

[BMIM]Cl performed as a better plasticizer than glycerol by inducing lower water absorption, lower glass transition temperature, higher electrical conductivity and higher elongation at break. As a consequence, the Young's modulus was lower for ionic liquid-plasticized TPS than for the glycerol-plasticized one.

Table 8.6 Main properties of thermoplastic starch plasticized by glycerol and 1-butyl-3-methylimidazolium chloride [BMIM]Cl.

Properties	TPS/glycerol 100/30	TPS/[BMIM]Cl 100/30
Water absorption (wt%)	20	13
Glass transition temperature (T_g) (°C)	−21	−13
Electrical conductivity (S cm^{-1})	$10^{-5.4}$	$10^{-4.6}$
Young's modulus (MPa)	8.3	0.5
Stress at break (%)	88	392

Sankri et al. 2010 [17]. Reproduced with permission of Elsevier.

8.2.2 Thermoplastics Blends

Polymer blending is an easy method to obtain polymers with modified properties and versatile applications. In fact, polymeric blend production was half of all plastic production in 2010 [18].

Different kinds of polymer blends exist, but, in this section, we discuss thermoplastic blend, focusing on oil-based and bio-derived thermoplastic polymers.

In general, a polymer blend is a mixture of two or more different polymers, which, in the molten state, are usually immiscible. As a consequence, the polymer in the lower amount assumes the geometric shape that is thermodynamically more stable, a sphere. In fact, considering the second law of thermodynamics, in order to mix polymers, the variation of free energy has to be negative.

$$\Delta G = \Delta H - T \Delta S$$

where ΔG is the variation of free energy, ΔH is the variation of enthalpy, T is the temperature, and ΔS is the entropy.

ΔH is negative just in case the affinity between polymers is higher than the affinity of the polymer itself: thus ΔH is generally positive. ΔS is inversely proportional to polymerization degree, which for polymers is generally very high. As a consequence, the variation of free energy is generally positive, causing immiscible blends between polymers [19].

In order to predict the solubility between two polymers, the Flory and Huggins theory is often used. They studied this question in the early 1940, and proposed a model based on the following equation:

$$\chi_{AB} = \frac{V_{ref}(\delta_A - \delta_B)^2}{RT}$$

where χ_{AB} is the Flory–Huggins interaction parameter, V_{ref} is a reference volume, generally 100 cm^3 mol^{-1}, δ is the solubility parameter of the polymer, R is the gas constant and T the temperature. The higher the χ_{AB}, the lower the miscibility between both polymers [20]. The solubility parameter was also deeply studied by Bicerano in his book 'Prediction of polymer properties' [21] because of its importance for compatibility and permeation prediction, bulk and solution properties of polymers [22]. Hildebrand theories [23, 24] provide definition of δ depending on the energy of total evaporation

$$\delta = \frac{\sqrt{\Delta H_{vap}}}{V}$$

in which ΔH_{vap} is the energy of evaporation and V is the molar volume of the solvent. This δ expression is valid only for regular solution (aliphatic and nonpolar solvents). Hansen [25] extends δ proposing a multidimensional solubility parameter

$$\delta = \sqrt{\delta_D^2 + \delta_H^2 + \delta_P^2}$$

where δ_D refers to nonspecific intermolecular interaction related to dispersion forces, δ_H refers to specific intermolecular interaction (such as hydrogen bonding, acid–base interaction) and δ_P refers to dipole–dipole forces.

One of the parameters that govern miscibility between polymers is the interfacial tension: the higher the interfacial tension, the higher the immiscibility between polymers [26].

Having high interfacial tension leads to particle formation, which could coalesce, decreasing mechanical properties. The general rule to achieve good blends is, like chemists say, 'similar dissolves similar'. Polymers with similar polarity are easier to blend than polymers with different polarities. Other important features are as follows:

- Specific group attractions: polymers with groups that could provide bonds or attraction between polymer chains are easier to blend.
- Molecular weight: polymers with low molecular weight are easier to mix, facilitating miscibility. As mentioned earlier, polymers with similar molecular weight are more miscible.
- Ratio: the mutual amount of polymers affects the possibility of mixing. A low amount of polymer could be soluble in another polymer, while a higher amount could not.
- Crystallinity: generally, if polymers in a blend crystallize, they will provide different crystalline phase, increasing the number of phases present. As a consequence, the mixing between polymers will occur with more difficulty.

During last decades, many studies have been conducted in order to reduce the amount of non-biodegradable charge. Thipmanee and Sane [27] analysed low-density polyethylene (LLDP)/TPS blend with zeolite 5A nanoparticles (Z) (Table 8.7), trying to improve the compatibility between the polymers.

The addition of the zeolite 5A allows obtaining both good mechanical and thermal properties of LLDPE/TPS blend. Zeolite 5A increases both tensile strength and elongation at break (Figure 8.4) and decreases both glass transition temperature and crystalline temperature (Table 8.8). Usually, zeolite increases T_c acting as a nucleating agent. In this study, zeolite allows an improvement in the miscibility of LLDPE/TPS blend; as a consequence, starch molecules depress the crystallization of PE during cooling. Decreases in T_c could be considered as a proof of zeolite work as coupling agent between LLDPE and TPS.

Because of their great importance and multitude of applications, PLA blends are deeply studied nowadays, in order to reduce PLA brittleness. Hence, soft and tough polymers can be blended to PLA, improving their properties and their applications. At the same time, the production of biodegradable blends with traditional polyolefins is also possible.

Table 8.7 Composition of polyethylene/thermoplastic starch blend with and without zeolite 5A in the amount of 1–3–5%.

Samples	LLDPE (wt%)	TPS (wt%)	Zeolite 5A (wt%)
PE	100	–	–
TPS	–	100	–
PE/TPS	70	30	–
PE/TPS/Z1	69.3	29.7	1
PE/TPS/Z3	67.9	29.1	3
PE/TPS/Z5	66.5	28.5	5

Thipmanee and Sane 2012 [27]. Reproduced with permission of John Wiley and Sons.

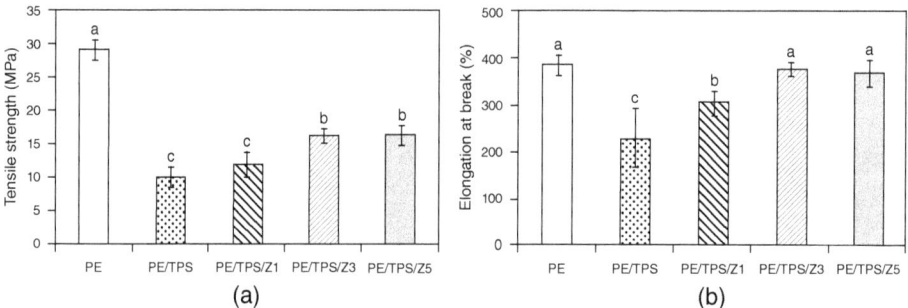

Figure 8.4 (a) Tensile strength of low-density polyethylene/thermoplastic starch with and without zeolite 5A. (b) Elongation at break of low-density polyethylene/thermoplastic starch blend with and without zeolite 5A. Thipmanee and Sane 2012 [27]. Reproduced with permission of John Wiley and Sons.

Table 8.8 Glass transition temperature and crystallization temperature of low-density polyethylene/thermoplastic starch blend with and without zeolite 5A.

Samples (numbers represent ratios)	T_g (°C)	T_c (°C)
PE 100	–	122.1 ± 0.2
TPS 100	51.6 ± 1.4	–
PE 70/TPS 30	47.4 ± 1.9	108.1 ± 0.2
PE 69.3/TPS 29.7/Z 1	43.0 ± 0.9	108.3 ± 0.3
PE 67.9/TPS 29.1/Z 3	44.2 ± 1.0	107.6 ± 0.5
PE 66.5/TPS 28.5/Z 5	43.4 ± 1.2	105.1 ± 1.1

Thipmanee and Sane 2012 [27]. Reproduced with permission of John Wiley and Sons.

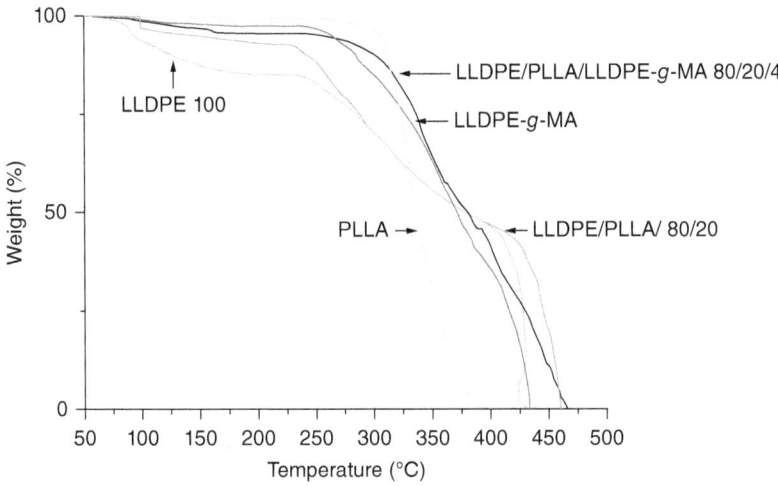

Figure 8.5 Thermal analysis results for low-density polyethylene/PLA blend with 4 phr of low-density polyethylene-graft-maleic anhydride. Singh et al. 2011 [28]. Reproduced with permission of Springer.

Singh et al. [28] analysed linear low-density polyethylene and PLA blends in order to develop a film with good mechanical properties and biodegradability under specific conditions. On facing the problem of immiscible blends, they decided to introduce low-density polyethylene-graft-maleic anhydride as compatibilizer. Analysing different ratios of polymer amount and compatibilizer, they stated that 80 wt% of LLPE, 20 wt% of PLLA and 4 phr of compatibilizer was the best blend in terms of mechanical properties, thermal stability and biodegradability. An example of their results is given for thermal analysis in Figure 8.5.

Interesting studies for medical applications have been done by Ploypetchara et al. [29] blending together different ratios of polypropylene and poly(lactic acid) with 3 wt% of polypropylene-grafted-maleic anhydride as compatibilizer.

Important properties of water vapour permeability were observed: PLA increased water vapour permeability of the blend, indicating that PP biodegradability might be improved.

Nayak [30] proposed a work based on TPS, blended with polybutylene adipate-co-terephthalate (PBAT) and nanoclay C30B at 3 wt% with and without the addition of maleic anhydride-grafted-polybutylene adipate-co-terephthalate (PBAT-g-MA) to optimize the blend. C30B is Cloisite 30B: natural montmorillonite organically modified with methyl tallow bis-2-hydroxy ethyl quaternary ammonium salt with cation exchange capacity (CEC) 90 meq/100 g. SEM analysis (Figure 8.6) displays an improved interface of TPS in PBAT thanks to PBAT-g-MA and addition of 3% of nanoclays, reducing size domains of TPS.

Both nanoclays and PBAT-g-MA increased the mechanical properties of PBAT–TPS blend, especially elongation at break. Another important effect of the improved compatibilization was the higher thermal stability of the blend thanks to both compatibilizer agents. A synergic effect of both methods of compatibilization was evident: both nanoclays and MA alone improved the properties of the blend, but their simultaneous interaction allowed to obtain a better blend (Table 8.9).

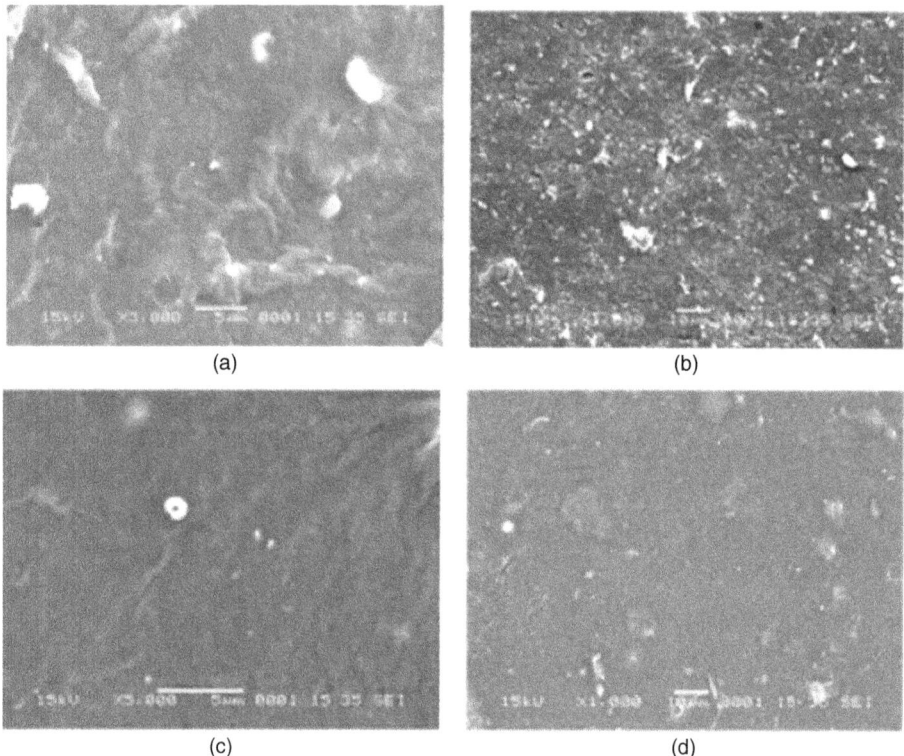

Figure 8.6 SEM analysis of (a) PBAT/TPS (b) PBAT-g-MA/TPS (c) PBAT/TPS/C30B (d) PBAT-g-MA/TPS/C30B. Nayak 2010 [30]. Reproduced with permission of Taylor and Francis.

Table 8.9 Mechanical and thermal properties of neat PBAT and PBAT/TPS blend modified with maleic anhydride and nanoclay C30B addition.

Samples (numbers represent ratios)	Tensile strength (MPa)	Elongation at break (%)	Degradation temperature (°C)
PBAT	11.8 ± 6.0	324.8 ± 10.1	412
PBAT 70/TPS 30	10.2 ± 2.3	392.4 ± 10.8	410
PBAT 70/TPS 30/C30B 3	11.2 ± 4.5	564.0 ± 12.1	417
PBAT-g-MA 70/TPS 30	10.5 ± 3.3	413.1 ± 11.3	418
PBAT-g-MA 70/TPS 30/C30B 3	11.8 ± 2.8	613.4 ± 11.6	423

Nayak 2010 [30]. Reproduced with permission of Taylor and Francis.

Leroy *et al.* [31] used 1-butyl-3-methylimidazoliumchloride ([BMIM]Cl) as plasticizer for starch–zein (main protein in corn) blend, comparing their results to glycerol plasticizer. The presence of [BMIM]Cl allowed to obtain a lower water uptake and better thermodynamical behaviour than samples with glycerol, resulting in an improved blend compatibility.

Another alternative to traditional compatibilizing methods was proposed by Vignon *et al.* [32] using γ irradiation exposure of polycarbonate (PC) and TPS (plasticized through glycerol) blends. The formulation tested was up to 20 wt% of TPS mixed with PC, evaluated by DSC, TGA, SEM and mechanical testing. All the analysis proved a better compatibilization between PC and TPS thanks to the recombination of the free radicals created during the irradiation: the T_g decreased, reduced thermal degradation. SEM analyses are consistent with the creation of an interphase (Figure 8.7), which is in agreement with improved mechanical properties (Figure 8.8). An optimization in irradiation dose and TPS percentages was needed to improve PC/TPS properties.

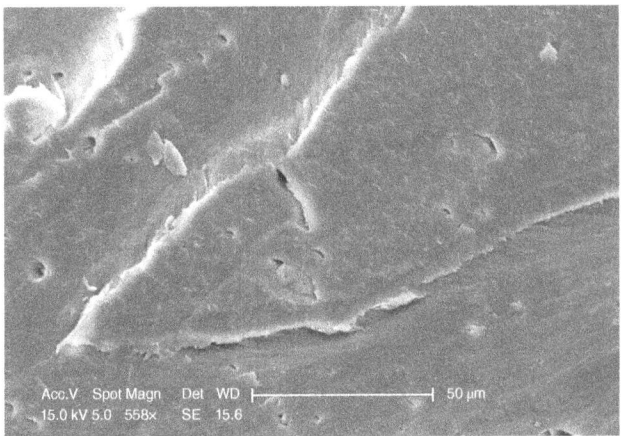

Figure 8.7 Image of 50 kGy irradiated Pc080/TPS20. Vignon *et al.* 2011 [32]. Reproduced with permission of John Wiley and Sons.

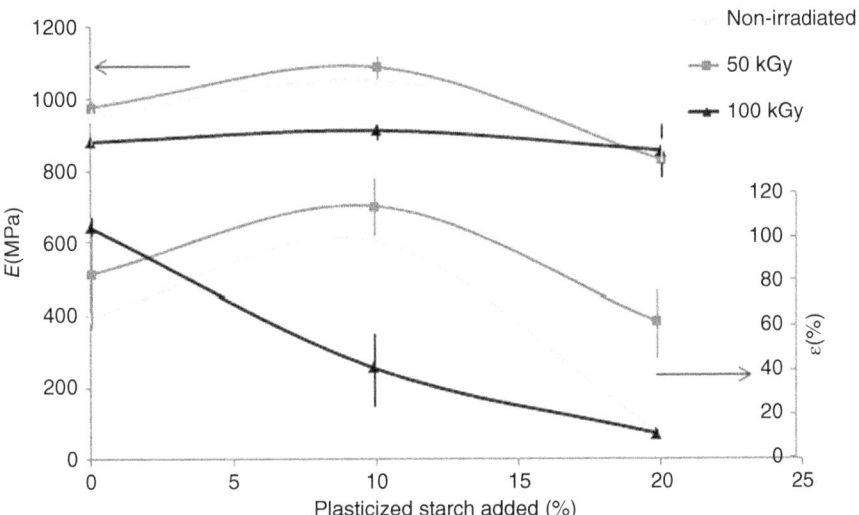

Figure 8.8 Effect of irradiation on the variation of Young's modulus for different amounts of starch contained in the blends. Vignon *et al.* 2011 [32]. Reproduced with permission of John Wiley and Sons.

These results pushed researchers into deep and innovative studies, proposing new thermoplastic materials, useful in many different applications. Much work has already been done in order to reduce plastic solid waste, but still much work has to be done: the important results obtained until now symbolize a step forward in facing environmental impact problems of plastic industries, hoping for a better future.

8.3 Thermoplastic Composites with Natural Fillers

During recent years, many researchers have focused their attention on natural fibres as fillers for thermoplastic composites. These kinds of fibres, in fact, are characterized by many advantages with respect to traditional glass and carbon fibres such as low cost, low density, comparable specific tensile properties, non-abrasive to the equipment, non-irritating the skin, reduced energy consumption, renewability, recyclability and biodegradability [33, 34]. Table 8.10 shows the main feature of natural fibres, in comparison to traditional glass fibres.

Some of the widespread applications for composites with natural fillers are aerospace, leisure, construction, sport, packaging and automotive. In fact, these industrial sectors have shown enormous interest in the development of new composite materials [35]. One of the first examples of using natural fibres instead of fuel-based ones is Trabant.

Trabant (Figure 8.9) is the first-known application of natural fibres in automotive industries, produced in 1958 in the German Democratic Republic. The main parts, manufactured with phenolic resin reinforced with cotton fibres, were roof, boot lid, bonnet, wings and doors [36].

Table 8.11 reports the mechanical properties of some natural fibres as compared to conventional fibres and polymers [37].

Natural fibres are a large family of fibres, including wood, cellulose, hemp, flax, jute, sisal, kenaf and many others [38]. In particular, wood–plastic composites (WPCs)

Table 8.10 Comparison between natural and glass fibres [can natural fibres replace glass?].

	Natural fibres	Glass fibres
Density	Low	Twice that of natural fibres
Cost	Low	Low, but higher than natural fibres
Renewability	Yes	No
Recyclability	Yes	No
Energy consumption	Low	High
Distribution	Wide	Wide
CO_2 neutral	Yes	No
Abrasion to machines	No	Yes
Health risk when inhaled	No	Yes
Disposal	Biodegradable	Non-biodegradable

Wambua *et al.* 2003 [34]. Reproduced with permission of Elsevier.

Figure 8.9 Trabant car, produced in 1958 in the German Democratic Republic with main parts made of composites with natural fillers. *Source:* https://en.wikipedia.org/wiki/Trabant#/media/File:Trabant_601_Mulhouse_FRA_001.JPG Used under CC BY 3.0, https://creativecommons.org/licenses/by-sa/3.0/.

Table 8.11 Mechanical properties of some natural fibres in comparison to traditional fibres and polymers.

	Density (g cm^{-3})	Elongation (%)	Tensile strength (MPa)	Young's modulus (GPa)
Fibres				
Cotton	1.5–1.6	7.0–8.0	278–800	5.5–12.6
Jute	1.3	1.5–1.8	393–773	26.5
Flax	1.5	2.7–3.2	345–1035	27.6
Hemps	1.5	1.6	690	70.0
Ramie	1.5	1.2–3.8	400–938	61.4–128.0
Sisal	1.5	2.0–2.5	511–635	9.4–22.0
Coir	1.2	30.0	175	4.0–6.0
Viscose (cord)		11.4	593	11.0
Soft wood (kraft)	1.5		1000	40.0
E-glass	2.5	2.5	2000–3500	70.0
S-glass	2.5	2.8	4570	86.0
Aramid (normal)	1.4	6–8	3000–3150	100–150
Carbon (standard)	1.8–2	1.4–1.8	4000	230.0–240.0
Polymers				
ABS	1.05	10	55	2.8
Polycarbonate	1.22	100	62	2.3

(Continued)

Table 8.11 (Continued)

	Density (g cm^{-3})	Elongation (%)	Tensile strength (MPa)	Young's modulus (GPa)
Polyetherimide	0		105	2.8
Nylon	1.12	29	66	3.5
Polyethylene (HDPE)	0.95	30	28	1.1–1.2
Polypropylene	0.9	200	35	1.5–2
Polystyrene (high impact)	1.05	15	35	3–3.5
Epoxy resin	1.3	3–4	32	4–6

Wambua et al. 2003 [34]. Reproduced with permission of Elsevier.

are one of the commonly known families of natural fibre–thermoplastic composites. Another interesting class of materials is characterized by thermoplastic composites with cellulosic fibres derived from wastes such as newspapers, additivated paper, paper board and paper mill sludge. These two classes of composites are described in detail.

The main drawback of composites with natural fillers is the poor compatibility between hydrophilic fibres and hydrophobic matrices: the different nature between filler and matrix causes the formation of fibre–fibre interactions instead of fibre–matrix ones. As a consequence, formation of agglomerates and lower dispersion of natural fibres is possible. These two problems, bad mixing and weak interface, will render low mechanical properties to a composite [39].

The most popular methods to improve fibre–matrix interface are the introduction of a coupling agent [40–43] as well as physical and chemical treatments [44, 45].

Natural fibres present drawbacks, such as interface problems with traditional polyolefins, low temperature processes, moisture absorption and swelling fibres, supply and demand cycles based on product availability and harvest yields, quality variations based on growing sites and seasonal factors, but their advantages are much more interesting and will lead to new materials taking care of our planet.

8.3.1 Wood–Plastic Composites

WPCs were born in Italy in the 1970s and then developed as commercial products in North America in the 1990s. By the start of the twenty-first century it spread to India, Singapore, Malaysia, Japan and China [46]. The introduction of WPCs in the decking market (Figure 8.10) was mainly responsible for its growth. A direct result of success in the decking market is that products are now being developed and introduced for new exterior applications such as railing, fencing, roofing and siding [47].

The WPCs are defined as thermoplastics reinforced with wood or other natural fibres, principally produced from commodity thermoplastics such as polyethylene, polyvinyl chloride or polypropylene [48]. The matrix chosen for WPC production is influenced by limited thermal stability of wood: only thermoplastics that melt or can be processed under 200 °C are commonly used [49]. These kinds of composites are generally manufactured by typical techniques of plastics industry: first extrusion compounding followed by profile extrusion or injection moulding. [50]. New production processes have been developed in order to improve the amount of fibre in composites avoiding the formation of agglomerates. An example is turbo-mixing process (Figure 8.11)

Figure 8.10 Example of wood–plastic composites application in decking market [7trust.com, green products].

Figure 8.11 Example of turbo mixer, developed by Valente *et al.* (University of Rome, La Sapienza) [ref. 51] to optimize and analyse the effect of compatibilizing agent in natural fibre–polymer composites. The turbo mixer works through kinetic energy developed by rotating blades as only heat source. Valente *et al.* 2016 [52]. Reproduced with permission of Elsevier.

[51, 52]: the only source of heat is generated by kinetic energy of rotating blades thanks to friction (Figure 8.9), using the same idea of working as screw extrusion, but in this case we can reach 3000–5000 rpm. After the charge is melted, the samples are produced through hot compression.

One of the biggest problems of WPCs, because of their different nature, is low compatibility between hydrophilic filler and hydrophobic matrix. Many researches have been conducted in order to face this problem. The introduction of a coupling agent is one of the most common methods adopted. A variety of coupling agents used in WPC exist such as acrylates, amides and imides, anhydrides, chlorotriazines and derivatives, epoxides, isocyanates, organic acids, monomers, polymers and copolymers, inorganic agents and organic–inorganic agents (silanes, titanates).

Polyolefins grafted with maleic anhydride as coupling agents have received attention because they allow improving mechanical properties of wood fibre–polyolefin composites. The maleic anhydride mechanism can be exposed through two phenomena: formation of bonds between the hydroxyl groups of natural fibres and carbonyl groups of maleic anhydride, and blending of polymer chains of coupling agent with the composite matrix [53]. Figure 8.12 shows an example of a polypropylene-grafted-maleic anhydride (MAPP) polyolefin, and Figure 8.13 shows the mechanism of coupling agent action.

The main effects of maleic anhydride are improving tensile and flexural modulus as well as tensile and flexural strength. However, an optimization of its concentration has to be performed [53–55].

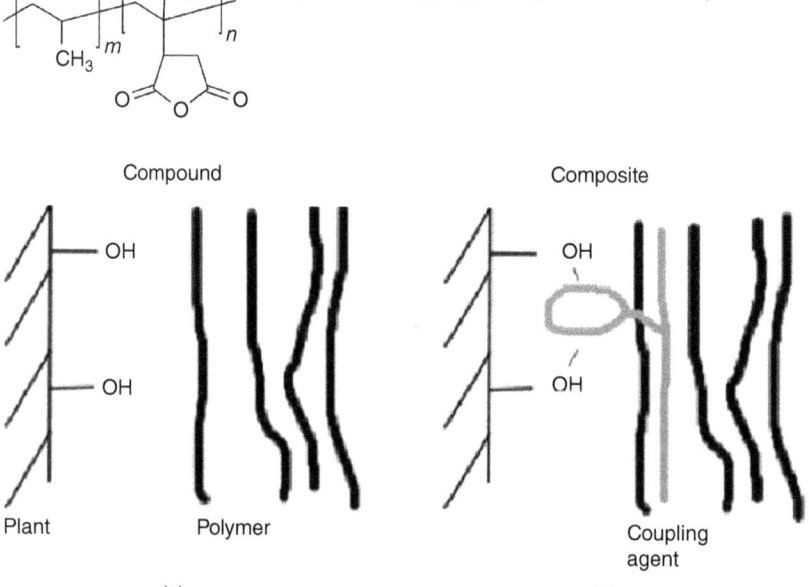

Figure 8.12 Polypropylene-grafted-maleic anhydride.

Figure 8.13 Working mechanism of polyolefin-grafted-maleic anhydride as coupling agent (elliptical shape is the maleic anhydride graft): (a) compound formation because of the weak interaction between plant fibres (hydrophilic) and polymers (hydrophobic); (b) composite formation thanks to the interaction of MAPP maleated graft with plant fibres and polymeric MAPP chains mixing with polymer matrix. Ashori 2008 [37]. Reproduced with permission of Elsevier.

Table 8.12 Properties of five MAPP with different maleic anhydride content (MA%), molecular weight ($\overline{M}_w$) and melt flow index (MFI).

MAPP	MA (%)	$\overline{M}_w$	MFI (g/10 min)	Providers
(A) Polybond 3150	0.5	46,000	20	Crompton Polybond
(B) Polybond 3200	1	42,000	104	Crompton Polybond
(C) G-3003	1.2	52,000	90	Eastman Chemical Products
(D) E-43	1.2	9,100	ND	Eastman Chemical Products
(E) Bondyram	0.8	66,000	90	Polyram

Kim *et al.* 2007 [56]. Reproduced with permission of Elsevier.

Kim *et al.* [56] analysed different maleic anhydride-grafted polypropylene (MAPP) and their influence on bio-flour-filled polypropylene composites. In particular, they studied rice husk flour (RHF) and wood flour (WF) as fillers (30 wt%), and they chose five different MAPP with different percentages of maleic anhydride (MA) graft, molecular weight ($\overline{M}_w$) and melt flow index (MFI) – displayed in Table 8.12. The amount of coupling agent was the 3 wt% with respect to the total weight of the composite (bio-flour and polypropylene).

An optimization of maleic anhydride percentages and $\overline{M}_w$ is needed in order to improve composite properties, as shown in Figures 8.14 and 8.15. Low MA content did not allow obtaining adequate interaction between bio-flour and polypropylene matrix. On the other hand, high MA content keeps the coupling agent near the hydrophilic surface, preventing a good interaction with the hydrophobic matrix. Low MAPP M_w

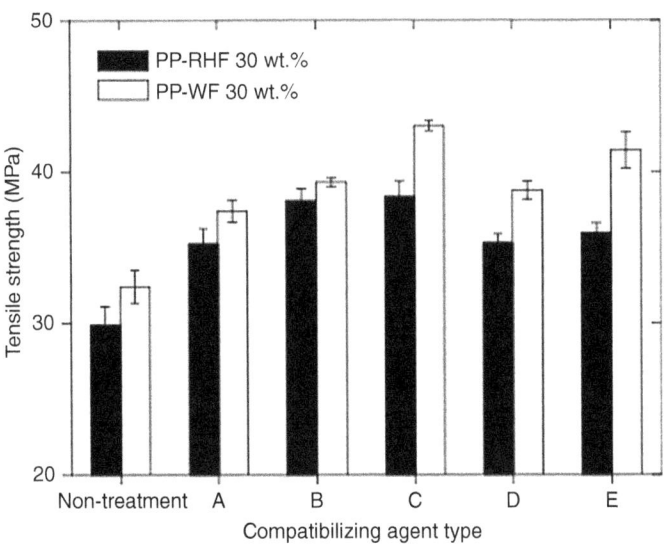

Figure 8.14 Tensile strength trend of polypropylene-rice husk flour (PP-RHF) neat and with five different compatibilizing agents. Kim *et al.* 2007 [56]. Reproduced with permission of Elsevier.

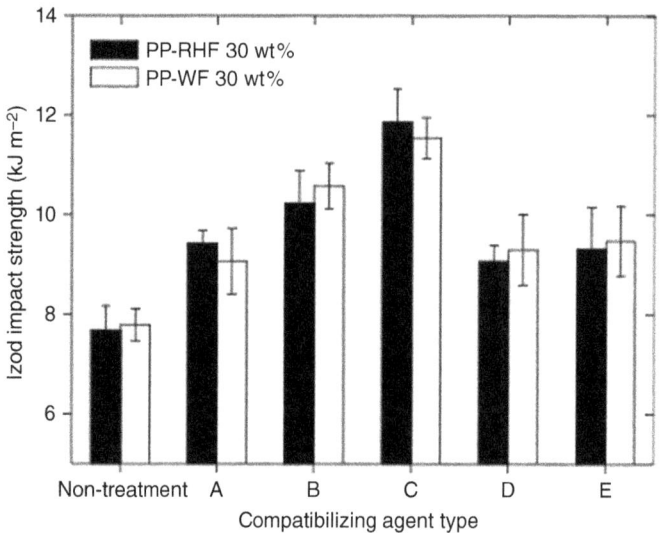

Figure 8.15 Izod impact strength trend of polypropylene-rice husk flour (PP-RHF) neat and with five different compatibilizing agents. Kim *et al*. 2007 [56]. Reproduced with permission of Elsevier.

does not allow a good mix with the polypropylene matrix, while a high M_w could provide interface problems. Therefore, G-3003 was considered as the best-optimized coupling agent, able to induce the best properties in the composite.

Selke and Wichman, using recycled HDPE from milk bottles, with 30 wt% of Aspen hardwood and 2 or 5 wt% of MAPP, reported another interesting work [57]. A good effect of MAPP addition with respect to neat HDPE was evident, and as literature revealed, a suitable MAPP percentages is needed in order to optimize the properties of the analysed composite. The evolution of elastic modulus is reported in Figure 8.16.

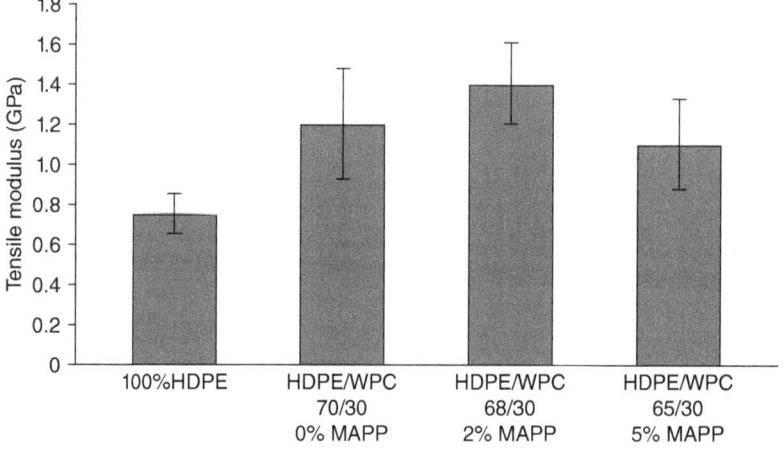

Figure 8.16 Tensile modulus trend of recycled HDPE from milk bottles, with 30 wt% of Aspen hardwood and 2 or 5 wt% of MAPP. Selke and Wichman 2004 [57]. Reproduced with permission of Elsevier.

Table 8.13 Thermal properties of neat high-density polyethylene (HDPE) in comparison to HDPE with 20–30–40–50–60 wt% of wood.

Samples (numbers represent ratios)	T_m (°C)	ΔC_p (J g^{-1} °C^{-1})	X (%)
HDPE 100	138.2	16.8	74.7
HDPE 80/Wood 20	142.0	14.0	76.6
HDPE 70/Wood 30	143.3	22.0	78.1
HDPE 60/Wood 40	140.0	10.0	78.5
HDPE 50/Wood 50	143.4	6.0	79.0
HDPE 40/Wood 60	140.9	4.0	86.1

Tazi *et al.* 2014 [58]. Reproduced with permission of John Wiley and Sons.

Important analyses were also conducted on the thermal stability of WPCs. Tazi *et al.* [58] studied the effect of wood and MAPE percentages on thermo-physical properties. They found that the addition of wood fillers and coupling agent (MAPE) to the polymer matrix resulted in an increase of the melting temperature of HDPE up to value higher than 141 °C; the same phenomenon was reported in the literature. Table 8.13 sums up their results.

Heat capacity (ΔC_p) decreases while the filler amount increases: wood fillers obstacle molecular chains mobility close to melting temperature. Moreover, wood fillers acted as nucleating agents, increasing crystallinity of the matrix, from 60% to 85%.

In order to improve wood-like effect, nailing and screwing properties and to reduce density, foam-like structure could be produced in WPC materials [59]. The most common ways to obtain foam are thermal decomposition of chemical blowing agents (CBAs), volatilization of physical agents (low boiling liquids), expansion of dissolved gas in a polymer and addition of gas-filled microspheres into a polymer system. The principles of a foam production are based on bubble nucleation, growth and stabilization.

Important properties of CBAs are a decomposition temperature suitable for the chosen polymer matrix, avoiding fast reaction during decomposition, easy agent mix and dispersion in the polymer, high gas yield and non-tool corrosivity. Exothermic and endothermic blowing agents [60] were studied by Mengeloglu and Matuana in PVC (70 wt%)/WF composite (30 wt%) [61]: they compared modified azocarbonamide (exothermic agent) and sodium bicarbonate (endothermic agent). Speaking about density, no big differences were noticed between the two CBAs; both of them provided lower densities compared to PVC/WF composite without CBAs. The average cell size instead was affected by the CBA type in particular azocarbonamide allowed obtaining lower cell size than sodium bicarbonate one. Some results are displayed in Figure 8.17 and Table 8.14.

Many parameters affected WPC foam properties, and Yoon *et al.* [62] studied the influence of N_2 amount (physical blowing agent), injection speed, weight reduction and mold temperature. Shrinkage of WPC foam seems to be affected by N_2 amount, weight reduction and mold temperature, while WPC foam deformation was influenced by

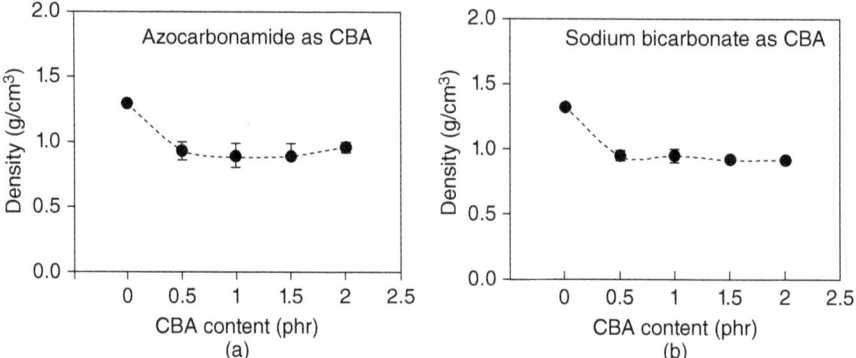

Figure 8.17 Influence of exothermic (a) and endothermic (b) blowing agents on PVC/wood flour density. Adhikary *et al.* 2012 [60] Reproduced with permission of Sage Publication.

Table 8.14 Influence of CBAs on average cell size.

CBA phr	Cell size
Azocarbonamide 0.5	44
Azocarbonamide 1	42
Azocarbonamide 2	41
Sodium bicarbonate 0.5	65
Sodium bicarbonate 1	65
Sodium bicarbonate 2	64

Adhikary *et al.* 2012 [60]. Reproduced with permission of Sage Publications.

injection speed. Moreover, cell density was conditioned by N_2 amount, injection speed and weight reduction. Some results are display in Tables 8.15 and 8.16.

WPCs are a well-established class of materials, with their known properties and their own market. However, improvement in order to increase WPC properties and environmental friendly character could still be done. The use of virgin wood for WPC materials could overcome its usual application; moreover, WPCs can be processed through traditional thermoplastics processes and post-processed with the same method of traditional wood.

8.3.2 Waste Paper as Filler in Thermoplastic Composites

Problems related to paper disposal has grown during last year, despite paper being the most recycled product in Europe and 54% of paper industry's raw materials come from waste paper and paperboard [63].

Paper cannot always be subjected to traditional recycling processes. In fact, approximately just 50% of all the cellulosic wastes is used in recycling processes, the rest has to be send to incinerators or landfilled as solid wastes [64].

In order to reduce the environmental impact of paper industries, the possibility of introducing different kinds of waste paper is opened.

Table 8.15 Influence of N_2 amount, injection speed, weight reduction and mold temperature on shrinkage, warpage, mechanical properties and cell density of WPC foams.

Dissolved gas amount (%)	Injection speed (ccm s^{-1})	Weight reduction (%)	Mold temperature (°C)	Shrinkage (%)	Warpage (%)	Yield stress (MPa)	Young's modulus (MPa)	Cell density (cells cm^{-3})
0.1	70	10	30	0.70	0.28	5.84	205	6.24E + 06
0.1	110	20	30	0.78	0.29	4.30	152	3.26E + 06
0.2	110	10	30	0.76	0.35	5.45	210	1.51E + 07
0.2	70	20	30	0.94	0.33	4.38	158	1.02E + 07
0.1	110	10	50	0.80	0.60	5.84	209	8.25E + 06
0.1	70	20	50	1.07	0.40	4.44	157	2.12E + 06
0.2	70	10	50	0.83	0.13	6.29	218	1.33E + 07
0.2	110	20	50	1.09	0.64	3.95	147	1.22E + 07

Yoon *et al.* 2009 [62]. Reproduced with permission of Taylor and Francis.

Table 8.16 Comparison of properties and cost between HDPE and WPC foam (weight reduction of 20%).

	HDPE	WPC foam (20%)	Difference
Density (g/cc)	0.95	0.88	7% reduced
Cost (CAD/kg)	1.5	0.9	40% reduced
Shrinkage (%)	3.5	0.87	68% reduced
Warpage (mm)	4.5	0.37	91% reduced
Yield stress (MPa)	4.898	5.059	3% increased
Young's modulus (MPa)	141.1	181.4	29% increased

Yoon *et al.* 2009 [62]. Reproduced with permission of Taylor and Francis.

Newspapers are one of the widespread sources of waste paper, and many studies have been conducted on it.

Sanadi *et al.* proposed one of the first studies on newspaper's fibre introduction in thermoplastic composites in 1994 [65]. They introduced 40 wt% of recycled old newspaper fibres (ONFs) in polypropylene, resulting in a tensile strength of 34.1 MPa, and unnotched Izod impact strength of 112 J m^{-1}. Thanks to the addition of MAPP (3% of the fibre weight), a 57 MPa tensile strength and a 212 J m^{-1} Izod impact strength were achieved. This study shows the good effect of newspaper introduction in polypropylene matrix, and the good effect of coupling agent. More recent works [66–70] not only validate the good results obtained by newspapers fibres but also proposed the possibility to substitute 20–30 wt% fibreglass (GF)–polypropylene composites by 40–50 wt% ONFs–polypropylene (PP) composites. This involves a reduction in the use of synthetic polymers, avoiding glass fibres, and exploiting the environmentally friendly character of natural fibres. Some interesting results about these works are summed up in Table 8.17.

Some interesting studies have been conducted on the introduction of paper sludge in thermoplastic composites. Sludge is an abundant waste source; about 300 kg of sludge is produced for each ton of recycled paper [71]. Properties and composition

Table 8.17 Mechanical properties of polyolefin–old newspaper composites compared to polyolefin–glass fibre composites.

Composites (numbers represent ratios)	Izod impact energy notched (J m^{-1})	Izod impact energy unnotched (J m^{-1})	Tensile modulus (GPa)	Tensile strength (MPa)	Publication
PP60/ONF40	20.8 ± 0.5	112 ± 18	4.42 ± 0.2	34.1 ± 1.1	[65]
PP60/ONF40/MAPP3	21.6 ± 0.1	212 ± 24		57.0 ± 0.7	[65]
PP80/ONF20			2.8 ± 0.1		[70]
PP70/ONF30			3.8 ± 0.1		[70]
PP60/ONF40			4.2 ± 0.1		[70]
PP50/ONF50			5.3 ± 0.2		[70]
PP80/GF20			4.1 ± 0.1		[70]
PP70/GF30			5.7 ± 0.1		[70]
PP60/GF40			7.7 ± 0.1		[70]
PP90/ONF10/MAPP5			2.7	44	[67]
PP80/ONF20/MAPP5			3.0	50	[67]
PP70/ONF30/MAPP5			3.3	58	[67]

PP – polypropylene, ONFs – old newspaper fibres, MAPP – polypropylene-grafted-maleic anhydride, GF – glass fibre.

of paper sludge depend on raw materials, the paper grade being manufactured, the amount of water used and the cleaning technique used for wastewater. Wood, cellulose fibres, lignin and organic binders are the main organic components, while kaolinite and calcium carbonate are the main inorganic ones [72]. Generally, paper sludge is disposed in landfills; as a consequence many researchers focused their attention on the potential use of this waste charge in composite materials.

Hamzeh, Ashori and Mirzaei have proposed an interesting study, introducing waste paper sludge in high-density polyethylene–WF composites [73]. They analysed different concentrations (20, 30, 40 and 60 wt%) of two sludge materials: one derived from wastewater treatment used in paper industries and the other ink-eliminated sludge, derived from the recycling treatment of waste paper. This study suggests that waste sludge can be used as reinforcing filler for thermoplastic composites: in fact, composites with paper sludge addition have shown better properties (except for some samples) in terms of flexural properties and water absorption uptake. Similar results were also obtained by Son [74], investigating the effect of different paper sludge's mixing ratio and types, as well as concentration of coupling agents on the physical and mechanical properties of paper sludge–thermoplastic composites. They analysed the dimensional stability of the composites, founding its increase with paper sludge content. From a mechanical point of view, they also confirmed the effect of paper sludge as reinforcement, regarding flexural strength. These two aspects can be seen in Figures 8.18 and 8.19.

The extraction process adopted to obtain fibres from waste paper directly affected the properties of the resulting composites. The choice depends on the application of the produced material. Valente et al. grinding waste paper with two different equipment (a mill working by impact shear and turbulence and a knife-mill), obtaining distinct mechanical properties, optimized for the applications [75].

Figure 8.18 Dimensional stability of polypropylene (PP), high-impact polypropylene (HIPP), high-density polyethylene (HDPE) and low-density polyethylene (LDPE) with different percentages of paper sludge. Hamzeh *et al.* 2011 [73]. Reproduced with permission fo Springer.

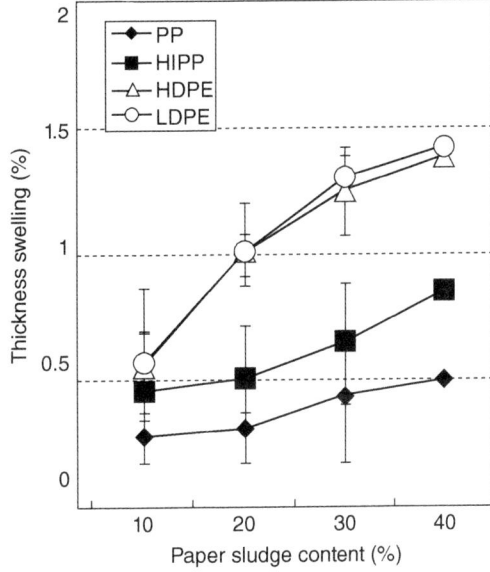

Figure 8.19 Flexural strength of polypropylene (PP), high-impact polypropylene (HIPP), high-density polyethylene (HDPE) and low-density polyethylene (LDPE) with different percentages of paper sludge. Hamzeh *et al.* 2011 [73]. Reproduced with permission fo Springer.

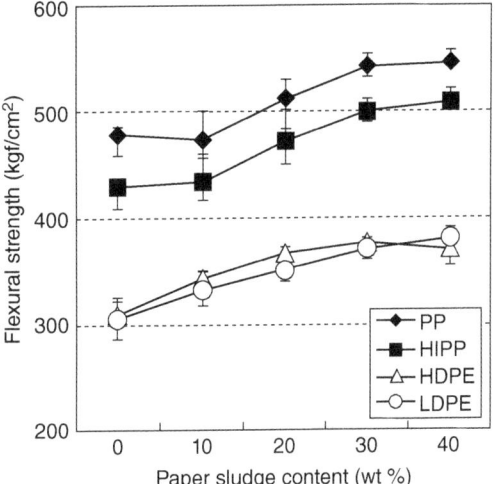

8.4 Conclusion

A rising part of the scientific community has chosen renewable materials as subject for their studies, trying to analyse the main features about this incredible field. In fact, during last decades, properties and applications of both natural fibres and bio-derived polymers are easily available, giving the basics for new projects and attractive challenges. The main reasons for the great interest in bio-derived materials are the renewability, recyclability and biodegradability: these are three themes deeply studied in these years, thanks to the improvement in environmental protection. Moreover, the performance of natural fibres could be comparable to traditional fibres (such as glass fibres), but a lot of studies must be done in order to potentially substitute traditional fibres with natural ones.

Oil-based polymers represent one of the most used materials in our daily life, thanks to their properties and costs. One of the first attempts to avoid petroleum sources was to produce these polymers from natural sources. As a consequence, bio-polyolefins have been developed, such as bio-polyethylene: the ethylene monomer is derived from ethanol, which can be produced by the fermentation of natural sources like sugar cane, corn or beet.

Generally, the hydrophobicity of oil-based polymers is one of the distinguishing features from bio-derived polymers: this hydrophobicity allows obtaining a good moisture resistance, preventing from bacterial attacks and reducing the water uptake.

Thus, bio-derived and conventional oil-based polymer blends could produce unique and interesting composites, mixing their properties and compensating for defects. Because of their different hydrophilic character, wide attention is given to compatibility between them. Many efforts have been focused on optimizing these blends, such as the introduction of coupling agents, ionic liquids and nanoparticles (zeolites, nanoclays, nanosilicas, etc.).

According to what has been said until now, many studies will be focused on these themes, trying to keep improving oil-based and bio-derived blends and composites, as these materials represent the future of our daily life.

8.5 Questions for Further Consideration

1. Will blend composed of oil-based and bio-derived polymers be matrices for WPC composites, improving in this way the bio-derived charge?
2. Could it be possible to totally replace virgin wood in the future?
3. Could cellulose fibres be the future?
4. Will the traditional industry evolve in an industry based on the valorization of wastes?
5. Are we ready for a new era of polymer materials?

References

1 W. Carothers, G. Dorough, and F. J. Van Natta, Studies of polymerization and ring formation. X. The reversible polymerization of six-membered cyclic esters, *J. Am. Chem. Soc.*, **54**, 761–772 (1932).
2 K. Hamad, M. Kaseem, H. W. Yang, F. Deri, and Y. G. Ko, Properties and medical applications of polylactic acid: A review, *eXPRESS Polym. Lett.*, **9** (5), 435–455 (2015).
3 M. Baiardo, G. Frisoni, M. Scandola, M. Rimelen, D. Lips, K. Ruffieux, and E. Wintermantel, Thermal and mechanical properties of plasticized poly(L-lactic acid), *J. Appl. Polym. Sci.*, **90**, 1731–1738 (2003).
4 O. Martin and L. Avérous, Poly(lactic acid): Plasticization and properties of biodegradable multiphase systems, *Polymer*, **42**(14), 6209–6219 (2001).
5 M. Maiza, M. T. Benaniba, and V. Massardier-Nageotte, Plasticizing effects of citrate esters on properties of poly(lactic acid), *J. Polym. Eng.*, **36**(4), 371–380, (2016).
6 B. K. Chen, T. Y. Wu, Y. M. Chang, and A. F. Chen, Ductile polylactic acid prepared with ionic liquids, *Chem. Eng. J.*, **215–216**, 886–893 (2013).

7 R. Saiah, R. Gattin, P. A. Srekumar, Properties and Biodegradation Nature of Thermoplastic Starch, *Thermoplastic Elastomers*, A. El-Sonbati (Ed.) InTech, pp. 57–78 (2012).

8 P. J. Halley, R. W. Truss, M. G. Markotsis, C. Chaleat, M. Russo, and A. Sargent, A review of biodegradable thermoplastic starch polymers, *Polym. Durab. Radiat. Eff.*, **978**, 287–300 (2007).

9 Y. Zhang, C. Rempel, and Q. Liu, Thermoplastic starch processing and characteristics-a review., *Crit. Rev. Food Sci. Nutr.*, **54**, 1353–1370 (2014).

10 E. M. Teixeira, A. L. Da Róz, A. J. F. Carvalho, and A. A. S. Curvelo, The effect of glycerol/sugar/water and sugar/water mixtures on the plasticization of thermoplastic cassava starch, *Carbohydr. Polym.*, **69**(4) 619–624 (2007).

11 O. Martin, E. Schwach, L. Averous, and Y. Couturier, Properties of biodegradable multilayer films based on plasticized wheat starch, *Starch - Stärke*, **53**(8), 372 (2001).

12 X. Ma, J. Yu, and J. F. Kennedy, Studies on the properties of natural fibers-reinforced thermoplastic starch composites, *Carbohydr. Polym.*, **62** (1), 19–24, (2005).

13 M. Huang, J. Yu, and X. Ma, High mechanical performance MMT-urea and formamide-plasticized thermoplastic cornstarch biodegradable nanocomposites, *Carbohydr. Polym.*, **63** (3) 393–399 (2006).

14 H. Li and M. A. Huneault, Comparison of sorbitol and glycerol as plasticizers for thermoplastic starch in TPS/PLA blends, *J. Appl. Polym. Sci.*, **119**, 2439–2448 (2011).

15 M. A. García, M..N. Martino, and N. E. Zaritzky, Microstructural characterization of plasticized starch-based films, *Starch/Staerke*, **52**(4), 118–124 (2000).

16 Y. Jiugao, W. Ning, and M. Xiaofei, The effects of citric acid on the properties of thermoplastic starch plasticized by glycerol, *Starch - Stärke*, **57** (10), 494–504 (2005).

17 A. Sankri, A. Arhaliass, I. Dez, A. C. Gaumont, Y. Grohens, D. Lourdin, I. Pillin, A. Rolland-Sabaté, and E. Leroy, Thermoplastic starch plasticized by an ionic liquid, *Carbohydr. Polym.*, **82** (2) 256–263 (2010).

18 J. Parameswaranpillai, S. Thomas, and Y. Grohens, Polymer Blends: State of the Art, New Challenges, and Opportunities, 1–6 (2015).

19 J. S. Higgins, J. E. G. Lipson, and R. P. White, A simple approach to polymer mixture miscibility, *Philos. Trans. A Math. Phys. Eng. Sci.*, **368**(1914), 1009–1025 (2010).

20 B. A. Miller-Chou, J. L. Koenig, A review of polymer dissolution, *Prog. Polym. Sci.*, **28** (8), 1223–1270 (2003).

21 J. Bicerano, *Prediction of Polymer Properties*. M. Dekker (Ed.), Marcel Dekker Inc., New York (2009).

22 X. Chen, C. Yuan, C. K. Y. Wong, G. Zhang, Molecular modeling of temperature dependence of solubility parameters for amorphous polymers, *J. Mol. Model.*, **18** (6), 2333–2341 (2012).

23 J. H. Hildebrand, Solubility, *J. Am. Chem. Soc.*, **5**, 45, (1916).

24 J. H. Hildebrand, Solubility. III. Relative values of internal pressures and their practical application, *J. Am. Chem. Soc.*, **38**, 1067–1080, (1919).

25 C. M. Hansen, *Hansen Solubility Parameters, a User's Handbook*, Taylor & Francis group, New York, (2007).

26 L. a. Utracki, Compatibilization of polymer blends, *Can. J. Chem. Eng.*, **80** (6), 1008–1016 (2002).

27 R. Thipmanee, A. Sane, Effect of zeolite 5A on compatibility and properties of linear low-density polyethylene/thermoplastic starch blend, *J. Appl. Polym. Sci.*, **126**, E251–E258 (2012).

28 G. Singh, H. Bhunia, A. Rajor, and V. Choudhary, Thermal properties and degradation characteristics of polylactide, linear low density polyethylene, and their blends, *Polym. Bull.*, **66** (7) 939–953 (2011).

29 N. Ploypetchara, P. Suppakul, D. Atong, and C. Pechyen, Blend of polypropylene/poly(lactic acid) for medical packaging application: Physicochemical, thermal, mechanical, and barrier properties, *Energy Procedia*, **56** (C) 201–210 (2014).

30 S. K. Nayak, Biodegradable PBAT/starch nanocomposites, *Polym. Plast. Technol. Eng.*, **February 2013**, 37–41 (2010).

31 E. Leroy, P. Jacquet, G. Coativy, A. L. Reguerre, and D. Lourdin, Compatibilization of starch-zein melt processed blends by an ionic liquid used as plasticizer, *Carbohydr. Polym.*, **89** (3) 955–963 (2012).

32 A. Vignon, A. Ayoub, and V. Massardier, The effect of γ-irradiation and molten medium on the structure and properties of polycarbonate and starch blends: A work oriented to the valorization of bio based polymers, *J. Appl. Polym. Sci.*, **127**, 4168–4176, (2011).

33 R. Malkapuram, V. Kumar, and Y. S. Negi, Recent development in natural fiber, *J. Reinf. Plast. Comp.*, **28**, 1169 (2009).

34 P. Wambua, J. Ivens, and I. Verpoest, Natural fibres: Can they replace glass in fibre reinforced plastics?, *Compos. Sci. Technol.*, **63** (9) 1259–1264, (2003).

35 H. Ku, H. Wang, N. Pattarachaiyakoop, and M. Trada, A review on the tensile properties of natural fiber reinforced polymer composites, *Composites, Part B*, **42** (4) 856–873, (2011).

36 B. Suddell, Industrial fibres: recent and current developments, *Symp. Nat. Fibres*, **44** (0) 71–82 (2008).

37 A. Ashori, Wood-plastic composites as promising green-composites for automotive industries!, *Bioresour. Technol.*, **99** (11) 4661–4667 (2008).

38 D. N. Saheb, J. P. Jog, and others, Natural fiber polymer composites: A review, *Adv. Polym. Technol.*, **18** (4), 351–363 (1999).

39 V. Mazzanti, F. Mollica, and N. El Kissi, *Rheological and Mechanical Characterization of Polypropylene-Based Wood Plastic Composites*, Wiley Online Libr., Wiley (2015).

40 A. L. Catto, B. V. Stefani, V. F. Ribeiro, R. Marlene, and C. Santana, Influence of coupling agent in compatibility of post-consumer HDPE in thermoplastic composites reinforced with eucalyptus fiber, *Mater. Res.*, **17**(1) 203–209 (2014).

41 Y. Lei, Q. Wu, F. Yao, and Y. Xu, Preparation and properties of recycled HDPE/natural fiber composites, *Composites, Part A*, **38**(7) 1664–1674 (2007).

42 T. J. Keener, R. K. Stuart, and T. K. Brown, Maleated coupling agents for natural fibre composites, *Composites, Part A*, **35**(3) 357–362 (2004).

43 J. R. Araùjo, W. R. Waldman, and M. A. De Paoli, Thermal properties of high density polyethylene composites with natural fibres: Coupling agent effect, *Polym. Degrad. Stab.*, **93**(10) 1770–1775 (2008).

44 M. M. Kabir, H. Wang, K. T. Lau, and F. Cardona, Chemical treatments on plant-based natural fibre reinforced polymer composites: An overview, *Composites, Part B*, **43** (7) 2883–2892 (2012).

45 X. Li, L. G. Tabil, and S. Panigrahi, Chemical treatments of natural fiber for use in natural fiber-reinforced composites: A review, *J. Polym. Environ.*, **15**(1) 25–33 (2007).
46 G. Pritchard, Two technologies merge: Wood plastic composites Geoff Pritchard describes how wood and resin are being, *Plast. Addit. Compd.*, **48** (6), 18–21 (2004).
47 N. M. Stark and L. M. Matuana, Characterization of weathered wood-plastic composite surfaces using FTIR spectroscopy, contact angle, and XPS, *Polym. Degrad. Stab.*, **92**(10) 1883–1890 (2007).
48 M. P. Wolcott, Wood-Plastic Composites, in *Encyclopedia of Materials - Science and Technology*, 9759–9763 (2001).
49 C. Clemons, Wood-plastic composites in the United States: The interfacing of two industries, *For. Prod. J.*, **52**(6) 10–18 (2002).
50 M. Hietala, J. Niinimäki, and K. Oksman, Processing of wood chip–plastic composites: Effect on wood particle size, microstructure and mechanical properties, *Plast., Rubber Compos.*, **40**(2) 49–56 (2011).
51 A. R. Sanadi and D. Caulfield, Thermoplastic polyolefins as formaldehyde free binders in highly filled lignocellulosic panel boards: Using glycerine as a processing aid in kenaf fiber polypropylene boards, *Mater. Res.*, **11**(4) 487–492 (2008).
52 M. Valente, J. Tirillò, A. Quitadamo, and C. Santulli, Paper fiber filled polymer. Mechanical evaluation and interfaces modification, *Composites, Part B*, **110**, 520–529, (2016).
53 M. Kazayawoko, J. J. Balatinecz, and L. M. Matuana, Surface modification and adhesion mechanisms in woodfiber-polypropylene composites, *J. Mater. Sci.*, **34**(24) 6189–6199, (1999).
54 N. Sombatsompop, C. Yotinwattanakumtorn, and C. Thongpin, Influence of type and concentration of maleic anhydride grafted polypropylene and impact modifiers on mechanical properties of PP/wood sawdust composites, *J. Appl. Polym. Sci.*, **97**(2) 475–484 (2005).
55 S. M. B. Nachtigall, G. S. Cerveira, and S. M. L. Rosa, New polymeric-coupling agent for polypropylene/wood-flour composites, *Polym. Test.*, **26**(5) 619–628 (2007).
56 H. S. Kim, B. H. Lee, S. W. Choi, S. Kim, and H. J. Kim, The effect of types of maleic anhydride-grafted polypropylene (MAPP) on the interfacial adhesion properties of bio-flour-filled polypropylene composites, *Composites, Part A*, **38** (6) 1473–1482 (2007).
57 S. E. Selke and I. Wichman, Wood fiber/polyolefin composites, *Composites, Part A*, **35** (3) 321–326 (2004).
58 M. Tazi, F. Erchiqui, F. Godard, H. Kaddami, and A. Ajji, Characterization of rheological and thermophysical properties of HDPE-wood composite, *J. Appl. Polym. Sci.*, **131**(13) 1–11 (2014).
59 A. K. Bledzki, O. Faruk, Injection moulded microcellular wood fibre-polypropylene composites, *Composites, Part A*, **37**(9), 1358–1367 (2006).
60 K. B. Adhikary, M. R. Islam, G. M. Rizvi, and C. B. Park, Effect of extrusion conditions on the surface quality, mechanical properties, and morphology of cellular wood flour/high-density polyethylene composite profiles, *J. Thermoplast. Compos. Mater.*, **26** (8) 1127–1144 (2012).
61 F. Mengeloglu, L. M. Matuana, Foaming of rigid PVC/wood-flour composites through a continuous extrusion process, *J. Vinyl Addit. Technol.*, 7(3), 142–148 (2001).

62 J. D. Yoon, T. Kuboki, P. U. Jung, J. Wang, and C. B. Park, Injection molding of wood–fiber/plastic composite foams, *Compos. Interfaces*, **16**, 797–811 (2009).
63 E. Pulp, *Key Statistics*, European Pulp and Paper Industry, (2012).
64 M. C. Flemings, B. Ilschner, E. J. Kramer, and S. Mahajan, Paper: Recycling and recycled materials, *Encycl. Mater. Sci. Technol.*, **6** (3) 6711–6720 (2001).
65 A. R. Sanadi, R. a. Young, C. Clemons, and R. M. Rowell, Recycled newspaper fibers as reinforcing fillers in thermoplastics: Part I-analysis of tensile and impact properties in polypropylene, *J. Reinf. Plast. Compos.*, **13**(1) 54–67 (1994).
66 I. Baroulaki, O. Karakasi, G. Pappa, P. A. Tarantili, D. Economides, and K. Magoulas, Preparation and study of plastic compounds containing polyolefins and post used newspaper fibers, *Composites, Part A*, **37**(10),1613–1625 (2006).
67 M. Prambauer, C. Paulik, and C. Burgstaller, The influence of paper type on the properties of structural paper - polypropylene composites, *Composites, Part A*, **74**, 107–113 (2015).
68 A. Serrano, F. X. Espinach, F. Julian, R. Del Rey, J. A. Mendez, and P. Mutje, Estimation of the interfacial shears strength, orientation factor and mean equivalent intrinsic tensile strength in old newspaper fiber/polypropylene composites, *Composites, Part B*, **50**, 232–238 (2013).
69 E. Franco-Marquès, J. A. Méndez, M. A. Pèlach, F. Vilaseca, J. Bayer, and P. Mutjé, Influence of coupling agents in the preparation of polypropylene composites reinforced with recycled fibers, *Chem. Eng. J.*, **166**(3) 1170–1178 (2011).
70 A. Serrano, F. X. Espinach, J. Tresserras, R. del Rey, N. Pellicer, and P. Mutje, Macro and micromechanics analysis of short fiber composites stiffness: The case of old newspaper fibers-polypropylene composites, *Mater. Des.*, **55**, 319–324 (2014).
71 S. A. Balwaik and S. P. Raut, Utilization of waste paper pulp by partial replacement of cement in concrete, *Int. J. Eng. Res. Ind. Appl.*, **1**(2) 300–309 (2011).
72 T. Kuokkanen, H. Nurmesniemi, R. Pöykiö, K. Kujala, J. Kaakinen, and M. Kuokkanen, Chemical and leaching properties of paper mill sludge, *Chem. Speciation Bioavailability*, **20**(2) 111–122 (2008).
73 Y. Hamzeh, A. Ashori, and B. Mirzaei, Effects of waste paper sludge on the physico-mechanical properties of high density polyethylene/wood flour composites, *J. Polym. Environ.*, **19** (1), 120–124 (2011).
74 J. Son, Physico-mechanical properties of paper sludge-thermoplastic polymer composites, *J. Thermoplast. Compos. Mater.*, **17**(6) 509–522 (2004).
75 M. Valente, J. Tirillò, and A. Quitadamo, Industrial paper recycling process: Suitable micronization for additive polymer application, *CSE J.*, **2**, 145–152 (2015).

Index

a
Applications   7, 21, 240

b
Biobased   95
Biomass composition   114
Biomass growth   11, 71

c
Catalyst   22
Cellulose   40, 183
Cell wall analysis   112, 116
Characterization   113, 117
Chemical industry   3, 10, 14, 18, 58
Coal   4, 8, 28, 66
Cracking biomass   181

d
Drugs   5, 23

e
Energy   3, 18, 63, 91, 95
Enzymatic polymerization   219
Ethanol   91

f
Fatty acid   226
Feedstock   30
Fermentation   92

Food   31
Fossil materials   2, 66,
Fourier transform infrared spectroscopy (FTIR)   121
Furan   229

g
Gas   28, 67, 82, 93, 94
Gel permeation chromatography (GPC)   123
Global production   6, 19, 22, 24, 229
Green house   26

h
Hemicellulose   43, 185
High-density polyethylene (HDPE)   240
High-performance liquid chromatography (HPLC)   115

i
Irradiation   205

l
Life-cycle analysis   96, 141
Lignin   45, 125, 197
Lignocellulose   15, 16, 39, 74, 183

m
Moisture   110

Introduction to Renewable Biomaterials: First Principles and Concepts, First Edition.
Edited by Ali S. Ayoub and Lucian A. Lucia.
© 2018 John Wiley & Sons Ltd. Published 2018 by John Wiley & Sons Ltd.

n

Natural gas 4, 66
Nuclear magnetic resonance (NMR) 126
Nutrition 24

o

Oil 4, 11, 240
Oil refinery 6, 8

p

Polyesters 221, 226
Polyethylenterephthalate (PET) 240
Polylactide (PLA) 240
Polymers 7, 24, 190
Polypropylene (PP) 240

r

Raman 122

s

Size of polymers 123
Spectroscopy 116
Starch 14, 15, 24, 55, 74
Sugar 13

t

Thermoplastic 240, 246
Transformation 17, 75, 85, 194

x

X-ray 118

www.ingramcontent.com/pod-product-compliance
Lightning Source LLC
LaVergne TN
LVHW081516060526
838200LV00005B/196